Springer Theses

Recognizing Outstanding Ph.D. Research

For further volumes:
http://www.springer.com/series/8790

Aims and Scope

The series "Springer Theses" brings together a selection of the very best Ph.D. theses from around the world and across the physical sciences. Nominated and endorsed by two recognized specialists, each published volume has been selected for its scientific excellence and the high impact of its contents for the pertinent field of research. For greater accessibility to non-specialists, the published versions include an extended introduction, as well as a foreword by the student's supervisor explaining the special relevance of the work for the field. As a whole, the series will provide a valuable resource both for newcomers to the research fields described, and for other scientists seeking detailed background information on special questions. Finally, it provides an accredited documentation of the valuable contributions made by today's younger generation of scientists.

Theses are accepted into the series by invited nomination only and must fulfill all of the following criteria

- They must be written in good English.
- The topic should fall within the confines of Chemistry, Physics, Earth Sciences, Engineering and related interdisciplinary fields such as Materials, Nanoscience, Chemical Engineering, Complex Systems and Biophysics.
- The work reported in the thesis must represent a significant scientific advance.
- If the thesis includes previously published material, permission to reproduce this must be gained from the respective copyright holder.
- They must have been examined and passed during the 12 months prior to nomination.
- Each thesis should include a foreword by the supervisor outlining the significance of its content.
- The theses should have a clearly defined structure including an introduction accessible to scientists not expert in that particular field.

Kohei Miyata

Highly Luminescent Lanthanide Complexes with Specific Coordination Structures

Doctoral Thesis accepted by
Hokkaido University, Sapporo, Japan

Author
Dr. Kohei Miyata
Hokkaido University
Sapporo
Japan

Supervisor
Prof. Yasuchika Hasegawa
Hokkaido University
Sapporo
Japan

ISSN 2190-5053 ISSN 2190-5061 (electronic)
ISBN 978-4-431-56398-3 ISBN 978-4-431-54944-4 (eBook)
DOI 10.1007/978-4-431-54944-4
Springer Tokyo Heidelberg New York Dordrecht London

Printed on acid-free paper

Springer is part of Springer Science+Business Media (www.springer.com)

Parts of this thesis have been published in the following journal articles:

K. Miyata, Y. Hasegawa, Y. Kuramochi, T. Nakagawa, T. Yokoo, T. Kawai, "Characteristic Structures and Photophysical Properties of Nine-Coordinated Europium(III) Complexes with Tandem-connected Phosphane Oxide Ligands", *Eur. J. Inorg. Chem.* **2009**, *32*, 4777.

K. Miyata, T. Nakagawa, R. Kawakami, Y. Kita, K. Sugimoto, T. Nakashima, T. Harada, T. Kawai, Y. Hasegawa, "Remarkable Luminescence Properties of Lanthanide Complexes with Asymmetric Dodecahedron Structures", *Chem. Eur. J.* **2011**, *17*, 521.

K. Miyata, T. Ohba, A. Kobayashi, M. Kato, T. Nakanishi, K. Fushimi, Y. Hasegawa, "Thermostable Organo-phosphor: Low-vibrational Coordination Polymers That Exhibit Different Intermolecular Interactions", *ChemPlusChem* **2012**, *77*, 277.

K. Miyata, T. Nakanishi, K. Fushimi, Y. Hasegawa, "Solvent-dependent Luminescence of Eu(III) Complexes with Bidentate Phosphine Oxide", *J. Photochem. Photobiol. A: Chem.* **2012**, *235*, 35.

K. Miyata, Y. Konno, T. Nakanishi, A. Kobayashi, M. Kato, K. Fushimi, Y. Hasegawa, "Chameleon Luminophore for Sensing Temperatures: Control of Metal-to-Metal and Back Energy Transfer in Lanthanide Coordination Polymers", *Angew. Chem. Int. Ed.* **2013**, *52*, 6413.

Supervisor's Foreword

There has been significant recent interest in the development of luminescent metal complexes for applications such as optical materials, organic light-emitting diodes (OLEDs), and fluorescent sensors. These luminescence properties are strongly dominated by the type of organic ligand and the coordination structure, which together control the ligand field. In particular, the ligand fields of transition-metal complexes are affected by the particular geometrical structure, and a large number of studies have reported the link between the structure and the luminescence properties of luminescent metal complexes.

In this thesis, luminescent lanthanide complexes with specific coordination structures are introduced by Dr. Kohei Miyata. Specific coordination structures result in lanthanide complexes with remarkable photophysical properties. This thesis provides academic studies of specific coordination structures (mono-capped square-antiprism, dodecahedron, and coordination-polymer structures) of luminescent lanthanide complexes. Dr. Miyata has also successfully prepared thermostable and thermosensing luminophores composed of lanthanide ions and characteristic organic ligands. I believe that his studies contribute to successful exploration of new science and technology in our future.

Lanthanide complexes with characteristic photophysical properties, narrow emission bands, and long emission lifetimes, have been regarded as attractive luminescent materials for use in electroluminescent (EL) devices, lasers, and luminescent biosensing applications. Luminescent lanthanide complexes with specific coordination structures may lead to the development of new fields in photophysical, coordination, and materials chemistry.

Hokkaido, January 2014 Prof. Yasuchika Hasegawa

Supervisor's Foreword

Acknowledgments

I am sincerely grateful to Prof. Yasuchika Hasegawa, Laboratory of Advanced Materials Chemistry, Division of Materials Chemistry, Graduate School of Engineering, Hokkaido University, for his kind guidance, valuable suggestions, and hearty encouragement throughout this work.

I would also like to express my gratitude to Assoc. Prof. Koji Fushimi and Asst. Prof. Takayuki Nakanishi, Graduate School of Engineering, Hokkaido University, for their kind help and useful advice.

I thank Prof. Tsuyoshi Kawai, Assoc. Prof. Takuya Nakashima, and Asst. Prof. Junpei Yuasa Graduate School of Materials Science, Nara Institute of Science and Technology (NAIST), for their helpful discussions and guidance in my master's thesis.

I acknowledge Prof. Masako Kato and Asst. Prof. Atsushi Kobayashi, Graduate School of Science, Hokkaido University, for their kind help in X-ray single-crystal analyses and valuable suggestions.

I also express my appreciation to Prof. Emer. Hideo Hirohara, Prof. Tsutomu Kumagai, Assoc. Prof. Yoshinori Inoue, and Asst. Prof. Munenori Takehara of the Department of Materials Science, University of Shiga Prefecture, for their kind support in my bachelor's thesis.

I gratefully acknowledge the financial support of the Japan Society for the Promotion of Science (Grant-in-Aid for JSPS fellows) and the Global COE Program "Catalysis as the Basis for Innovation in Materials Science" from the Ministry of Education, Culture, Sports, Science and Technology, Japan.

Finally, I am particularly grateful to my family, Kiichi Tanaka, Matsue Tanaka, Toshio Miyata, Atsuko Miyata, and Kazuna Miyata, for their constant understanding and support.

Contents

Chapter 1
General Introduction

1.1 Luminescence of Trivalent Lanthanide Ion

The luminescence properties of lanthanide (rare-earth) compounds have been fascinating many researchers for decades [1–9]. An attractive feature of luminescent lanthanide compounds is their line-like emission, which results in a high color purity of the emitted light. The trivalent ions of the lanthanide series are characterized by a gradual filling of the 4f orbitals, from $4f^0$ (for La(III)) to $4f^{14}$ (Lu(III)) as summarized in Table 1.1. One of the most interesting features of these ions is their photoluminescence. Several lanthanide ions show luminescence in the visible or near-infrared spectral regions upon irradiation with ultraviolet (UV) radiation. The color of the emitted light depends on the lanthanide ion. For instance, Eu(III) emits red light, Tb(III) green light, Sm(III) orange light, and Tm(III) blue light. Yb(III), Nd(III), and Er(III) are well-known for their near-infrared luminescence, and Gd(III) emits in the UV region. Therefore, lanthanide(III) compounds are the popular luminescent materials for application in electroluminescent (EL) devices, optical amplifiers, lasers, bio-sensing, and so on.

The lanthanides in the stable (III) oxidation state are simply characterized by an incompletely filled 4f shell. The 4f orbital is shielded from the surroundings by the filled $5s^2$ and $5p^2$ orbitals. Therefore, the influence of the host media on the optical transition within the $4f^n$ configuration is small. The energy levels of the trivalent Eu ion are given in Fig. 1.1. The energy levels are actually split by the geometric field (crystal or ligand fields). As a matter of fact, the splitting is very small due to the shielding by $5s^2$ and $5p^2$ electrons, whereas the geometric field strength in the case of the transition metal ions (d^n) is characteristically several tens of thousands of cm^{-1}; for the lanthanide(III) ions (f^n), it amounts to several hundreds of cm^{-1}.

In a configurational coordinate diagram, these level appear as parallel parabolas (small off-set, Fig. 1.2), because the 4f electrons are well shielded from their surroundings. Emission transitions yield, therefore, sharp lines in the spectra. The emission of light radiation for lanthanide(III) ions comes mainly from the electric dipole transition. Transitions that occur in the 4f inner shell of free ions are forbidden because they do not correspond to a change of parity. However, the

K. Miyata, *Highly Luminescent Lanthanide Complexes with Specific Coordination Structures*, Springer Theses,
DOI: 10.1007/978-4-431-54944-4_1, © Springer Japan 2014

Table 1.1 Electronic structure of the trivalent lanthanide ions

Element	Symbol	Atomic number (Z)	Configuration Ln(III)	Ground state Ln(III)
Lanthanum	La	57	[Xe]	$^1S_0$
Cerium	Ce	58	$[Xe]4f^1$	$^2F_{5/2}$
Praseodymium	Pr	59	$[Xe]4f^2$	$^3H_4$
Neodymium	Nd	60	$[Xe]4f^3$	$^4I_{9/2}$
Promethium	Pm	61	$[Xe]4f^4$	$^5I_4$
Samarium	Sm	62	$[Xe]4f^5$	$^6H_{5/2}$
Europium	Eu	63	$[Xe]4f^6$	$^7F_0$
Gadolinium	Gd	64	$[Xe]4f^7$	$^8S_{7/2}$
Terbium	Tb	65	$[Xe]4f^8$	$^7F_6$
Dysprosium	Dy	66	$[Xe]4f^9$	$^6H_{15/2}$
Holmium	Ho	67	$[Xe]4f^{10}$	$^5I_8$
Erbium	Er	68	$[Xe]4f^{11}$	$^4I_{15/2}$
Thulium	Tm	69	$[Xe]4f^{12}$	$^3H_6$
Ytterbium	Yb	70	$[Xe]4f^{13}$	$^2F_{7/2}$
Lutetium	Lu	71	$[Xe]4f^{14}$	$^1S_0$

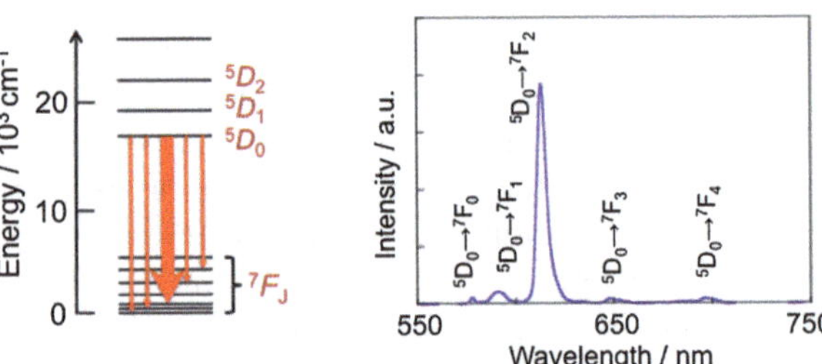

Fig. 1.1 The energy levels and emission spectra of Eu(III) ions

transitions that are forbidden by odd parity become partially allowed by mixing 4f and 5d states through the ligand field. Since the parity does not change dramatically in such a transition, the lifetime of the excited state is long (>1 μs).

1.2 Luminescent Lanthanide Complex

The 1930s and early 1940s witnessed the first spectroscopic studies of lanthanide ions in solution. Freed et al. found that the relative intensities of the absorption lines of Eu(III) were different in various solvents [10], and Weissman discovered that complexes of Eu(III) with certain ultraviolet absorbing ligands were highly luminescent when excited with ultraviolet light [11]. Since Eu(III) itself has only a few very weak absorption bands, the solution of this ion does not exhibit very bright luminescence. Obviously, certain organic ligands can serve to photosensitize the luminescence of lanthanide ions. It was also found that lanthanide ions

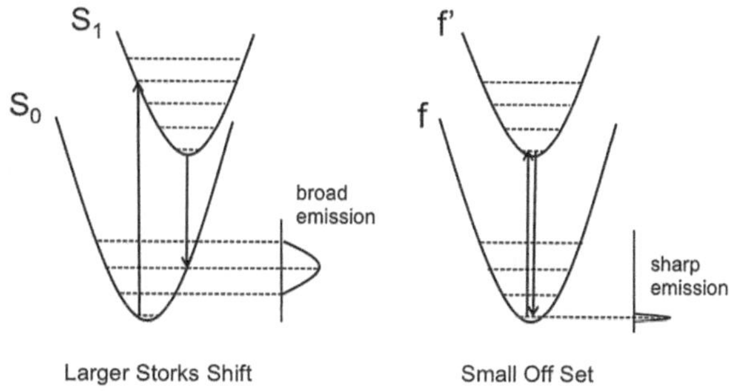

Fig. 1.2 Comparison of the emission process of lanthanide(III) ions with that of general organic compounds or metal complexes

quench the fluorescence of organic ligands [12]. Those studies can be considered as the starting points for the research and development of luminescent lanthanide complexes. It took, however, about 20 years before photochemists took a serious interest in these materials. At the 1990s, the field received a new impulse when it became clear that lanthanide complexes could find widespread use in medical diagnostics [13]. In the second half of that decade, lanthanide ions also started to receive the attention of the growing group of supramolecular chemists.

Luminescent lanthanide complexes consist of a lanthanide ion encapsulated in a ligand. Generally speaking, the ligand can be tailored to contain built-in functionalities that give the overall complex desired properties. Parameters such as solubility, electrochemical activity, binding affinity for other molecular building blocks, responsively to external stimuli (such as the presence of certain ions) etc. are all influenced by the overall structure of the ligand. In most cases, the ligand contains a light-absorbing group in the form of an organic chromophore as illustrated in Fig. 1.3a [14–21]. Such a group is generally referred to as the antenna chromophore, in analogy to the light-harvesting center in photo-synthetic reaction center. The photonic energy absorbed by this antenna can be transferred to the encapsulated lanthanide ion, thus circumventing the photo-excitation bottleneck posed by the small absorption of lanthanide ions.

Indirect excitation, in other words, sensitization has to be used and proceeds in three steps. First, light is absorbed by the immediate environment of the Ln(III) ion through the attached organic ligands. Energy is then transferred onto one or several excited states of the metal ion, and finally, the metal ion emits light. A multitude of organic ligands bearing aromatic chromophores has been proposed for this purpose, derived, for instance, from bipyridine, terpyridine, terphenylene, quinoline, or substituted phenyl and naphthyl groups. Although often discussed and modeled in terms of a simple ligand (S_1) → ligand (T_1) → Ln* energy flow which can be optimized by adjusting the energy gap between the lowest ligand triplet state and the Ln(III) emitting level [22–24], sensitization of trivalent lanthanide ions is an exceedingly complex process involving numerous rate constants (Fig. 1.3b) [25, 26]. The characteristic luminescent properties of lanthanide complexes have been extensively studied [27–69].

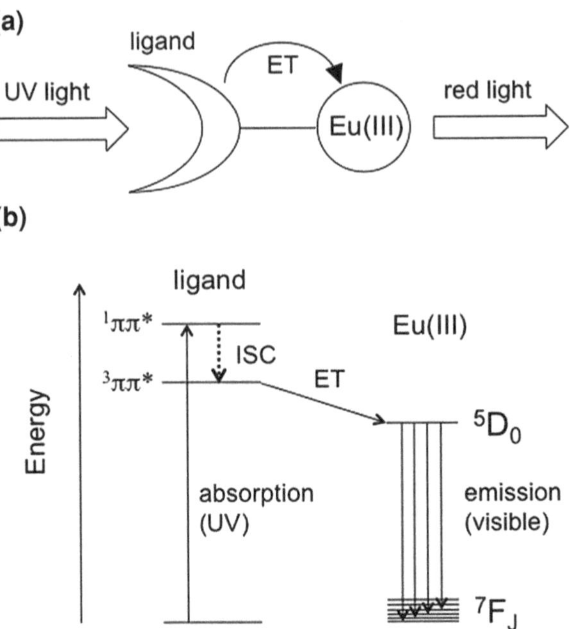

Fig. 1.3 **a** A schematic representation of the antenna effect occurring in monometallic Eu(III) complexes possessing aromatic light-harvesting chromophores and **b** associated energy level diagram (ISC = intersystem crossing, ET = energy transfer)

The author describes the strategies for design of strong-luminescent lanthanide complexes from the next section.

1.3 Molecular Design for Enhancement of Luminescence of Lanthanide Complex

1.3.1 Suppression of Vibrational Relaxation

The energy transfer process via vibrational excitation has been considered as a dominant process that quenches the excited states of lanthanide ion. For instance, in the case of Nd(III), the difficulty in the enhancement of luminescence of Nd(III) in liquid systems is explained as being due to the radiationless relaxation of the emitting level ($^4F_{3/2}$) via vibrational excitation of the liquid matrix, which is made up of ligand and solvent molecules (Fig. 1.4a). The vibrational transition probability is proportional to the Frank–Condon factor, i.e., the overlap integrals between the energy gap and the vibrational energy. In terms of the energy gap theory, the rate constant for the radiationless transition W_{RT} is given by the following expression:

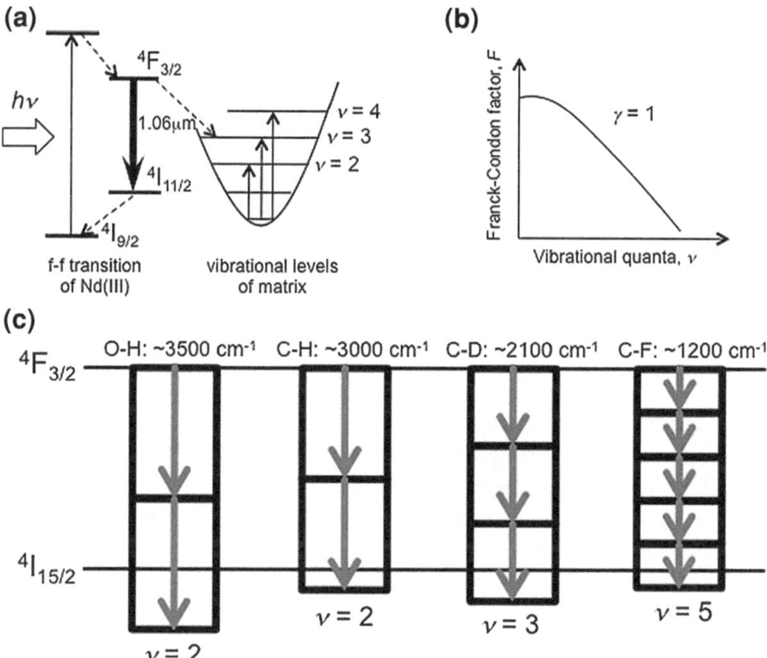

Fig. 1.4 **a** The radiationless transition via vibrational relaxation, **b** Franck–Condon function, **c** the vibrational energy levels and matching the electronic energy gap of Nd(III) ion

$$W_{RT} = \frac{2\pi\rho}{\hbar} J^2 F \quad (1.1)$$

where ρ is the density of states, J is the coupling constant between the electronic wave functions due to nuclear motion, and F the Frank–Condon factor. The factor F should be estimated quantitatively when the approximation using the undistorted oscillator model is adopted. In the model, F is given by

$$F = \frac{\exp(-\gamma)\gamma^{v}}{v!} \quad (1.2)$$

If γ is assumed to be 1, the Frank–Condon factor decreases with increasing vibrational quantum number v, as shown in Fig. 1.4b. The energy gap of the crucial radiative transition in Nd(III) ($^4F_{3/2}$–$^4I_{15/2}$: 5400 cm^{-1}) matches well the vibrations of the C–H and O–H bonds (5900 and 6900 cm^{-1}, respectively) with vibrational quantum number $v = 2$, and vibrational excitation leads to effective quenching of the excited state of Nd(III). In contrast, C–D, C–O, C–C, and C–F bonds give larger vibrational quanta ($v = 3$, 5 Fig. 1.4c). In theoretical calculations, the Frank–Condon factors of C–H and C–F are determined to be 0.18 and 0.0031, respectively. These facts suggest that if organic ligands surrounding

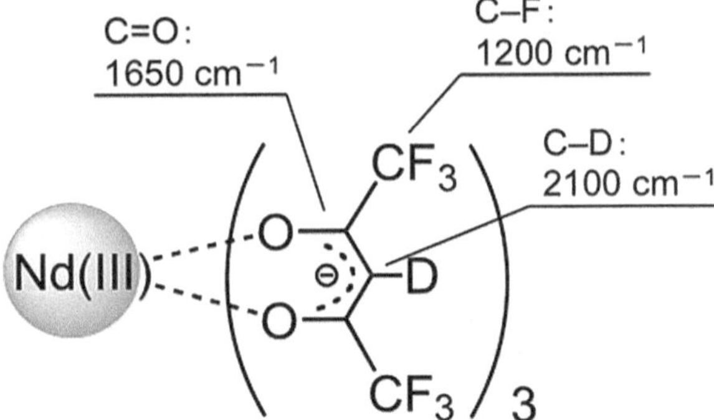

Fig. 1.5 Chemical structure of Nd(hfa-D)$_3$

Nd(III) ion can be constructed with C–D and C–F bonds instead of C–H bonds, the radiationless transition through vibrational excitation should be suppressed (Fig. 1.5). This theoretical understanding led to the first successful utilization of deuterated hexafluoroacetylacetone (hfa-D) as an effective ligand, which enables the resulting Nd(III) complex to exhibit near-infrared luminescence. The absence of C–H and O–H vibrations should make it possible to observe luminescence from Nd(III) in an organic solvent [70].

1.3.2 Control of Coordination Structure

The luminescence properties of lanthanide complexes are mainly derived from the electric transitions in 4f orbitals. Generally, 4f orbitals of lanthanide are shielded from direct perturbations by outer filled 5s and 5p shells [71]. The electric dipole transitions from the 4f inner shell of lanthanide ions are intrinsically forbidden because of their odd parity. However, they can be partially allowed upon mixing 4f and 5d states through the ligand field effects. The partially allowed transitions and electrically shielded nature of the excited states promote long emission lifetimes and characteristic sharp emission bands. Radiative rates of lanthanide complexes are linked to their geometric structures. If there is no inversion symmetry at the lanthanide ion sites, uneven ligand field components can mix with opposite-parity states in $4f^n$-configuration levels. Electric dipole transitions of lanthanide complexes are then no longer strictly forbidden in the ligand fields, resulting in faster radiative rates and higher emission quantum yields. Richardson and Reid have estimated the transition intensity parameters of lanthanide complexes from the ligand field [72, 73]. Binnemans has proposed to evaluate the transition intensity by using Judd-Ofelt analysis [14]. Since these studies, it has been widely accepted that the radiative transition probability between 4f orbitals is enhanced by reducing the coordination structure's geometrical symmetry [74–82].

In order to prepare intensely luminescent lanthanide complexes, a large radiative rate constant is necessary based on reducing the geometrical symmetry.

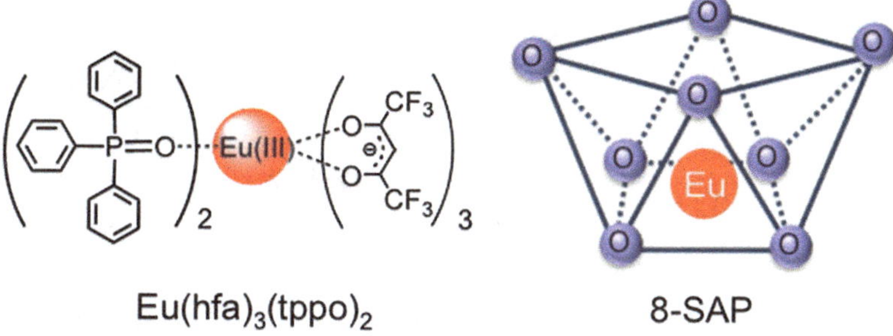

Fig. 1.6 Chemical and coordination structure of $Eu(hfa)_3(tppo)_2$

Lanthanide complexes with odd parity can be created using certain geometric and coordination structures. Hasegawa has reported on Eu(III) complexes with low-vibrational frequency (LVF) hfa (hexafluoroacetylacetone) and tppo (triphenyl-phosphine oxide) ligands, $Eu(hfa)_3(tppo)_2$, for which the chemical structure is shown in Fig. 1.6 [83]. The coordination structure composed of LVF phosphine oxides ($P = O$: 1125 cm^{-1}) and hfa ligands are expected to provide the Eu(III) complex with a high emission quantum yield and a relatively small nonradiative rate constant. The coordination geometry of $Eu(hfa)_3(tppo)_2$ is categorized as a square antiprism (8-SAP) which has no inversion center. The characteristic 8-SAP structure is composed of three hfa and two tppo ligands, which leads to a reduction of the geometrical symmetry of the Eu(III) complex and consequently $Eu(hfa)_3(tppo)_2$ shows a high emission quantum yield ($\Phi = 65$ %) and a fast radiative rate constant.

1.4 Lanthanide Complex with Thermostability

Lanthanide complexes have been also recognized as attractive compounds for fabricating lanthanide-doped polymers which are applicable to luminescent displays, plastic optical devices, and thin-film lasers [84–86]. In order to achieve the effective luminescence of lanthanide ion in polymer matrices, lanthanide complexes are required to possess high thermal stability. Luminescent plastics including lanthanide complexes are manufactured by mixing plasticizers and polymers at extremely high temperature. For example, polycarbonate, which is one of the major plastics for industrial products, is processed at 290 °C [87]. The thermostability of lanthanide complexes should be stable above 300 °C for optic applications such as lightings, displays, and optoelectronic devices. With this criterion in mind, Manseki has developed thermostable tetranuclear Eu(III) complexes, $[Eu_4(\mu\text{-}O)(L)_{10}]$ (L = 2-hydroxy-4-octyloxybenzophenone), having effective photosensitized luminescence as illustrated in Fig. 1.7 [88]. Differential scanning calorimetry (DSC) measurements indicated that the decomposition points

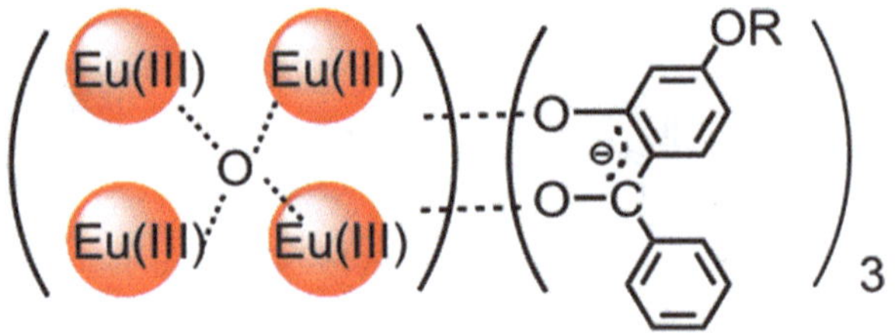

Fig. 1.7 Chemical structure of $[Eu_4(\mu\text{-}O)(L)_{10}]$ (L = 2-hydroxy-4-octyloxy benzophenone)

of tetranuclear Eu(III) complexes were 309 °C. Although many types of luminescent organic dyes are generally decomposed at temperatures under 200 °C, this disadvantage can be overcome by using polynuclear complex and coordination polymer (Sect. 1.5).

1.5 Lanthanide Coordination Polymer

Lanthanide coordination polymers and metal-organic frameworks composed of lanthanide ions and organic linker ligands have attracted considerable attention. The one-dimensional alternating sequence of metal ions and organic ligands exhibits remarkable characteristics as novel organic materials with (1) ease of preparation, (2) various structures, and (3) unique physical properties by the combination of metal ions and organic ligands [89–94]. Although lanthanide coordination polymers have been an active research field for about a decade [95–98], most of the attention has been paid to the structural characterization of these compounds. More recently, luminescence studies of the coordination polymers have been reported [99–106].

One of the significant characteristics of coordination polymers is their high thermal stability as well as polynuclear complexes. For instance, Carlos has reported novel three-dimensional lanthanide-organic frameworks with 2, 5-pyridinedicarboxylic acid [107]. Reddy has developed thermostable Eu(III) coordination polymers with 4-(dipyridin-2-yl)aminobenzoate ligands [63]. However, their emission quantum yields have been relatively low, because previous lanthanide coordination polymers contain numerous C–H and O–H bonds close to the metal center, which lead to the radiationless transition via vibrational relaxations as introduced in Sect. 1.3 [108–110]. To enhance emission quantum yields of lanthanide coordination polymers, the coordination sites of the organic linker ligands should be composed of low-vibrational frequency (LVF) modes for the suppression of vibrational relaxation. Luminescent coordination polymer with both high thermostability and strong-luminescent property is required as a novel luminophore in the field of future opto-electronic devices.

1.6 Objectives

As mentioned above, design of coordination structures of lanthanide complexes enable precise manipulations of the electron transitions in f orbitals. The geometrical, vibrational, and steric structures of lanthanide complexes are directly linked to luminescence properties such as emission quantum yield, radiative, and nonradiative rate constants. The control of the coordination structures of lanthanide complexes is expected to provide desirable emission properties. Lanthanide polynuclear complexes and coordination polymers have been also studied from the viewpoint of thermostability of complexes. However, the correlation between coordination structures and photophysical properties of lanthanide complexes has been scarcely investigated, and the optimum coordination environment for strong-luminescent properties has been still unclear. Additionally, lanthanide coordination polymer with both high thermostability and strong-luminescent properties has not been previously reported.

In this thesis, the control of coordination structures and functionalization of luminescent lanthanide(III) complexes with LVF hfa and phosphine oxide ligands were investigated. In particular, the author constructed (1) lanthanide coordination polymer with thermostability and thermosensing ability and (2) lanthanide complexes with novel asymmetric coordination structures to improve their emission quantum yields.

1.7 Contents of this Thesis

This thesis consists of a total of seven chapters. In this chapter, the previous studies are overviewed and the significance and objectives in this study are described. In Chap. 2, lanthanide coordination polymers composed of Eu(III) ions and phosphine oxide linkers are proposed to improve thermostability of lanthanide compounds. The characteristic structures, thermogravimetric analyses and remarkable photophysical properties of coordination polymers are also demonstrated. Chapter 3 presents a novel temperature-sensing material, chameleon luminophore, that has a high thermostability and thermosensing ability based on the results of Chap. 2. It is composed of color-changing luminescent coordination polymers containing Eu(III) and Tb(III) ions. The dual emission bands of Eu(III) and Tb(III) ions are expected to enable more accurate temperature measurements than previous lanthanide complexes. Temperature sensing dyes with thermostable structures and dual sensing units are demonstrated for the first time. In Chap. 4, novel nona-coordinated Eu(III) complexes with three kinds of low-vibrational frequency triphosphine oxides are reported. The characteristic luminescence properties of nona-coodinated Eu(III) complexes are elucidated in terms of geometrical, vibrational, and chemical structures. In Chap. 5, Eu(III) and Sm(III) complexes with novel asymmetric structures composed of oxo-linked bidentate phosphine oxides are reported. From the results of the structural and photophysical measurements, remarkably strong luminescence properties of lanthanide complexes with characteristic trigonal dodecahedron structures (8-TDH) are demonstrated for the first time. Based on the findings of Chap. 5, solvent-dependent luminescence of octa-coordinated Eu(III) complexes are described in Chap. 6. The relationships between photophysical properties and coordination structures in various organic media are discussed. Finally, all the results and achievements in this study are summarized in Chap. 7.

References

1. J.-C.G. Bünzli, *Luminescent Probes. In Lanthanide Probes in Life, Chemical and Earth Sciences: Theory and Practice* (Elsevier, Amsterdam, 1989)
2. A.J. Kenyon, Prog. Quantum Electron. **26**, 225 (2002)
3. G. Blasse, B.C. Grabmaier, *Luminescent Materials* (Spinger-Verlag, Berlin, 1994)
4. G. Blasse, Prog. Solid State Chem. **18**, 79 (1988)
5. M. Elbanowski, B. Makowsaka, J. Photochem. Photobiol. A **99**, 85 (1996)
6. J.-C.G. Bünzli, C. Piguet, Chem. Soc. Rev. **2005**, 34 (1048)
7. Y. Hasegawa, Y. Wada, S. Yanagida, J. Photochem. Photobiol. C **5**, 183 (2004)
8. B.M. Tissue, Chem. Mater. **10**, 2837 (1998)
9. K. Binnemans, Chem. Rev. **109**, 4283 (2009)
10. S. Freed, S.I. Weissman, F.E. Fortress, H.F. Jacobson, J. Chem. Phys. **7**, 824 (1939)
11. S.I. Weissman, J. Chem. Phys. **10**, 214 (1942)
12. P. Yuster, S.I. Weissman, J. Chem. Phys. **17**, 1182 (1949)
13. I. Hemmila, J. Alloy Compd. **225**, 480 (1995)
14. K. Binneman, R. Van Deun, C. Gorller-Walrand, S.R. Collinson, F. Martin, D.W. Bruce, C. Wickleder, Phys. Chem. Chem. Phys. **2**, 3753 (2000)
15. F.R. Goncalves e Silva, R. Longo, O.L. Malta, C. Piguet, J.-C.G. Bünzli, Phys. Chem. Chem. Phys.**2**, 5400 (2000)
16. D.M. Epstein, L.L. Chappell, H. Khalili, R.M. Supkowski, W.D. Horrocks Jr, J.R. Morrow, Inorg. Chem. **39**, 2130 (2000)
17. N. Fatin-Rouge, E. Toth, D. Perret, R.H. Backer, A.E. Merbach, J.-C.G. Bünzli, J. Am. Chem. Soc. **122**, 10810 (2000)
18. J.J. Lessmann, W.D. Horrocks Jr, Inorg. Chem. **39**, 3114 (2000)
19. M.D. McGehee, M.A. Diaz-Garcia, F. Hide, R. Gupta, E.K. Miller, D. Moses, A.J. Heeger, Appl. Phys. Lett. **72**, 1536 (1998)
20. P.J. Skinner, A. Beeby, R.S. Dickins, D. Parker, S. Aime, M. Botta, J. Chem. Soc. Perkin 2 **7**, 1329 (2000)
21. H. Tsukube, M. Hosokubo, M. Wada, S. Shinoda, H. Tamiaki, Inorg. Chem. **40**, 740 (2001)
22. M. Latva, H. Takalob, V.M. Mukkala, C. Matachescu, J.C. Rodriguez-Ubis, J. Kankare, J. Lumin. **75**, 149 (1997)
23. F. Gutierrrez, C. Tedeschi, L. Maron, J.P. Daudey, R. Poteau, J. Azema, P. Tisnes, C. Picard, Dalton Trans. 1334 (2004)
24. R.D. Archer, H. Chen, L.C. Thompson, Inorg. Chem. **37**, 2809 (1998)
25. F.R. Goncalves e Silva, O.L. Malta, C. Reinhard, H.U. Gudel, C. Piguet, J.E. Moser, J.-C.G. Bünzli, J. Phys. Chem. A **106**, 1670 (2002)
26. G.F. De Sa, O.L. Malta, C. De Mello Donega, A.M. Simas, R.L. Longo, P.A. Santa-Cruz, E.F. Da Silva Jr, Coord. Chem. Rev. **196**, 165 (2000)
27. R. Pal, D. Parker, Chem. Commun. 474 (2007)

28. G.S. Kottas, M. Mehlstäubl, R. Fröhlich, L. De Cola, Eur. J. Inrog. Chem. **22**, 3465 (2007)
29. S. Petoud, G. Muller, E.G. Moore, J. Xu, J. Sokolnicki, J.P. Riehl, U.N. Le, S.M. Cohen, K.N. Raymond, J. Am. Chem. Soc. **129**, 77 (2007)
30. J.P. Leonard, P. Jensen, T. McCabe, J.E. O'Brien, R.D. Peacock, P.E. Kruger, T. Gunnlaugsson, J. Am. Chem. Soc. **129**, 10986 (2007)
31. A. De Bettencourt-Dias, S. Viswanathan, A. Rollett, J. Am. Chem. Soc. **129**, 15436 (2007)
32. X.Y. Chen, X. Yang, B.J. Holliday, J. Am. Chem. Soc. **130**, 1546 (2008)
33. E.G. Moore, J. Xu, C.J. Jocher, I. Castro-Rodriguez, K.N. Raymond, Inorg. Chem. **47**, 3105 (2008)
34. O. Moudam, B.C. Rowan, M. Alamiry, P. Richardson, B.S. Richards, A.C. Jones, N. Robertson, Chem. Commun. 6649 (2009)
35. M. Osawa, M. Hoshino, T. Wada, F. Hayashi, S. Osanai, J. Phys. Chem. A **113**, 10895 (2009)
36. D.P. Li, C.H. Li, J. Wang, L.C. Kang, T. Wu, Y.Z. Li, X.Z. You, Eur. J. Inorg. Chem. **2009**, 4844 (2009)
37. N.M. Shavaleev, S.V. Eliseeva, R. Scopelliti, J.-C.G. Bünzli, Chem. Eur. J. **15**, 10790 (2009)
38. G.E. Kiefer, M. Woods, Inorg. Chem. **48**, 11767 (2009)
39. K.A. White, D.A. Chengelis, K.A. Gogick, J. Stehman, N.L. Rosi, S. Petoud, J. Am. Chem. Soc. **131**, 18069 (2009)
40. D.B. A. Raj, S. Biju, M.L.P. Reddy, Dalton Trans. 7519 (2009)
41. B. McMahon, P. Mauer, C.P. McCoy, T.C. Lee, T. Gunnlaugsson, J. Am. Chem. Soc. **131**, 17542 (2009)
42. M.-H. Ha-Thi, J.A. Delaire, V. Michelet, I. Leray, J. Phys. Chem. A **114**, 3264 (2010)
43. J.-C.G. Bünzli, Chem. Rev. **110**, 2729 (2010)
44. M. Tropiano, N.L. Kilah, M. Morten, H.R.J.J. Davis, P.D. Beer, S. Faulkner, J. Am. Chem. Soc. **133**, 11847 (2011)
45. D. Sykes, M.D. Ward, Chem. Commun. **47**, 2279 (2011)
46. M. Sturzbecher-Hoehne, C.N.P. Leung, A. D'Aléo, B. Kullgren, A.-L. Prigent, D.K. Shuh, K.N. Raymond, R.J. Abergel, Dalton Trans. **40**, 8340 (2011)
47. C.M. Andolina, J.R. Morrow, Eur. J. Inorg. Chem. **2011**, 154 (2011)
48. J. Andres, A.-S. Chauvin, Inorg. Chem. **50**, 10082 (2011)
49. Q.-B. Bo, H.-Y. Wang, D.-Q. Wang, Z.-W. Zhang, J.-L. Miao, G.-X. Sun, Inorg. Chem. **50**, 10163 (2011)
50. S.V. Eliseeva, D.N. Pleshkov, K.A. Lyssenko, L.S. Lepnev, J.-C.G. Bünzli, N.P. Kuzmina, Inorg. Chem. **50**, 5137 (2011)
51. J. An, C.M. Shade, D.A. Chengelis-Czegan, S. Petoud, N.L. Rosi, J. Am. Chem. Soc. **133**, 1220 (2011)
52. D.J. Lewis, P.B. Glover, M.C. Solomons, Z. Pikramenou, J. Am. Chem. Soc. **2011**, 133 (1033)
53. M. Varlan, B.A. Blight, S. Wang, Chem. Commun. **48**, 12059 (2012)
54. D.G. Smith, B.K. McMahon, R. Pal, D. Parker, Chem. Commun. **48**, 8520 (2012)
55. R.M. Edkins, D. Sykes, A. Beeby, M.D. Ward, Chem. Commun. **48**, 9977 (2012)
56. A. Nonat, M. Regueiro-Figueroa, D. Esteban-Gomez, A. de Blas, T. Rodríguez-Blas, C. Platas-Iglesias, L.J. Charbonnière, Chem. Eur. J. **18**, 8163 (2012)
57. S. Nadella, P.M. Selvakumar, E. Suresh, P.S. Subramanian, M. Albrecht, M. Giese, R. Frohlich, Chem. Eur. J. **18**, 16784 (2012)
58. D.G. Smith, R. Pal, D. Parker, Chem. Eur. J. **18**, 11604 (2012)
59. E.S. Andreiadis, D. Imbert, J. Pécaut, R. Demadrille, M. Mazzanti, Dalton Trans. **41**, 1268 (2012)
60. D.J. Lewis, F. Moretta, A.T. Holloway, Z. Pikramenou, Dalton Trans. **41**, 13138 (2012)
61. Y.-A. Li, S.-K. Ren, Q.-K. Liu, J.-P. Ma, X. Chen, H. Zhu, Y.-B. Dong, Inorg. Chem. **51**, 9629 (2012)
62. S. Mohapatra, S. Adhikari, H. Riju, T.K. Maji, Inorg. Chem. **51**, 4891 (2012)

63. A.R. Ramya, D. Sharma, S. Natarajan, M.L.P. Reddy, Inorg. Chem. **51**, 8818 (2012)
64. A. de Bettencourt-Dias, P.S. Barber, S. Bauer, J. Am. Chem. Soc. **134**, 6987 (2012)
65. M.O. Rodrigues, J.D.L. Dutra, L.A.O. Nunes, G.F. de Sá, W.M. de Azevedo, P. Silva, F.A.A. Paz, R.O. Freire, S.A. Júnior, J. Phys. Chem. C **116**, 19951 (2012)
66. K.-N.T. Hua, J. Xu, E.E. Quiroz, S. Lopez, A.J. Ingram, V.A. Johnson, A.R. Tisch, A. de Bettencourt-Dias, D.A. Straus, G. Muller, Inorg. Chem. **51**, 647 (2012)
67. A. Ablet, S.-M. Li, W. Cao, X.-J. Zheng, W.-T. Wong, L.-P. Jin, Chem. Asian J. **8**, 95 (2013)
68. J. Xu, L. Jia, N. Jin, Y. Ma, X. Liu, W. Wu, W. Liu, Y. Tang, F. Zhou, Chem. Eur. J. **19**, 4556 (2013)
69. N. Wartenberg, O. Raccurt, E. Bourgeat-Lami, D. Imbert, M. Mazzanti, Chem. Eur. J. **19**, 3477 (2013)
70. Y. Hasegawa, K. Murakoshi, Y. Wada, S. Yanagida, J. Kim, N. Nakashima, T. Yamanaka, Chem. Phys. Lett. **248**, 8 (1996)
71. V.S. Sastri, J.-C.G. Bünzli, V.R. Rao, G.V.S. Rayudu, J.R. Perumareddi, *In Modern Aspects of Rare Earth and Their Complexes* (Elsevier, New York, 2003)
72. E.M. Stephens, M.F. Reid, F.S. Richardson, Inorg. Chem. **23**, 4611 (1984)
73. M.T. Devlin, E.M. Stephens, M.F. Reid, F.S. Richardson, Inorg. Chem. **26**, 1208 (1987)
74. S.F. Mason, R.D. Peacock, B. Stewart, Chem. Phys. Lett. **29**, 149 (1974)
75. S.F. Mason, J. Indian Chem. Soc. **63**, 73 (1986)
76. A.F. Kirby, F.S. Richardson, J. Phys. Chem. **87**, 2544 (1983)
77. M. Montalti, L. Prodi, N. Zaccheroni, L. Charbonnière, L. Douce, R. Ziessel, J. Am. Chem. Soc. **123**, 12694 (2001)
78. K. Driesen, P. Lenaerts, K. Binnemans, C. Görller-Walrand, Phys. Chem. Chem. Phys. **4**, 552 (2002)
79. W. Liu, T. Jiao, Y. Li, Q. Liu, M. Tan, H. Wang, L. Wang, J. Am. Chem. Soc. **126**, 2280 (2004)
80. J.P. Cross, M. Lauz, P.D. Badger, S. Petoud, J. Am. Chem. Soc. **126**, 16278 (2004)
81. P. Nockemann, B. Thijs, N. Postelmans, K.V. Hecke, L.V. Meervelt, K. Binnemans, J. Am. Chem. Soc. **128**, 13658 (2006)
82. A. Wada, M. Watanabe, Y. Yamanoi, T. Nankawa, K. Namiki, M. Yamasaki, M. Murata, H. Nishihara, Bull. Chem. Soc. Jpn **80**, 335 (2007)
83. Y. Hasegawa, M. Yamamuro, Y. Wada, N. Kanehisa, Y. Kai, S. Yanagida, J. Phys. Chem. A **107**, 1697 (2003)
84. T. Jüstel, H. Nikol, C. Ronda, Angew. Chem. Int. Ed. **37**, 3084 (1998)
85. J. Kido, Y. Okamoto, Chem. Rev. **102**, 2357 (2002)
86. J. Yu, L. Zhou, H. Zhang, Y. Zheng, H. Li, R. Deng, Z. Peng, Z. Li, Inorg. Chem. **44**, 1611 (2005)
87. E.S. Wilks, *Industrial Polymer Handbook* (Wiley-VCH, Weinheim, 2000), p. 291
88. K. Manseki, Y. Hasegawa, Y. Wada, S. Yanagida, J. Lumin. **111**, 183 (2005)
89. H. Li, M. Eddaoudi, M. O'Keeffe, O.M. Yaghi, Nature **402**, 276 (1999)
90. B. Moulton, M.J. Zaworotko, Chem. Rev. **101**, 1629 (2001)
91. S.L. James, Chem. Soc. Rev. **32**, 276 (2003)
92. S. Kitagawa, R. Kitaura, S. Noro, Angew. Chem. Int. Ed. **43**, 2334 (2004)
93. I.G. Georgiev, L.R. MacGillivray, Chem. Soc. Rev. **36**, 1239 (2007)
94. R. Cao, D.F. Sun, Y.C. Liang, M.C. Hong, K. Tatsumi, Q. Shi, Inorg. Chem. **41**, 2087 (2002)
95. L. Pan, X.Y. Huang, J. Li, Y.G. Wu, N.W. Zheng, Angew. Chem. Int. Ed. **39**, 527 (2000)
96. L. Pan, K.M. Adams, H.E. Hernandez, X.T. Wang, C. Zheng, Y. Hattori, K. Kaneko, J. Am. Chem. Soc. **125**, 3062 (2003)
97. D. L. Long, A. J. Blake, N. R. Champness, M. Schroder, Chem. Commun. 1369 (2000)
98. O. Guillou, C. Daiguebonne, Lanthanide-containing coordination polymers. In *Handbook on the Physics and Chemistry of Rare Earths*, Vol. 34, Chapter 221, 359, ed by K.A. Gschneidner Jr., J.-C.G. Bünzli, V. Pescharsky (Elsevier, Amsterdam, 2004), p. 359

99. J. Rocha, L.D. Carlos, Curr. Opin. Solid State Mater. Sci. **7**, 199 (2003)
100. M.D. Allendorf, C.A. Bauer, R.K. Bhakta, R.J.T. Houk, Chem. Soc. Rev. **38**, 1330 (2009)
101. M. Eddaoudi, D.B. Moler, H.L. Li, B.L. Chen, T.M. Reineke, M. O'Keeffe, O.M. Yaghi, Acc. Chem. Res. **34**, 319 (2001)
102. A.K. Cheetham, C.N.R. Rao, R.K. Feller, Chem. Commun. 4780 (2006)
103. C. Daiguebonne, N. Kerbellec, K. Bernot, Y. Gerault, A. Deluzet, O. Guillou, Inorg. Chem. **45**, 5399 (2006)
104. X. Guo, G. Zhu, F. Sun, Z. Li, X. Zhao, X. Li, H. Wang, S. Qiu, Inorg. Chem. **45**, 2581 (2006)
105. L. Pan, N. Zheng, Y. Wu, S. Han, R. Yang, X. Huang, J. Li, Inorg. Chem. **40**, 828 (2001)
106. X.P. Yang, R.A. Jones, J.H. Rivers, R.P.J. Lai, Dalton Trans. 3936 (2007)
107. L.D. Carlos, R.A.S. Ferreira, V. de Zea, Bermudez, B. Julian-Lopez, P. Escribano. Chem. Soc. Rev. **40**, 536 (2011)
108. Y. Hasegawa, Y. Kimura, K. Murakoshi, Y. Wada, J. Kim, N. Nakashima, T. Yamanaka, S. Yanagida, J. Phys. Chem. **100**, 10201 (1996)
109. Y. Hasegawa, K. Murakoshi, Y. Wada, J. Kim, N. Nakashima, T. Yamanaka, S. Yanagida, Chem. Phys. Lett. **260**, 173 (1996)
110. G. Stain, E. Würzberg, J. Chem. Phys. **62**, 208 (1975)

Chapter 2
Luminescence Properties of Thermostable Lanthanide Coordination Polymers with Intermolecular Interactions

2.1 Introduction

Luminophores with high thermal stability are promising candidates as active materials for electroluminescent (EL) devices, lasers and luminescent biosensing applications. As described in Chap. 1, lanthanide polynuclear complexes, coordination polymers, and metal-organic frameworks have been widely studied from the viewpoint of thermostable structure [1–7]. For example, Marchetti has reported a thermostable coordination polymer with Eu(III) ions and 4-acyl-pyrazolone ligands [8]. Wang has also reported that the decomposition point of a lanthanide coordination polymer with glutaric acid is over 300 °C [9]. However, their emission quantum yields have been extremely low, because their coordination polymers are attributed to the nonradiative transition via vibrational relaxation of high-vibrational frequency O–H bonds in polymer structure [10, 11]. Luminescent coordination polymer with both high thermostability and emission quantum efficiency is required as a novel luminophore in the field of future opto-electronic devices.

Here, the author considers that the introduction of low-vibrational frequency (LVF) ligands as a linker part in the polymer chains would result in the preparation of a lanthanide coordination polymer with strong luminescence properties. Strongly luminescent lanthanide complexes composed of LVF hfa and bidentate phosphine oxide ligands have been described in former chapters. The author also proposes that the introduction of aromatic aryl groups in the linker part of lanthanide coordination polymer is effective for the construction of thermostable luminophores with intermolecular interactions, such as CH/F, π–π, and CH/π interactions.

In this chapter, the author reports on novel coordination polymers composed of Eu(III) ion and 4 types of aryl units; $[Eu(hfa)_3(dpb)]_n$, $[Eu(hfa)_3(dpbp)]_n$, $[Eu(hfa)_3(dpbt)]_n$, and $[Eu(hfa)_3(dppcz)]_n$ (dpb: 1,4-bis(diphenylphosphoryl) benzene, dpbp: 4,4′-bis(diphenylphosphoryl)biphenyl, dpbt: 4,4′-bis(diphenylphos phoryl)bithiophene, dppcz: 3,6-bis(diphenylphosphoryl)-9-phenylcarbazole) as shown in Fig. 2.1. Their thermal stabilities and luminescence properties are characterized by

K. Miyata, *Highly Luminescent Lanthanide Complexes with Specific Coordination Structures*, Springer Theses,
DOI: 10.1007/978-4-431-54944-4_2, © Springer Japan 2014

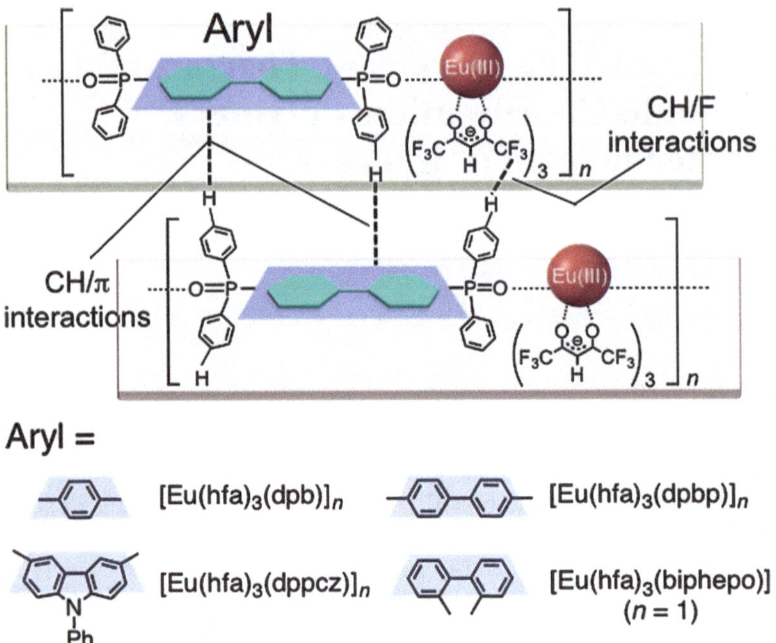

Fig. 2.1 Chemical structures of Eu(III) coordination polymers developed in this study

the thermogravimetric analyses (TGA and DSC), emission quantum yields, emission lifetimes, photosensitized energy transfer efficiency, and the nonradiative rate constants. In particular, $[Eu(hfa)_3(dppcz)]_n$ has both thermal stability (decomposition point =300 °C) and a high emission quantum yield (Φ_{Ln} = 83 % in the solid state). Novel luminophores with heat durability are expected to open up new fields of material chemistry.

2.2 Experimental Section

2.2.1 General

The starting materials, 1,4-difluorobenzene (TCI, >95 %), 4,4′-dibromobiphenyl (TCI, >98 %), 2,2′-bithiophene (TCI, >98 %), 3,6-dibromo-9-phenylcarbazole (TCI, >98 %) were used as received. All other chemicals were reagent grade and were used without further purification. All reactions involving air- and moisture-sensitive reagents were carried out under an argon atmosphere using dried solvents. Synthetic routes of organic ligands are shown in Scheme 2.1. The preparation method of $[Eu(hfa)_3(H_2O)_2]$ was described in Chap 5. Infrared spectra were recorded on a JASCO FT/IR–350 spectrometer. ^{1}H NMR spectra were recorded on a

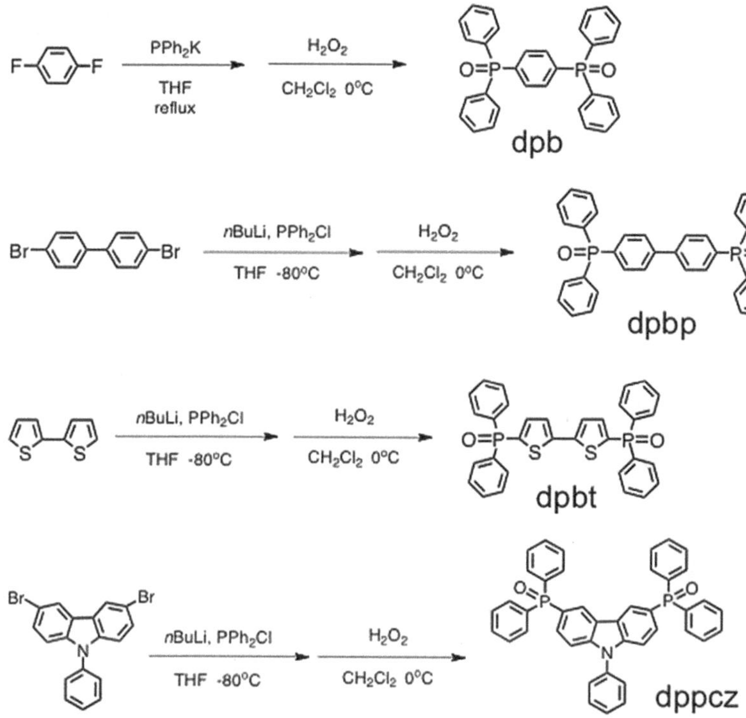

Scheme 2.1 Synthetic routes of organic ligands of Eu(III) coordination polymers

JEOL JNM-EX270 (270 MHz). ^{1}H NMR chemical shifts were determined by tetramethylsilane (TMS) as an internal standard. Elemental analysis and mass spectrometry were performed at the Instrumental Analysis Division in Hokkaido University. Thermogravimetric analysis was performed on a Rigaku TermoEvo TG8120 analyzer in an argon atmosphere at a heating rate of 1 °C min^{-1}. DSC measurement was performed on a MAC DSC3220 at a heating rate of 1 °C min^{-1}.

2.2.2 Syntheses

2.2.2.1 Preparation of 1,4-bis(diphenylphosphoryl)benzene (dpb)

1,4-bis(diphenylphosphoryl)benzene was synthesized according to the published procedure [12]. A solution of 1,4-difluorobenzene (0.80 mL, 8.0 mmol) was added dropwise to a solution of potassium diphenylphosphide (40 mL, 0.5 M THF, 20 mmol). The mixture was allowed to stir for 1 h and then was brought to reflux for 12 h. THF was removed under reduced pressure, and methanol (*ca.* 40 mL) was added. The mixture was then heated to reflux for 30 min. A gray precipitate was formed, after which the methanol was decanted off. The obtained gray solid and dichloromethane (20 mL) were placed in a flask. The solution was cooled to 0 °C and then 30 % H_2O_2 aqueous solution (5 mL) was added to it. The reaction mixture was stirred for 2 h. The product was extracted with dichloromethane, the extracts washed with brine for three times and dried over anhydrous $MgSO_4$. The solvent was evaporated to afford a white powder. Recrystallization from dichloromethane for gave white crystals of the titled compound.

Yield: 2.5 g (66 %). ^{1}H NMR (270 MHz, $CDCl_3$, 25 °C) δ 7.48–7.78 (m, 24H; P-C_6H_5, C_6H_4) ppm. ESI–Mass (*m/z*) = 479.1 $[M+H]^+$. Anal. Calcd for $C_{30}H_{24}O_2P_2$: C, 75.31; H, 5.06. Found: C, 74.86; H, 5.11.

2.2.2.2 Preparation of 4,4′-bis(diphenylphosphoryl)biphenyl (dpbp)

4,4′-bis(diphenylphosphoryl)biphenyl was synthesized according to the published procedure [13–15]. A solution of *n*-BuLi (9.3 mL, 1.6 M hexane, 15 mmol), was added dropwise to a solution of 4,4′-dibromobiphenyl (1.9 g, 6.0 mmol) in dry THF (30 mL) at –80 °C. The addition was completed in *ca.* 15 min during which time a yellow precipitate was formed. The mixture was allowed to stir for 3 h at –10 °C, after which a PPh_2Cl (2.7 mL, 15 mmol) was added dropwise at –80 °C. The mixture was gradually brought to room temperature, and stirred for 14 h. The product was extracted with ethyl acetate, the extracts washed with brine for three times and dried over anhydrous $MgSO_4$. The solvent was evaporated, and resulting residue was washed with acetone and ethanol for several times. The obtained white solid and dichloromethane (40 mL) were placed in a flask. The solution was cooled to 0 °C and then 30 % H_2O_2 aqueous solution (5 mL) was added to it. The reaction mixture was stirred for 2 h. The product was extracted with dichloromethane, the extracts washed with brine for three times and dried over anhydrous $MgSO_4$. The solvent was evaporated to afford a white powder. Recrystallization from dichloromethane gave white crystals of the titled compound.

Yield: 1.7 g (54 %). IR (KBr): 1120 (st, P=O) cm^{-1}. ^{1}H NMR (270 MHz, $CDCl_3$, 25 °C) δ 7.67–7.80 (m, 16H; P-C_6H_5, C_6H_4), 7.45–7.60 (m, 12H; P-C_6H_5, C_6H_4) ppm. ESI–Mass (*m/z*) = 555.2 $[M+H]^+$. Anal. Calcd for $C_{36}H_{28}O_2P_2$: C, 77.97; H, 5.09. Found: C, 77.49; H, 5.20.

2.2.2.3 Preparation of 4,4′-bis(diphenylphosphoryl)bithiophene (dpbt)

A solution of *n*-BuLi (13 mL, 1.6 M hexane, 20 mmol), was added dropwise to a solution of 2,2′-bithiophene (1.2 g, 7.2 mmol) in dry THF (20 mL) at –80 °C. The addition was completed in *ca.* 15 min during which time a yellow precipitate was formed. The mixture was allowed to stir for 3 h at –10 °C, after which a PPh_2Cl (3.7 mL, 20 mmol) was added dropwise at –80 °C. The mixture was gradually brought to room temperature, and stirred for 18 h. The product was extracted with ethyl acetate, the extracts washed with brine for three times and dried over anhydrous $MgSO_4$. The solvent was evaporated, and resulting residue was washed with methanol for several times. The obtained yellow solid and dichloromethane (40 mL) were placed in a flask. The solution was cooled to 0 °C and then 30 % H_2O_2 aqueous solution (10 mL) was added to it. The reaction mixture was stirred for 2 h. The product was extracted with dichloromethane, the extracts washed with brine for three times and dried over anhydrous $MgSO_4$. The solvent was evaporated to afford a yellow powder. Recrystallization from dichloromethane gave yellow crystals of the titled compound.

Yield: 1.4 g (31 %). IR (KBr): 1122 (st, P=O) cm^{-1}. ^1H NMR (270 MHz, $CDCl_3$, 25 °C) δ 7.45–7.79 (m, 20H; P-C_6H_5), 7.33–7.37 (m, 2H; C_4H_2S), 7.24–7.27 (m, 2H; C_4H_2S) ppm. ESI–Mass (m/z) = 567.1 $[M+H]^+$. Anal. Calcd for $C_{32}H_{24}O_2P_2S_2$: C, 67.83; H, 4.27. Found: C, 67.13; H, 4.40.

2.2.2.4 Preparation of 3,6-bis(diphenylphosphoryl)-9-phenylcarbazole (dppcz)

A solution of n-BuLi (8.8 mL, 1.6 M hexane, 14 mmol), was added dropwise to a solution of 3,6-dibromo-9-phenylcarbazole (2.4 g, 6.0 mmol) in dry THF (30 mL) at –80 °C. The addition was completed in *ca.* 10 min during which time a white yellow precipitate was formed. The mixture was allowed to stir for 2 h at –10 °C, after which a PPh_2Cl (2.6 mL, 14 mmol) was added dropwise at –80 °C. The mixture was gradually brought to room temperature, and stirred for 18 h to give a white precipitate. The precipitate was filtered, washed with methanol for several times, and dried in vacuo. The obtained white powder and dichloromethane (40 mL) were placed in a flask. The solution was cooled to 0 °C and then 30 % H_2O_2 aqueous solution (8 mL) was added to it. The reaction mixture was stirred for 2 h. The product was extracted with dichloromethane, the extracts washed with brine for three times and dried over anhydrous $MgSO_4$. The solvent was evaporated to afford a white powder. Recrystallization from dichloromethane/hexane gave colorless crystals of the titled compound. Yield: 2.0 g (53 %). IR (KBr): 1122 (st, P=O) cm^{-1}. ^1H NMR (270 MHz, $CDCl_3$, 25 °C) δ 8.43–8.47 (d, J = 10.8 Hz, 2H; P-C_6H_5), 7.63–7.76 (m, 11H; C_4H_2S), 7.43–7.60 (m, 18H; C_4H_2S) ppm. ESI–Mass (m/z) = 644.2 $[M+H]^+$. Anal. Calcd for $C_{43}H_{31}NO_2P_2$: C, 78.37; H, 4.85; N, 2.18. Found: C, 78.42; H, 5.00; N, 2.18.

2.2.2.5 General Procedure for the Preparation of Eu(III) Coordination Polymers

Phosphine oxide ligand (1 equiv.) and $[Eu(hfa)_3(H_2O)_2]$ (1 equiv.) were dissolved in chloroform–methanol. The solution was refluxed while stirring for 8 h, and the reaction mixture was concentrated to dryness. A single crystal suitable for X-ray structural determination of Eu(III) coordination polymer was obtained by the diffusion method of methanol–chloroform solution at room temperature.

[Eu(hfa)$_3$(dpb)]$_n$. Yield: 65 mg (42 %; for monomer). IR (KBr): 1652 (st, C=O), 1256–1145 (st, C–F), 1128 (st, P=O) cm^{-1}. ESI–Mass (*m/z*) = 1045.05 [Eu(hfa)$_2$(dpb)]$^+$, 2297.18 [Eu$_2$(hfa)$_5$(dpb)$_2$]$^+$. Anal. Calcd for [C$_{45}$H$_{27}$Eu-F$_{18}$O$_8$P$_2$]$_n$: C, 43.18; H, 2.17. Found: C, 43.12; H 2.28.

[Eu(hfa)$_3$(dpbp)]$_n$. Yield: 98 mg (67 %; for monomer). IR (KBr): 1653 (st, C=O), 1255–1145 (st, C–F), 1127 (st, P=O) cm^{-1}. ESI–Mass (*m/z*) = 1120.08 [Eu(hfa)$_2$(dpbp)]$^+$, 2447.15 [Eu$_2$(hfa)$_5$(dpbp)$_2$]$^+$. Anal. Calcd for [C$_{51}$H$_{31}$Eu-F$_{18}$O$_8$P$_2$]$_n$: C, 46.14; H, 2.35. Found: C, 45.59; H 2.49.

[Eu(hfa)$_3$(dpbt)]$_n$. Yield: 160 mg (68 %; for monomer). IR (KBr): 1651 (st, C=O), 1254–1145 (st, C–F), 1128 (st, P=O) cm^{-1}. ESI–Mass (*m/z*) = 1133.00 [Eu(hfa)$_2$(dpbt)]$^+$, 2473.02 [Eu$_2$(hfa)$_5$(dpbt)$_2$]$^+$. Anal. Calcd for [C$_{47}$H$_{27}$Eu-F$_{18}$O$_8$P$_2$S$_2$]$_n$: C, 42.14; H, 2.03. Found: C, 42.67; H 2.12.

[Eu(hfa)$_3$(dppcz)]$_n$. Yield: 110 mg (50 %; for monomer). IR (KBr): 1652 (st, C=O), 1256–1145 (st, C–F), 1128 (st, P=O) cm^{-1}. ESI–Mass (*m/z*) = 1210.13, [Eu(hfa)$_2$(dppcz)]$^+$, 1853.34, [Eu(hfa)$_2$(dppcz)$_2$]$^+$. Anal. Calcd for [C$_{57}$H$_{34}$EuF$_{18}$N O$_8$P$_2$]$_n$: C, 48.32; H, 2.42; N, 0.99. Found: C, 47.95; H, 2.77; N, 1.06.

2.2.3 Crystallography

Colorless single crystals of Eu(III) coordination polymers were mounted on a glass fiber using paraffin oil. All measurements were made on a Rigaku Mercury CCD area detector with graphite monochromated Mo-Kα radiation. All structures were solved by direct methods (SIR 2004) [16] and expanded using Fourier techniques. The non-hydrogen atoms were refined anisotropically. Hydrogen atoms were refined using the riding model. All calculations were performed using the Crystal Structure crystallographic software package except for refinement, which was performed using SHELXL-97 [17]. The author confirmed the CIF data using the check CIF/PLATON service.

2.2.4 Optical Measurements

Emission spectra of Eu(III) coordination polymers were measured on a JASCO F-6300-H spectrometer and corrected for the response of the detector system. Emission lifetimes (τ_{obs}) were measured using the third harmonics (355 nm) of a Q-switched Nd:YAG laser (Spectra Physics, INDI-50, fwhm = 5 ns, λ = 1064 nm) and a photomultiplier (Hamamatsu photonics, R5108, response time $\leq$1.1 ns). The Nd:YAG laser response was monitored with a digital oscilloscope (Sony Tektronix, TDS3052, 500 MHz) synchronized to the single-pulse excitation. Emission lifetimes were determined from the slope of logarithmic plots of the decay profiles. The emission quantum yields excited at 380 nm (Φ_{tot}) were estimated using JASCO F-6300-H spectrometer attached with JASCO ILF-533 integrating sphere unit

Fig. 2.2 JASCO ILF-533 integrating sphere unit

(ϕ = 100 mm, Fig. 2.2). The wavelength dependences of the detector response and the beam intensity of Xe light source for each spectrum were calibrated using a standard light source.

The photophysical properties of the Eu(III) coordination polymers in the solid state were investigated from estimation of the 4f–4f emission quantum yields (Φ_{Ln}), and the radiative (k_r) and nonradiative (k_{nr}) rate constants from the radiative (τ_{rad}) and observed lifetimes (τ_{obs}). The radiative lifetime is defined as an ideal emission lifetime without nonradiative processes. The radiative and observed lifetimes are expressed by

$$\tau_{rad} = \frac{1}{k_r} \tag{2.1}$$

$$\tau_{obs} = \frac{1}{k_r + k_{nr}} \tag{2.2}$$

Φ_{Ln}, k_r, and k_{nr} of the Eu(III) compounds are given by

$$\Phi_{Ln} = \frac{k_r}{k_r + k_{nr}} = \frac{\tau_{obs}}{\tau_{rad}} \tag{2.3}$$

$$\frac{1}{\tau_{rad}} = A_{MD,0} n^3 \left(\frac{I_{tot}}{I_{MD}} \right) \tag{2.4}$$

$$k_r = \frac{1}{\tau_{rad}} \tag{2.5}$$

$$k_{nr} = \frac{1}{\tau_{obs}} - \frac{1}{\tau_{rad}} \tag{2.6}$$

where $A_{MD,0}$ is the spontaneous emission probability for the $^5D_0$–$^7F_1$ transition in vacuo (14.65 s^{-1}), n is the refractive index of the medium (an average index of refraction equal to 1.5 was employed [18]), and (I_{tot}/I_{MD}) is the ratio of the total area of the corrected Eu(III) emission spectrum to the area of the $^5D_0$–$^7F_1$ band.

2.3 Results and Discussion

2.3.1 Coordination and Network Structures

Single crystals of Eu(III) coordination polymers with phosphine oxides, $[Eu(hfa)_3(dpb)]_n$, $[Eu(hfa)_3(dpbp)]_n$, $[Eu(hfa)_3(dpbt)]_n$, and $[Eu(hfa)_3(dppcz)]_n$ were successfully prepared for X-ray single crystal analyses by the diffusion method using methanol–chloroform solutions. The resulting crystal data are summarized in Fig. 2.3 and Table 2.1. The ORTEP views show that the phosphine oxide ligand acts as a bidentate bridge between lanthanide ions in one-dimensional polymeric chains. The coordination sites of $[Eu(hfa)_3(dpb)]_n$, $[Eu(hfa)_3(dpbp)]_n$, $[Eu(hfa)_3(dpbt)]_n$, and $[Eu(hfa)_3(dppcz)]_n$ comprise three hfa ligands and two phosphine oxide units. Their structures are categorized as an octa-coordinated square antiprism with no inverted center in the crystal field, so that enhancement of their luminescence properties is expected with allowed electronic transition probabilities [19, 20].

Selected bond lengths between Eu(III) ions and oxygen atoms in coordination sites (Eu–O and Eu–Eu distances) are summarized in Table 2.2. The Eu–Eu distances of $[Eu(hfa)_3(dpb)]_n$, $[Eu(hfa)_3(dpbp)]_n$, $[Eu(hfa)_3(dpbt)]_n$, and $[Eu(hfa)_3(dppcz)]_n$ were determined 9.91, 10.36, 10.60, and 11.76 Å, respectively. These distances may be longer than the critical distance for nonradiative dipole-dipole energy transfer between Eu(III) ions. Average Eu–O distances between Eu(III) ions and oxygen atoms of phosphine oxide ligands in $[Eu(hfa)_3(dpb)]_n$, $[Eu(hfa)_3(dpbp)]_n$, $[Eu(hfa)_3(dpbt)]_n$ and $[Eu(hfa)_3(dppcz)]_n$ were estimated to be 2.31, 2.32, 2.31, and 2.30 Å, respectively, which are shorter than the Eu–O distances of hfa ligands in $[Eu(hfa)_3(dpb)]_n$ (2.41 Å), $[Eu(hfa)_3(dpbp)]_n$ (2.42 Å), $[Eu(hfa)_3(dpbt)]_n$ (2.41 Å), and $[Eu(hfa)_3(dppcz)]_n$ (2.42 Å). These results indicate that the coordination ability of phosphine oxide is stronger than that of the hfa ligands. The strong coordination ability of phosphine oxide in Eu(III) complexes has been reported. The tight-coordination of phosphine oxide molecules with low vibrational frequency (P=O: 1120 cm^{-1}) prevents coordination of high-vibrational molecules such as water, which results in enhancement of their luminescence properties.

X-ray analyses also reveal intermolecular interactions between one-dimensional polymeric chains (Fig. 2.4). Two CH/F interactions in one unit were identified for $[Eu(hfa)_3(dpb)]_n$, $[Eu(hfa)_3(dpbp)]_n$, $[Eu(hfa)_3(dpbt)]_n$, and $[Eu(hfa)_3(dppcz)]_n$. Remarkable CH/π interactions in the polymer chains of $[Eu(hfa)_3(dpbp)]_n$,

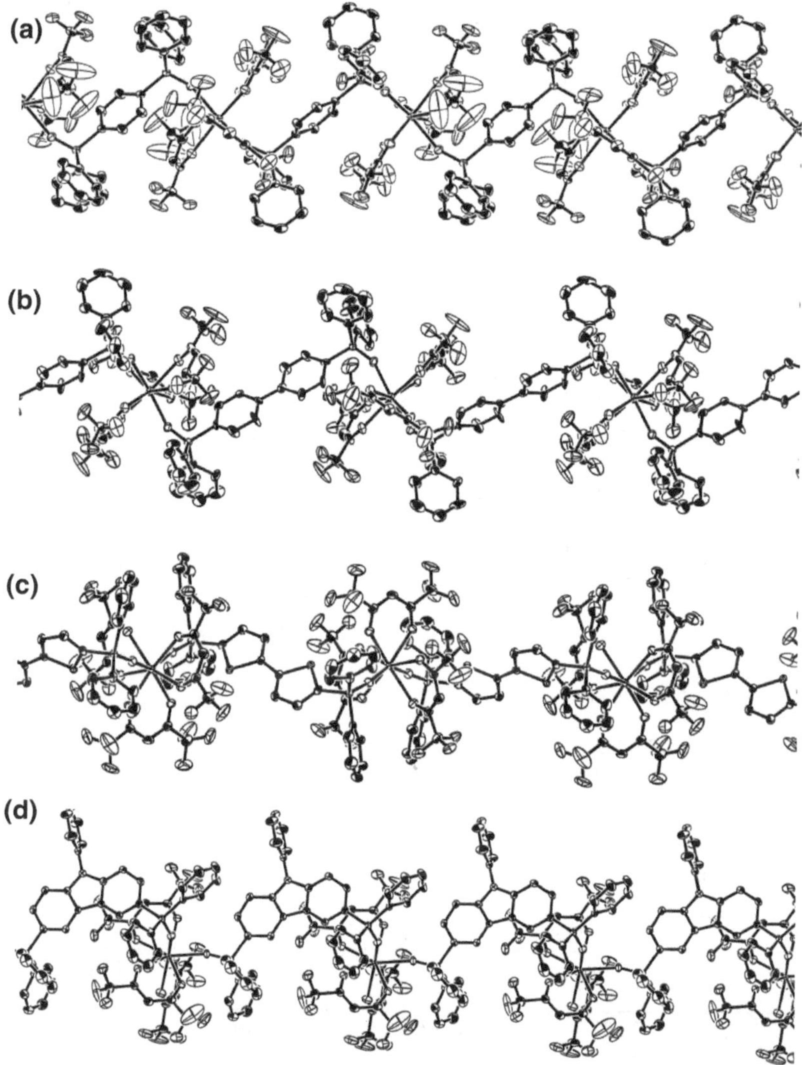

Fig. 2.3 ORTEP drawings (showing 50 % probability displacement ellipsoids) of **a** $[Eu(hfa)_3(dpb)]_n$, **b** $[Eu(hfa)_3(dpbp)]_n$, **c** $[Eu(hfa)_3(dpbt)]_n$, and **d** $[Eu(hfa)_3(dppcz)]_n$

$[Eu(hfa)_3(dpbt)]_n$, and $[Eu(hfa)_3(dppcz)]_n$ were observed. The author considers that tight-binding structures composed of Eu(III) ions and phosphine oxide result in the excellent thermogravimetric and emission properties.

Table 2.1 Crystal data of Eu(III) coordination polymers

	$[Eu(hfa)_3(dpb)]_n$	$[Eu(hfa)_3(dpbp)]_n$	$[Eu(hfa)_3(dpbt)]_n$	$[Eu(hfa)_3(dppcz)]_n$
Chemical formula	$C_{45}H_{27}F_{18}O_8P_2$ Eu	$C_{52}H_{31}F_{18}O_8P_2$ Eu	$C_{47}H_{27}F_{18}O_8P_2S_2$ Eu	$C_{57}H_{34}F_{18}NO_8P_2$ Eu
Formula weight	1251.58	1339.69	1339.72	1416.78
Crystal color, habit	Colorless, prism	Colorless, prism	Colorless, prism	Colorless, prism
Crystal system	Triclinic	Monoclinic	Monoclinic	Monoclinic
Space group	*P*-1 (#2)	*C*2/*c* (#15)	*C*2/*c* (#15)	$P2_1/n$ (#14)
a/Å	12.610(3)	23.477(4)	22.786(2)	12.3580(14)
b/Å	13.622(3)	13.367(2)	13.6381(8)	21.170(2)
c/Å	15.282(3)	17.168(3)	17.0846(11)	22.482(3)
α/deg	71.928(8)			
β/deg	79.279(9)	97.5749(8)	103.0178(15)	99.6974(5)
γ/deg	81.346(9)			
$V/Å^3$	2439.9(9)	5340.5(16)	5172.8(7)	5797.6(11)
Z	2	4	4	4
d_{calc}/g cm^{-3}	1.703	1.666	1.720	1.623
T/°C	-123 ± 1	-123 ± 1	-123 ± 1	-123 ± 1
μ (Mo Kα)/cm^{-1}	14.676	13.472	14.685	12.465
Max 2θ/deg	55.0	55.0	55.0	55.0
No. of measured reflections	19890	21091	20459	45925
No. of unique reflections	10984	6099	5928	13215
R ($I > 2\sigma$(I))[a]	0.0335	0.0266	0.0258	0.0319
R_w ($I > 2\sigma$(I))[b]	0.0949	0.0668	0.0689	0.0796

[a] $R = \Sigma \, ||F_o| - |F_c|| / \Sigma \, |F_o|$
[b] $R_w = [(\Sigma \, w \, (|F_o| - |F_c|)^2 / \Sigma \, w \, F_o^2)]^{1/2}$

Table 2.2 Selected bond lengths (Å) of Eu(III) coordination polymers

Bond	$[Eu(hfa)_3(dpb)]_n$	$[Eu(hfa)_3(dpbp)]_n$	$[Eu(hfa)_3(dpbt)]_n$	$[Eu(hfa)_3(dppcz)]_n$
Eu–O1 (P=O)	2.307	2.319	2.314	2.321
Eu–O2 (P=O)	2.312	2.319	2.314	2.284
Average bond length	2.310	2.319	2.314	2.303
Eu–O3 (C=O)	2.457	2.416	2.407	2.420
Eu–O4 (C=O)	2.362	2.423	2.413	2.404
Eu–O5 (C=O)	2.407	2.411	2.407	2.408
Eu–O6 (C=O)	2.409	2.411	2.413	2.446
Eu–O7 (C=O)	2.424	2.416	2.398	2.392
Eu–O8 (C=O)	2.399	2.423	2.398	2.445
Average bond length	2.410	2.417	2.406	2.419
Eu–Eu	9.910	10.367	10.602	11.755

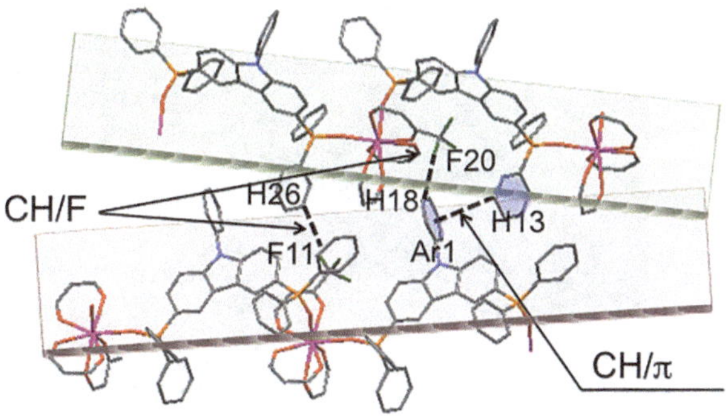

Fig. 2.4 Crystal structures of $[Eu(hfa)_3(dppcz)]_n$ focused on the intermolecular interactions between one-dimensional polymeric chains. *Dashed lines* denote CH/F and CH/π interactions. Two CH/F interactions ($d_{CH/F}$ = 2.62 Å (H18–F20), 2.74 Å (H26–F11)) and one CH/π interaction ($d_{CH/\pi}$ = 3.07 Å (H13–Ar1)) are displayed

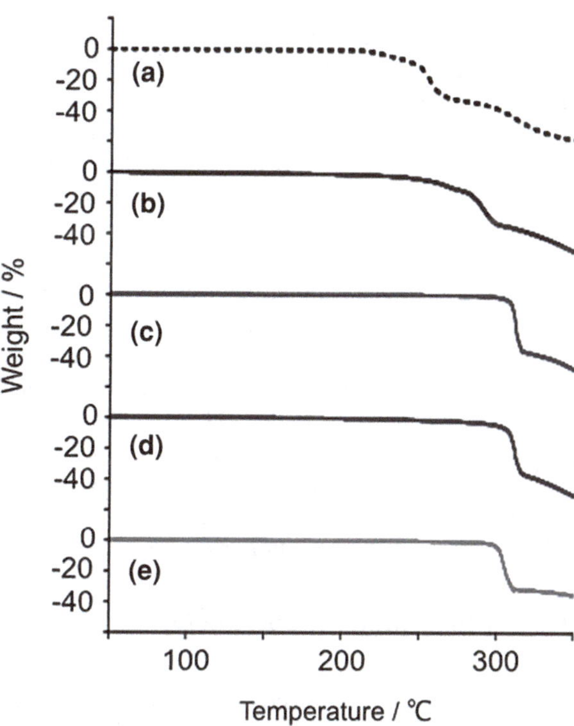

Fig. 2.5 TGA thermograms of **a** $[Eu(hfa)_3(biphepo)]$, **b** $[Eu(hfa)_3(dpb)]_n$, **c** $[Eu(hfa)_3(dpbp)]_n$, **d** $[Eu(hfa)_3(dpbt)]_n$, and **e** $[Eu(hfa)_3(dppcz)]_n$ under an argon atmosphere

2.3.2 Thermal Analyses

To estimate the thermal stability of Eu(III) coordination polymers, TGA and DSC were conducted. Decomposition points from the TGA thermograms are 261, 308, and 300 °C for $[Eu(hfa)_3(dpb)]_n$, $[Eu(hfa)_3(dpbp)]_n$, and $[Eu(hfa)_3(dppcz)]_n$, respectively (Fig. 2.5). In contrast, the decomposition point of the reported Eu(III) complex, $[Eu(hfa)_3(biphepo)]$ (biphepo: 2,2′-bis(diphenylphosphoryl)biphenyl),

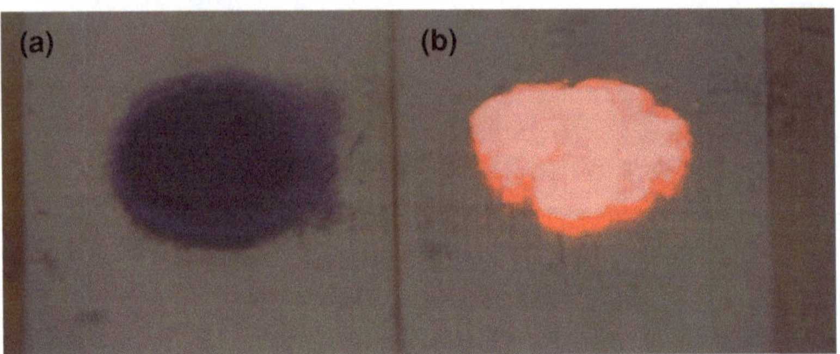

Fig. 2.6 Photographs of **a** [Eu(hfa)$_3$(biphepo)] and **b** [Eu(hfa)$_3$(dpbp)]$_n$ after heating at 300 °C for 10 min under UV irradiation

was found to be 230 °C. The high thermal stability of [Eu(hfa)$_3$(dpbp)]$_n$ and [Eu(hfa)$_3$(dppcz)]$_n$ might be due to their tight-binding structures supported by a combination of CH/F and CH/π interactions (Fig. 2.4). The binding energies of CH/F (hydrogen bond) and CH/π interactions are generally known to be 10–30 and 2–10 kJ mol^{-1}, respectively [21]. The author considers that a combination of both CH/F and CH/π interactions in coordination polymers is effective for the construction of thermostable luminophore compounds. As displayed in Fig. 2.6, [Eu(hfa)$_3$(dpbp)]$_n$ after heating at 300 °C for 10 min exhibits brilliant red luminescence under UV irradiation (excitation at 365 nm), whereas [Eu(hfa)$_3$(biphepo)] does not emit photons owing to thermal decomposition at 230 °C. The enhanced photostability of organic molecules containing hydrogen bonds has been reported [22]. In the author's system, high photostability of [Eu(hfa)$_3$(dpbp)]$_n$ has been also observed under UV irradiation. The DSC thermograms of [Eu(hfa)$_3$(dpb)]$_n$, [Eu(hfa)$_3$ (dpbp)]$_n$, and [Eu(hfa)$_3$(dppcz)]$_n$ show that exothermic peaks based on the decomposition points and endothermic peaks corresponding to melting points were observed (Fig. 2.7). Thus, the Eu(III) coordination polymers exhibit thermal behavior similar to polymer compounds.

2.3.3 Photophysical Properties

Steady-state emission spectra of the Eu(III) coordination polymers in the solid state are shown in Fig. 2.8 (right). Emission bands were observed at around 578, 591, 613, 650, and 698 nm, and are attributed to the f–f transitions of $^5D_0$–^7F_J with $J = 0, 1, 2, 3$, and 4, respectively. The emission band at 613 nm ($^5D_0$–$^7F_2$) is due to electric dipole transitions, which is strongly dependent on the coordination geometry. Their time-resolved emission profiles revealed single-exponential decays with millisecond scale lifetimes. The observed emission lifetimes from $^5D_0$ excited level (τ_{obs}) were determined from the slopes of logarithmic decay profiles. The emission lifetimes of [Eu(hfa)$_3$(dpb)]$_n$, [Eu(hfa)$_3$(dpbp)]$_n$, and [Eu(hfa)$_3$ (dppcz)]$_n$ were determined to be 0.93, 0.85, and 0.93 ms, respectively (Fig. 2.9).

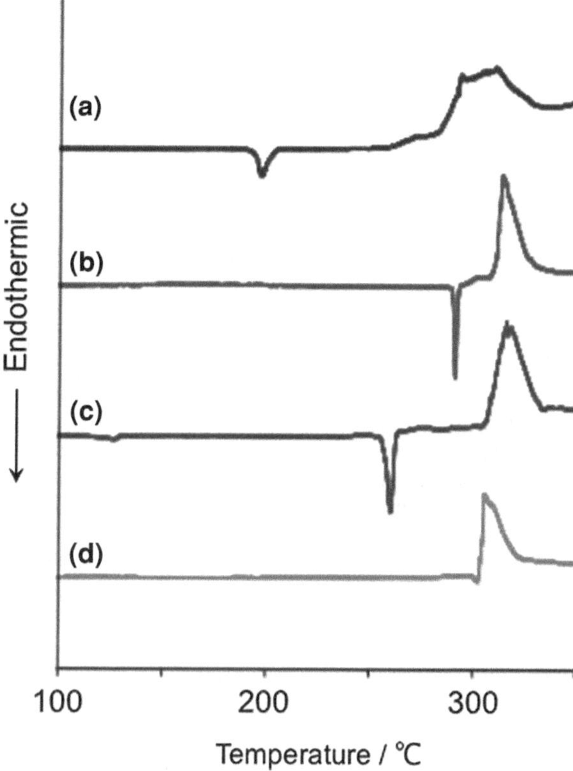

Fig. 2.7 DSC thermograms of **a** $[Eu(hfa)_3(dpb)]_n$, **b** $[Eu(hfa)_3(dpbp)]_n$, **c** $[Eu(hfa)_3(dpbt)]_n$, and **d** $[Eu(hfa)_3(dppcz)]_n$

The photophysical properties of the Eu(III) coordination polymers in the solid state are summarized in Table 2.3. The 4f–4f emission quantum yields (Φ_{Ln}) [18, 23, 24] for $[Eu(hfa)_3(dpb)]_n$, $[Eu(hfa)_3(dpbp)]_n$, and $[Eu(hfa)_3(dppcz)]_n$ were estimated to be 70, 72, and 83 %, respectively. These quantum yields are approximately four times larger than that for $[Eu(hfa)_3(H_2O)_2]$ [25, 26], and similar to that reported for $[Eu(hfa)_3(biphepo)]$ ($\Phi_{Ln} = 73$ % in the solid state), which have been reported as high luminescent Eu(III) complexes. According to the photosensitized energy transfer efficiency (η_{sens}) calculated from Φ_{Ln} (excited at 465 nm) and Φ_{tot} (excited at 380 nm), $[Eu(hfa)_3\ (dppcz)]_n$ are approximately five times larger than that for reported $[Eu(hfa)_3(H_2O)_2]$. Since absorption band at 380 nm is assigned to a π–π^* transition of hfa ligand (Fig. 2.8 left), these results suggest that the photosensitized energy transfer from hfa ligands to Eu(III) ions is enhanced in the coordination polymers, in particular, $[Eu(hfa)_3(dppcz)]_n$.

The nonradiative rate constants (k_{nr}) for the Eu(III) coordination polymers (1.8–$3.3 \times 10^2\ s^{-1}$) were approximately ten times smaller than that for $[Eu(hfa)_3(H_2O)_2]$ ($3.7 \times 10^3\ s^{-1}$). The smaller k_{nr} for the Eu(III) coordination polymers is attributed to the structural rigidity and low-vibrational frequencies in the crystal lattice. Our research group has previously reported that the nonradiative

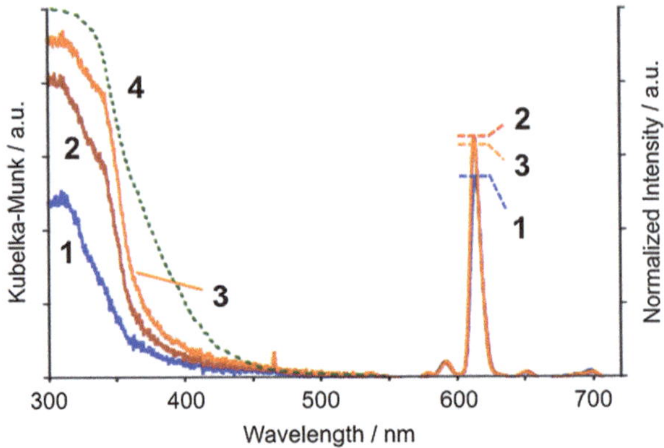

Fig. 2.8 Left: Diffuse-reflectance absorption spectra of $[Eu(hfa)_3(dpb)]_n$ (*line 1*), $[Eu(hfa)_3(dpbp)]_n$ (*line 2*), $[Eu(hfa)_3(dppcz)]_n$ (*line 3*), and $[Eu(hfa)_3(H_2O)_2]$ (*dot line 4*) in the solid state. The absorption bands at 310 nm are attributed to a π-π* transition of the hfa lilgands. The small bands at 465 nm are assigned to the $^7F_{0-5}D_2$ transition in the Eu(III) ion. Right: Emission spectra of $[Eu(hfa)_3(dpbp)]_n$, $[Eu(hfa)_3(dppcz)]_n$, and $[Eu(hfa)_3(dpb)]_n$ in the solid state. Excitation wavelength is 465 nm. The spectra were normalized with respect to the magnetic dipole transition ($^5D_0$–$^7F_1$)

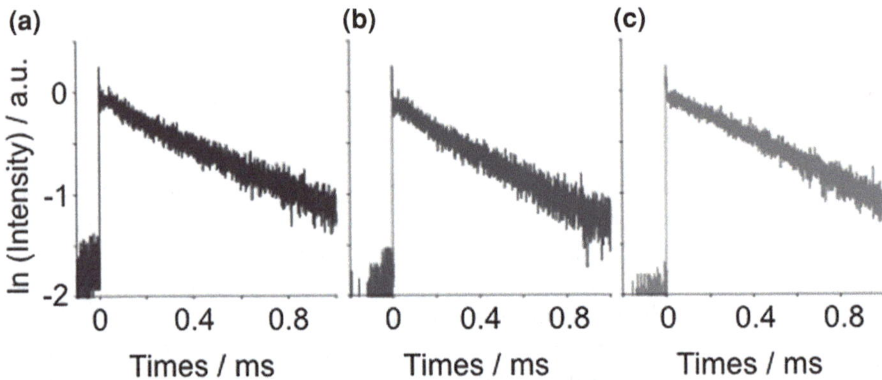

Fig. 2.9 The decay profiles of **a** $[Eu(hfa)_3(dpb)]_n$, **b** $[Eu(hfa)_3(dpbp)]_n$, and **c** $[Eu(hfa)_3(dppcz)]_n$ in the solid state

transitions of lanthanide complexes are caused by the high-vibrational frequency of O–H bonds, such as in water and alcohol. k_{nr} for [Eu(hfa)$_3$(biphepo)] is similar to those of $[Eu(hfa)_3(dpb)]_n$, $[Eu(hfa)_3(dpbp)]_n$, and $[Eu(hfa)_3(dppcz)]_n$. We consider that introduction of the LVF phosphine oxide ligand as a linker part in the polymer chains is effective for preparation of luminophores with high thermal stabilities and emission quantum yields.

Table 2.3 Photophysical and thermal properties of Eu(III) compounds in the solid state

Compound	interaction[a]	Decomp. point/°C[b]	Photophysical properties						
			τ_{obs} / ms[c]	τ_{rad} / ms[d]	Φ_{Ln} /%[d]	Φ_{tot} /%[e]	η_{sens} / %[f]	k_r /s^{-1} [d]	k_{nr} /s^{-1} [d]
$[Eu(hfa)_3(H_2O)_2]$	–	220[g]	0.22[h]	1.1[h]	19[h]	2.6[h]	13[h]	8.8×10^2	3.7×10^3
$[Eu(hfa)_3(biphepo)]$	–	230	0.94	1.3	73	21	29	7.8×10^2	2.8×10^2
$[Eu(hfa)_3(dpb)]_n$	CH/F	261	0.93	1.3	70	31	44	7.5×10^2	3.3×10^2
$[Eu(hfa)_3(dpbp)]_n$	CH/F, CH/π	308	0.85	1.2	72	29	40	8.5×10^2	3.2×10^2
$[Eu(hfa)_3(dppcz)]_n$	CH/F, CH/π	300	0.93	1.1	83	53	64	8.9×10^2	1.8×10^2

[a] From X-ray single crystal analyses (Fig. 2.4)
[b] From TGA thermograms (Fig. 2.5)
[c] Emission lifetime (τ_{obs}) of the Eu(III) coordination polymer were measured by excitation at 355 nm (Nd:YAG 3ω)
[d] Calculation methods are described in Experimental section
[e] Total emission quantum yield (excitation at 380 nm)
[f] Photosensitized energy transfer efficiency $\eta_{sens} = \Phi_{tot} / \Phi_{Ln}$
[g] From Ref. [18]
[h] From Ref. [25]

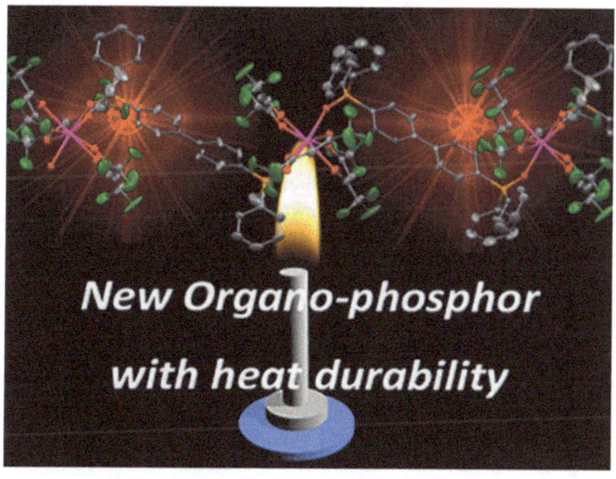

Fig. 2.10 The conceptual diagram in Chap. 2

2.3.4 Conclusions

The thermostable luminophores composed of Eu(III) coordination polymers were successfully synthesized (Fig. 2.10). In particular, $[Eu(hfa)_3(dppcz)]_n$ exhibits both high emission quantum yields (Φ_{Ln} = 83 %) and remarkable thermal stability (decomposition point =300 °C) due to a tight-binding structure composed of Eu(III) ions and low-vibrational phosphine oxide, although many types of luminescent organic dyes are generally decomposed at temperatures under 200 °C. The emission quantum yields of the author's coordination polymers are similar to those of strong-luminescent coordination polymers in former chapters. Reported new luminophores are expected to employ in optics applications such as luminescent plastics, displays, and opto-electronic devices.

References

1. A. Thirumurugan, S.K. Pati, M.A. Greenc, S. Natarajan, J. Mater. Chem. **13**, 2937 (2003)
2. C. Daiguebonne, N. Kerbellec, O. Guillou, J.-C.G. Bünzli, F. Gumy, L. Catala, T. Mallah, N. Audebrand, Y. Geraut, K. Bernot, G. Calvez, Inorg. Chem. **47**, 3700 (2008)
3. K. Binnemans, Chem. Rev. **109**, 4283 (2009)
4. C. Marchal, Y. Filinchuk, X.-Y. Chen, D. Imbert, M. Mazzanti, Chem. Eur. J. **15**, 5273 (2009)
5. T.K. Prasad, M.V. Rajasekharan, Inorg. Chem. **48**, 11543 (2009)
6. J. Rocha, L.D. Carlos, F.A. Almeida Paz, D. Ananias, Chem. Soc. Rev. **40**, 926 (2011)
7. S.V. Eliseeva, D.N. Pleshkov, K.A. Lyssenko, L.S. Lepnev, J.-C.G. Bünzli, N.P. Kuzmina, Inorg. Chem. **49**, 9300 (2010)
8. F. Marchetti, C. Pettinari, A. Pizzabiocca, A.A. Drozdov, S.I. Troyanov, C.O. Zhuravlev, S.N. Semenov, Y.A. Belousov, I.G. Timokhin, Inorg. Chim. Acta **363**, 4038 (2010)
9. C. Wang, Y. Xing, Z. Li, J. Li, X. Zeng, M. Ge, S. Niu, J. Mol. Struct. **931**, 76 (2009)
10. Y. Hasegawa, Y. Kimura, K. Murakoshi, Y. Wada, J. Kim, N. Nakashima, T. Yamanaka, S. Yanagida, J. Phys. Chem. **100**, 10201 (1996)
11. Y. Hasegawa, K. Murakoshi, Y. Wada, J. Kim, N. Nakashima, T. Yamanaka, S. Yanagida, Chem. Phys. Lett. **260**, 173 (1996)
12. P.W. Miller, M. Nieuwenhuyzen, J.P.H. Charmant, S.L. James, Inorg. Chem. **47**, 8367 (2008)
13. J.S. Field, R.J. Haines, E.I. Lakoba, M.H. Sosabowski, J. Chem. Soc. Perkin Trans. **1**, 3352 (2001)
14. R.D. Myrex, C.S. Colbert, G.M. Gray, Organometallics **23**, 409 (2004)
15. I.O. Koshevoy, A.J. Karttunen, S.P. Tunik, M. Haukka, S.I. Selivanov, A.S. Melnikov, P.Y. Serdobintsev, M.A. Khodorkovskiy, T.A. Pakkanen, Inorg. Chem. **47**, 9478 (2008)
16. M.C. Burla, R. Caliandro, M. Camalli, B. Carrozzini, G.L. Cascarano, L.D. Caro, C. Giacovazzo, G. Polidori, R. Spagana, J. Appl. Crystallogr. **38**, 381 (2005)
17. G.M. Sheldrick, Acta Crystallogr. Sect. A: Found. Crystallogr **64**, 112 (2008)
18. R. Pavithran, N.S. Saleesh Kumar, S. Biju, M.L. P. Reddy, S., Jr. Alves, R.O. Freire, Inorg. Chem. **45**, 2184 (2006)
19. Y. Hasegawa, M. Yamamuro, Y. Wada, N. Kanehisa, Y. Kai, S. Yanagida, J. Phys. Chem. A **107**, 1697 (2003)
20. R.B. King, J. Am. Chem. Soc. **91**, 7211 (1969)
21. G.R. Desiraju, T. Steiner, *The Weak Hydrogen Bond in Structural Chemistry and Biology* (Oxford University Press, Oxford 1999)
22. P. Slavíček, M. Fárník, Phys. Chem. Chem. Phys. **13**, 12123 (2011)
23. M.H.V. Werts, R.T.F. Jukes, J.W. Verhoeven, Phys. Chem. Chem. Phys. **4**, 1542 (2002)
24. J.-C.G. Bünzli, S.V. Eliseeva, In Springer *Series on Fluorescence. Lanthanide Luminescence: Photophysical, Analytical and Biological Aspects*, Vol. 8 ed. by P. Hänninen, H. Härmä, (Springer Verlag: Berlin, 2010)
25. S.V. Eliseeva, M. Ryazanov, F. Gumy, S.I. Troyanov, L.S. Lepnev, J.-C.G. Bünzli, N.P. Kuzmina, Eur. J. Inorg. Chem. **23** 4809 (2006)
26. S.V. Eliseeva, D.N. Pleshkov, K.A. Lyssenko, L.S. Lepnev, J.-C.G. Bünzli, N.P. Kuzmina, Inorg. Chem. **50**, 5137 (2011)

Chapter 3
Chameleon Luminophore for a Wide Range Temperature-Sensor Composed of Lanthanide Coordination Polymers

3.1 Introduction

In the fields of fluid dynamics, aeronautical engineering, environment engineering, and energy technology, it is critical to accurately measure the physical parameters of a material surface [1]. Optoelectronic devices have generally been employed as temperature and pressure sensors [2]. However, their sensing area is limited to a single point on a surface. There is a need to measure entire surfaces and obtain multidimensional data for mapping surfaces. There are high expectations that materials for surface measurements, such as temperature and pressure-sensitive dyes, will overcome this intrinsic limitation of optoelectronic devices.

The author proposes a new idea to employ the temperature-dependent energy transfer process in lanthanide complexes for thermosensing in this chapter. First, the author seeks to design temperature-sensitive dyes using luminescent lanthanide complexes. Lanthanide complexes exhibit characteristic luminescence with narrow emission bands and long emission lifetimes as described in Chap. 1 [3–11], which make them suitable for use in sensing devices. In 2003, Amao reported the first temperature-sensitive dye that employed Eu(III) complex in a polymer film [12]. Khalil demonstrated the high performance of Eu(III) complex for temperature-sensitive paint (temperature sensitivity: 4.42 % $°C^{-1}$) [13]. Katagiri has reported Tb(III) complex, $Tb(hfa)_3(H_2O)_2$ (hfa: hexafluoroacetylacetonato), that is suitable as a temperature-sensing probe since it exhibits effective back energy transfer (BEnT) from the emitting level of the Tb(III) ion to the excited triplet state of the hfa ligand [14]. Since BEnT depends on the energy barrier of the process, the emission intensity varies with temperature.

To improve the thermosensing performance, it is necessary to develop a thermostable structure for high temperature sensing and to implement a dual sensing unit for a high sensing ability. First, the author focused on a lanthanide coordination polymer to produce a thermostable structure based on the findings of Chap. 2. Introducing Tb(III) ion and hfa ligands to coordination polymer frameworks will produce a Tb(III) coordination polymer that can be used as a temperature-sensing probe. The triplet state of hfa (22,000 cm^{-1}) is very close to the emitting level of the

K. Miyata, *Highly Luminescent Lanthanide Complexes with Specific Coordination Structures*, Springer Theses,
DOI: 10.1007/978-4-431-54944-4_3, © Springer Japan 2014

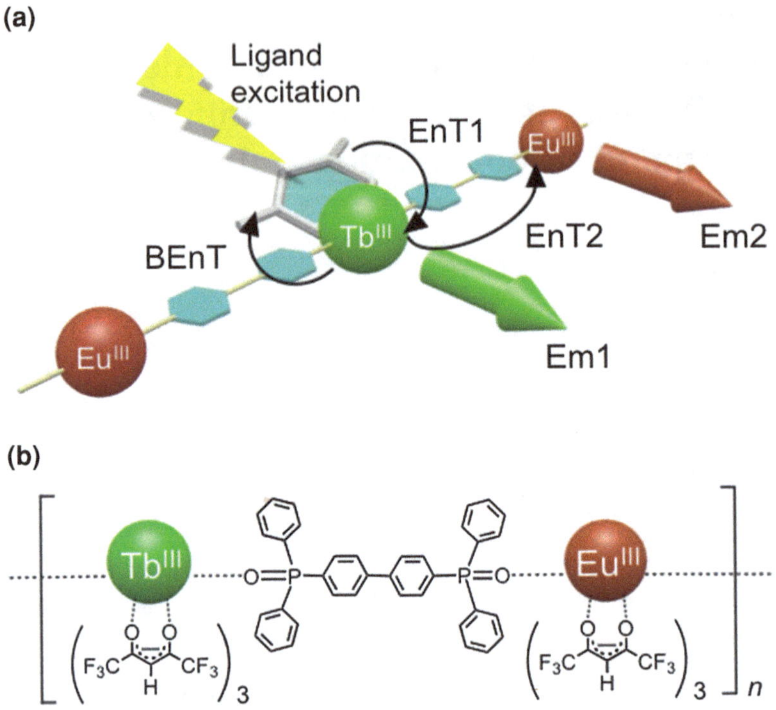

Fig. 3.1 **a** A schematic representation of energy transfer processes of Eu(III) and Tb(III) coordination polymer (*EnT* energy transfer, *BEnT* back energy transfer, *Em* emission). **b** Chemical structure of $[Ln(hfa)_3(dpbp)]_n$ (Ln = Eu, Tb)

Tb(III) ion (20,500 cm^{-1}), resulting in effective EnT1 and BEnT and thus high-performance thermosensing dyes (Fig. 3.1a and 3.2). The author also selected low-vibrational frequency phosphine oxide as the linking part in the Tb(III) coordination polymer because lanthanide complexes with high emission quantum yields composed of hfa and bidentate phosphine oxide ligands were described in former chapters. Second, the author attempted to impart ratiometric temperature sensing by using luminescent Eu(III) and Tb(III) ions in coordination polymer frameworks to realize a high thermosensing ability. Two independent emission bands are expected to enable more accurate thermal measurements than a previous single lanthanide complex [15–18]. Their emission intensities may be also dependent on the temperature based on the energy transfer from Tb(III) to Eu(III), EnT2 (Fig. 3.1a). Temperature-sensitive dyes with a thermostable structure and dual sensing units have not been previously reported.

In this chapter, the author reports a novel thermosensing dye that consists of a lanthanide coordination polymer; $[Ln(hfa)_3(dpbp)]_n$ (dpbp: 4,4′-bis(diphenylphosphoryl)biphenyl, Ln = Eu, Tb), as shown in Fig. 3.1b. Its thermal stability and luminescence properties are characterized by thermogravimetric analysis

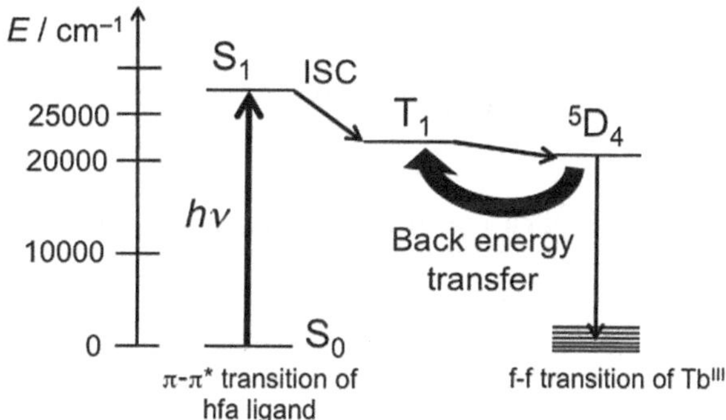

Fig. 3.2 Energy diagram of Tb(III) complexes and back energy transfer processes

(TGA) and by emission quantum yield and emission lifetime measurements. This coordination polymer has a high emission quantum yield ($\Phi_{tot} = 40\ \%$ for $[Tb(hfa)_3(dpbp)]_n$ at room temperature) and a temperature sensitivity over a wide temperature range of 200–500 K. The thermosensing dye composed of Tb(III) and Eu(III) coordination polymers is expected to open up new fields in the molecular chemistry of temperature-sensing materials.

3.2 Experimental Section

3.2.1 General

All chemicals were reagent grade and were used without further purification. Infrared spectra were recorded on a JASCO FT/IR–350 spectrometer. Mass spectra were measured using a JEOL JMS–T100LP. Elemental analyses were performed on a Yanaco CHN corder MT-6. Inductively coupled plasma (ICP) emission spectroscopy was performed on a Shimadzu ICPE-9000. Thermogravimetric analyses were performed on a MAC TG-DTA 2000 in an argon atmosphere at a heating rate of 1 °C min^{-1}.

3.2.2 Syntheses

3.2.2.1 Preparation of $[Tb(hfa)_3(dpbp)]_n$

Phosphine oxide ligand (70 mg, 0.13 mmol) and $Tb(hfa)_3(H_2O)_2$ (120 mg, 0.14 mmol) were dissolved in methanol (15 mL). The solution was refluxed while stirring for 2 h to give a white precipitate. The precipitate was filtered, washed with methanol and chloroform for several times, and dried in vacuo.

Yield: 105 mg (63 %). IR (KBr): 1653 (st, C=O), 1253–1142 (st, C–F), 1125 (st, P=O) cm^{-1}. ESI-Mass (*m/z*) = 1127.05 $[Tb(hfa)_2(dpbp)]^+$. Anal. Calcd. for $[C_{51}H_{31}F_{18}O_8\ P_2Tb]_n$: C, 45.90; H, 2.34. Found: C, 45.76; H 2.48.

3.2.2.2 Preparation of $[Tb_{0.99}Eu_{0.01}(hfa)_3(dpbp)]_n$

Phosphine oxide ligand (1.0 eq), $Tb(hfa)_3(H_2O)_2$ (0.99 eq), and $Eu(hfa)_3(H_2O)_2$ (0.01 eq) were dissolved in methanol. The solution was refluxed while stirring for 2 h to give a white precipitate. The precipitate was filtered, washed with methanol and chloroform for several times, and dried in vacuo.

To determine a Eu:Tb ratios of $[Tb_xEu_{1-x}(hfa)_3(dpbp)]_n$, ICP-AES was conducted. ICP samples were prepared by thoroughly drying $[Tb_xEu_{1-x}(hfa)_3(dpbp)]_n$, then digesting the samples to concentrated (60 %) HNO_3 and diluting to 1.2 % HNO_3. Calibration standards were made by diluting 1000 ppm standards of both Eu and Tb (purchased from Kanto Chemical) over the range of 0.0–1.0 ppm for each. Calibration curves for high and low concentrations were made for three emission wavelengths because instrument response was not linear over the broad range of concentrations required to accurately determine small x values. Three emission wavelengths for each Eu (381.967, 412.970, and 420.505 nm) and Tb (350.917, 367.635, and 384.873 nm) were measured. Concentrations were calculated from the emission intensities for the three Eu and three Tb wavelengths for each sample were converted to concentrations using the appropriate calibration curves and averaged. These analyses revealed a ratio of Eu:Tb = 0.9:99.1 (calcd. 1:99).

3.2.3 Optical Measurements

Emission spectra were recorded on a JASCO F-6300-H spectrometer and corrected for the response of the detector system. Emission lifetimes (τ_{obs}) were measured using the third harmonics (355 nm) of a Q-switched Nd:YAG laser (Spectra Physics, INDI-50, fwhm = 5 ns, λ = 1064 nm) and a photomultiplier (Hamamatsu photonics, R5108, response time $\leq$ 1.1 ns). The Nd:YAG laser response was monitored with a digital oscilloscope (Sony Tektronix, TDS3052, 500 MHz) synchronized to the single-pulse excitation. Emission lifetimes were determined from the slope of logarithmic plots of the decay profiles. The emission quantum yields of $[Tb(hfa)_3(dpbp)]_n$ excited at 380 nm (Φ_{tot}) were estimated using JASCO F-6300-H spectrometer attached with JASCO ILF-533 integrating sphere unit (ϕ = 100 mm). The wavelength dependences of the detector response and the beam intensity of Xe light source for each spectrum were calibrated using a standard light source. Emission spectra at low temperature (77–300 K) were measured with a nitrogen bath cryostat (Oxford Instruments, Optistat DN) and a temperature controller (Oxford, Instruments, ITC 502S). As displayed in Fig. 3.3, emission

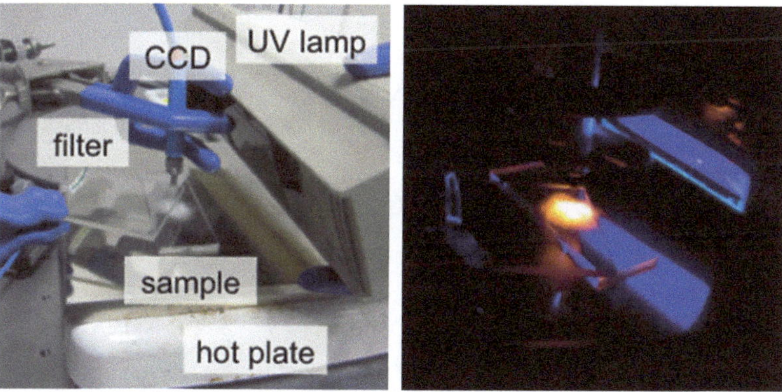

Fig. 3.3 Experimental setup of optical measurements at high temperature

spectra at high temperature (300–500 K) were measured using an electrical hot plate equipped with a digital temperature indicator, and an Ocean Optics multi-channel Analyzer USB4000 corrected for the response of the detector system.

3.3 Results and Discussion

3.3.1 Syntheses and Characterization of Eu/Tb Coordination Polymer

$[Ln(hfa)_3(dpbp)]_n$ and $[Tb_{0.99}Eu_{0.01}(hfa)_3(dpbp)]_n$ (Ln = Eu, Tb) were synthesized by complexing phosphine oxide ligands (dpbp) and $Ln(hfa)_3(H_2O)_2$ in methanol for 2 h. The structure of $[Eu(hfa)_3(dpbp)]_n$ was determined by X-ray single crystal analyses as described in Chap. 2. Phosphine oxide ligands acted as a bidentate bridge between lanthanide ions in one-dimensional polymeric chains. The coordination sites of $[Eu(hfa)_3(dpbp)]_n$ comprised three hfa ligands and two phosphine oxide units. This coordination structure was categorized as an octa-coordinated square antiprism. The powder XRD spectra of $[Eu(hfa)_3(dpbp)]_n$, $[Tb(hfa)_3(dpbp)]_n$, and $[Tb_{0.99}Eu_{0.01}(hfa)_3(dpbp)]_n$ are shown in Fig. 3.4. The XRD patterns of $[Tb(hfa)_3(dpbp)]_n$ and $[Tb_{0.99}Eu_{0.01}(hfa)_3(dpbp)]_n$ are the same as that of $[Eu(hfa)_3(dpbp)]_n$, because of their ionic radii (Eu^{III}: 0.95 Å, Tb^{III}: 0.92 Å). These results indicate that the coordination structures of $[Tb(hfa)_3(dpbp)]_n$ and $[Tb_{0.99}Eu_{0.01}(hfa)_3(dpbp)]_n$ are much similar to that of $[Eu(hfa)_3(dpbp)]_n$.

To evaluate the thermal stability of lanthanide coordination polymers, the author conducted thermogravimetric analyses (TGA) in an argon atmosphere at a heating rate of 1 °C min^{-1} (Fig. 3.5). Their TGA thermograms revealed that $[Eu(hfa)_3(dpbp)]_n$ and $[Tb(hfa)_3(dpbp)]_n$ have decomposition points of 308 and 317 °C, respectively (Table 3.1). In contrast, the decomposition point of Tb(III)

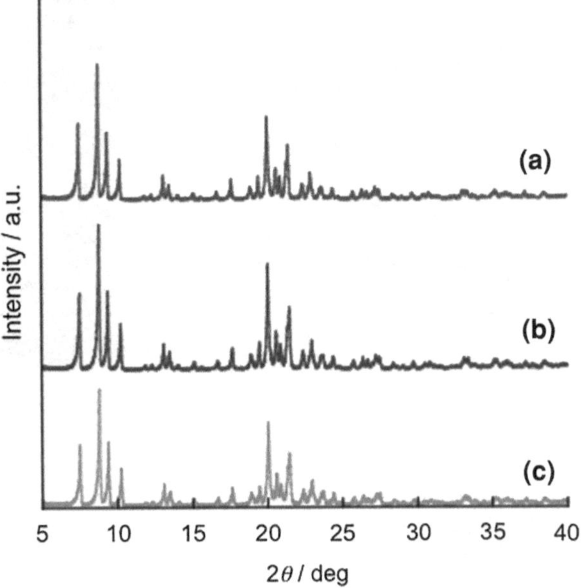

Fig. 3.4 Powder XRD patterns of **a** $[Eu(hfa)_3(dpbp)]_n$, **b** $[Tb(hfa)_3(dpbp)]_n$, and **c** $[Tb_{0.99}Eu_{0.01}(hfa)_3(dpbp)]_n$ coordination polymers

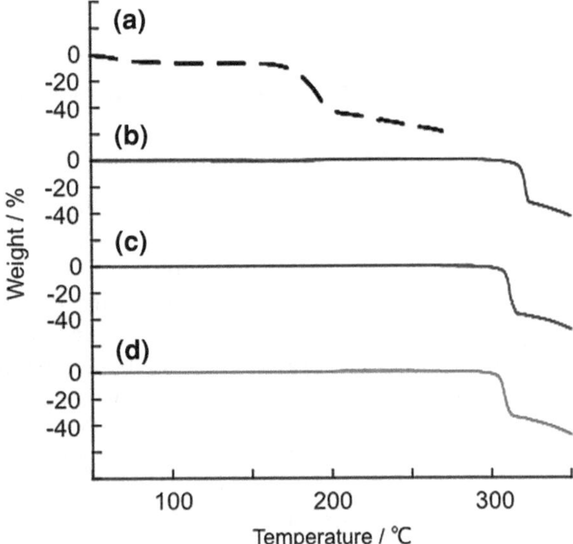

Fig. 3.5 TGA thermograms of **a** $Tb(hfa)_3(H_2O)_2$, **b** $[Tb(hfa)_3(dpbp)]_n$, **c** $[Eu(hfa)_3(dpbp)]_n$, and **d** $[Tb_{0.99}Eu_{0.01}(hfa)_3(dpbp)]_n$ in an argon atmosphere at a heating rate of 1 °C min^{-1}

complex, $Tb(hfa)_3(H_2O)_2$ was found to be 177 °C. The high thermal stability of $[Eu(hfa)_3(dpbp)]_n$ and $[Tb(hfa)_3(dpbp)]_n$ might be due to their tight-binding structures in coordination polymer units as described in Chap. 2 [19].

Table 3.1 The emission and thermal properties of Ln(III) compounds in the solid state

	Decomp. point/ °C	Φ_{tot}/%[a]	τ_{obs}/ms[a,b]	E_a/kJ mol^{-1} [c]	$\Delta G^{\ddagger}$/ kJ mol^{-1}	Temp. sensitivity/ % °C^{-1} [d]
$Tb(NO_3)_3$	–	–	0.86	–	–	<0.03
$Tb(hfa)_3(H_2O)_2$	177	27[e]	0.37	33	57	0.70
$[Tb(hfa)_3(dpbp)]_n$	317	40[e]	0.35	24	58	0.64
$[Eu(hfa)_3(dpbp)]_n$	308[g]	72[f,g]	0.85[g]	–	–	<0.05

[a] At room temperature. [b] Emission lifetimes (τ_{obs}) were measured by excitation at 355 nm (Nd:YAG 3ω). [c] From the slope of the Arrhenius plot (Fig. 3.7). [d] In the temperature range of 200–300 K. [e] Excited at 380 nm. [f] Excited at 465 nm. [g] From Chap. 2

Temperature-dependent emission spectra of $Tb(NO_3)_3$, $[Tb(hfa)_3(dpbp)]_n$, and $[Eu(hfa)_3(dpbp)]_n$ in the solid state are shown in Fig. 3.6. The emission intensities of $[Tb(hfa)_3(dpbp)]_n$ decreased dramatically with increasing temperature. In contrast, changes of the emission intensities of $Tb(NO_3)_3$ and $[Eu(hfa)_3(dpbp)]_n$ were not observed. The energy gaps between the emitting level of the lanthanide ion and the excited triplet state of the hfa ligand in $[Tb(hfa)_3(dpbp)]_n$ and $[Eu(hfa)_3(dpbp)]_n$ are 1700 and 4900 cm^{-1}, respectively [14, 20]. When the energy gap is less than 1850 cm^{-1}, the BEnT from lanthanide ion to the ligands is enhanced [21]. Therefore, the combination of Tb(III) ions with hfa ligands should lead to a condition for enhancement of BEnT, resulting in a high-temperature sensitivity.

The temperature-dependence of the BEnT rate is expected to follow an Arrhenius type equation with an energy barrier E_a. To analyze the BEnT mechanism in detail, the author attempted to estimate the back energy transfer rates (k_{back}) using kinetic analysis. k_{back} is assumed to obey the following Arrhenius type equation:

$$\ln k_{back} = \ln\left(\frac{1}{\tau_{obs}} - \frac{1}{\tau_{77K}}\right) = \ln A - \frac{E_a}{RT} \tag{3.1}$$

where τ_{obs}, τ_{77K}, A, E_a, R, T are emission lifetime, emission lifetime at 77 K, frequency factor, activation energy, gas constant, and temperature, respectively. This assumption is supported by the findings of Blasse and Grabmaier [22]. In order to estimate E_a using Eq. (3.1), the author measured the temperature-dependences of the emission lifetimes of $Tb(hfa)_3(H_2O)_2$ and $[Tb(hfa)_3(dpbp)]_n$. Figure 3.7 shows the temperature-dependences of the emission lifetimes and the Arrhenius plots for k_{back} and Table 3.1 lists the calculated activation energy E_a and the activation Gibbs function $\Delta G^{\ddagger}$ values. The author found that E_a and $\Delta G^{\ddagger}$ of $[Tb(hfa)_3(dpbp)]_n$ are similar to those of $Tb(hfa)_3(H_2O)_2$. These results indicate that phosphine oxide ligands might not affect to the back energy transfer rate.

The emission and thermosensing properties of lanthanide coordination polymers in the solid state are also summarized in Table 3.1. The emission quantum yields (Φ_{tot}) for $Tb(hfa)_3(H_2O)_2$ and $[Tb(hfa)_3(dpbp)]_n$ were estimated to be 27 and

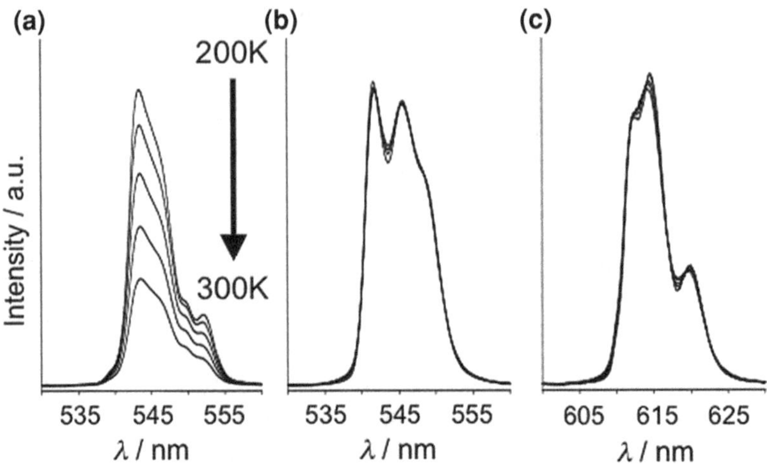

Fig. 3.6 Temperature-dependent emission spectra of **a** $[Tb(hfa)_3(dpbp)]_n$. **b** $Tb(NO_3)_3$, and **c** $[Eu(hfa)_3(dpbp)]_n$ in the solid state in the temperature range of 200–300 K (λ_{ex} = 380 nm)

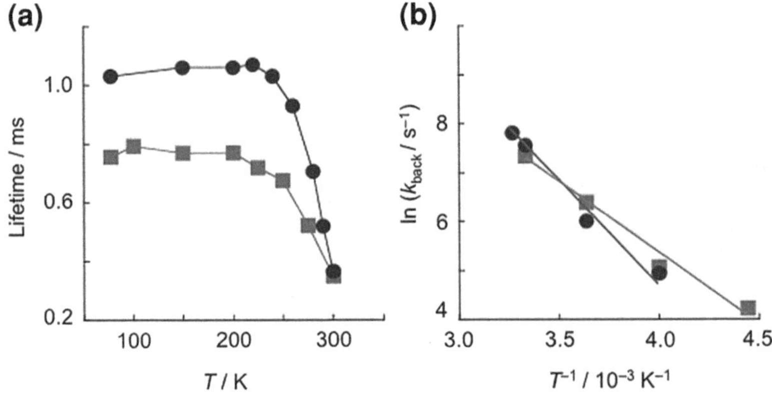

Fig. 3.7 **a** Temperature-dependences of emission lifetimes and **b** Arrhenius plots for back energy transfer rate constants of $Tb(hfa)_3(H_2O)_2$ (●) and $[Tb(hfa)_3(dpbp)]_n$ (■)

40 % at room temperature, respectively. The quantum yields of $[Tb(hfa)_3(dpbp)]_n$ are approximately 1.5 times larger than that of $Tb(hfa)_3(H_2O)_2$. The phosphine oxide ligands should contribute to the enhanced emission of the Tb(III) coordination polymer since they should suppress vibrational relaxation. $[Tb(hfa)_3(dpbp)]_n$ is the best dye in the view point of high-performance of thermosensing dyes.

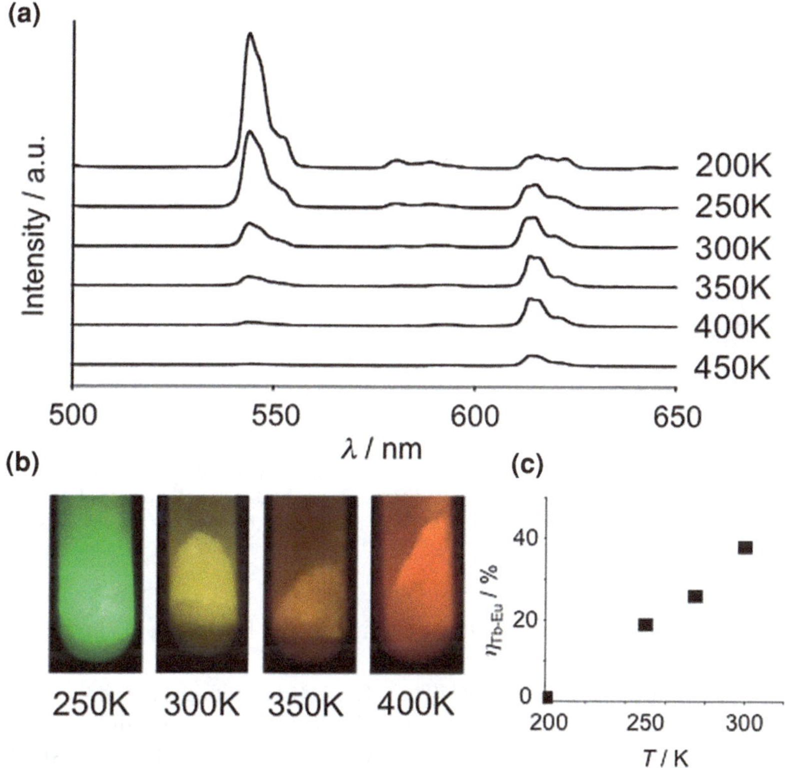

Fig. 3.8 **a** Temperature-dependent emission spectra of $[Tb_{0.99}Eu_{0.01}(hfa)_3(dpbp)]_n$ in the solid state in the temperature range of 200–450 K ($\lambda_{ex} = 380$ nm). **b** Color pictures of the samples under UV (365 nm) irradiation. **c** Temperature-dependence of the energy transfer efficiency ($\eta_{Tb\text{-}Eu}$) of $[Tb_{0.99}Eu_{0.01}(hfa)_3(dpbp)]_n$

3.3.2 Temperature-Dependent Photosensitized Luminescence of Eu/Tb Coordination Polymer

In order to investigate the ratiometric temperature-sensing properties of coordination polymers, $[Tb_{0.99}Eu_{0.01}(hfa)_3(dpbp)]_n$ containing Tb(III) and Eu(III) ions was prepared. Temperature-dependent emission spectra of $[Tb_{0.99}Eu_{0.01}(hfa)_3(dpbp)]_n$ in the solid state in the temperature range of 200–450 K are shown in Fig. 3.8a, and color pictures of the samples are displayed in Fig. 3.8b. The characteristic emission bands at 543 and 613 nm are attributed to the f–f transitions of Tb(III) ($^5D_4$–$^7F_5$) and Eu(III) ($^5D_0$–$^7F_2$), respectively. The emission intensities at 543 nm dramatically decreased with increasing temperature. In contrast, the emission intensities at 613 nm slightly increased. $[Tb_{0.99}Eu_{0.01}(hfa)_3(dpbp)]_n$ exhibits brilliant green, yellow, orange, and red photoluminescence under UV irradiation (365 nm) at 250, 300, 350, and 400 K, respectively. The

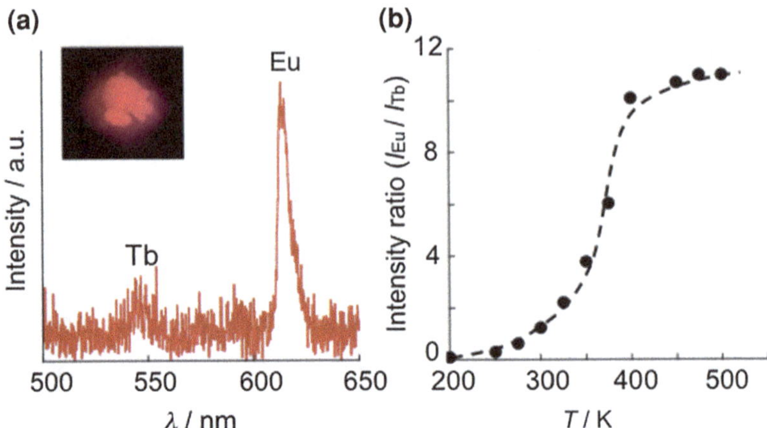

Fig. 3.9 **a** Emission spectra of $[Tb_{0.99}Eu_{0.01}(hfa)_3(dpbp)]_n$ in the solid state at 500 K ($\lambda_{ex} = 365$ nm). *Photograph* emission of $[Tb_{0.99}Eu_{0.01}(hfa)_3(dpbp)]_n$ under UV (365 nm) irradiation at 500 K. **b** The relationship between the ratio of the two emission intensities (I_{Eu}/I_{Tb}) and temperature

author achieved color tuning of the coordination polymers in response to temperature changes.

The temperature sensitivity for $[Tb_{0.99}Eu_{0.01}(hfa)_3(dpbp)]_n$ (0.83 % °C^{-1}) is higher than that for $[Tb(hfa)_3(dpbp)]_n$ (0.64 % °C^{-1}). This result indicates that the energy is transferred to both the excited triplet state of the hfa ligands (BEnT) and to the Eu(III) ion from the emitting level of the Tb(III) ion. The energy transfer from Tb(III) to Eu(III) strongly depends on the temperature, giving rise to temperature-dependent ratios of the luminescence intensities of the Tb(III) emission bands and the Eu(III) emission bands. The energy transfer can be explained by the Förster mechanism. The energy transfer efficiency from the Tb(III) to Eu(III) ion ($\eta_{Tb\text{-}Eu}$) is calculated from the following equation [23]:

$$\eta_{Tb-Eu} = 1 - \left(\frac{\tau_{obs}}{\tau_{Tb}}\right) \tag{3.2}$$

where τ_{obs} and τ_{Tb} are the emission lifetimes of $[Tb_{0.99}Eu_{0.01}(hfa)_3(dpbp)]_n$ and $[Tb(hfa)_3(dpbp)]_n$, respectively. The relation between $\eta_{Tb\text{-}Eu}$ and the temperature (200–300 K) are shown in Fig. 3.8c. The values of $\eta_{Tb\text{-}Eu}$ at 200, 250, 275, and 300 K are estimated to be 1, 19, 26, and 38 %, respectively. $\eta_{Tb\text{-}Eu}$ is increased with increasing temperature. The author considered that the energy transfer occurs both to the excited triplet state of hfa ligands and to the Eu(III) ion. $[Tb_{0.99}Eu_{0.01}(hfa)_3(dpbp)]_n$ exhibits dual emission under UV irradiation even at 500 K as shown in Fig. 3.9a. It is also possible to monitor the temperature changes by using the ratios of the luminescence intensities of the Tb(III) emission bands and the Eu(III) emission bands (Fig. 3.9b). The author also demonstrated that the

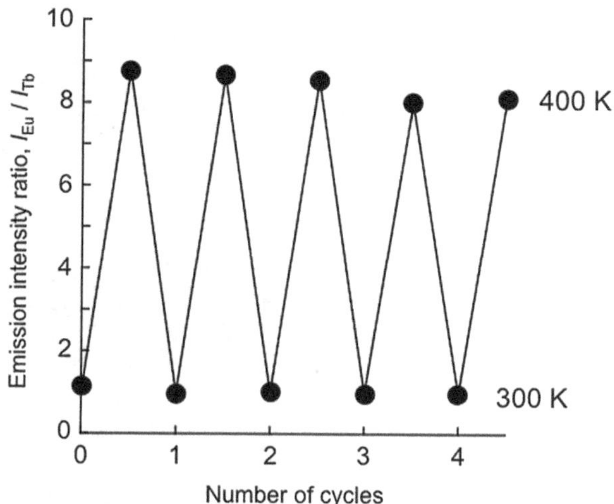

Fig. 3.10 The reversible changes of emission intensity ratio of Eu(III) and Tb(III) (I_{Eu}/I_{Tb}) of $[Tb_{0.99}Eu_{0.01}(hfa)_3(dpbp)]_n$ by the alternative thermo-cycles in the range of 300 and 400 K

luminescent reversibly undergo repeated thermo-cycles. The reversible changes of emission intensity ratio of Eu(III) and Tb(III) (I_{Eu}/I_{Tb}) of $[Tb_{0.99}Eu_{0.01}(hfa)_3(dpbp)]_n$ are observed by the alternative thermo-cycles in the range of 300 and 400 K (Fig. 3.10). The changes in the emission intensity ratio are stably repeated between the I_{Eu}/I_{Tb} values of 1.0 (300 K, yellow emission) and 8.7 (400 K, red emission). Thus, the author successfully synthesized an effective luminophore with a wide temperature sensing range of 200 to 500 K.

3.4 Conclusions

A novel thermosensing dye; "chameleon luminophore" (Fig. 3.11) composed of a lanthanide coordination polymer was successfully synthesized. This coordination polymer is thermally stable and exhibits a high emission quantum yield (Φ_{tot} = 40 % for $[Tb(hfa)_3(dpbp)]_n$ at room temperature) and a temperature sensitivity over a wide temperature range of 200–500 K. The results obtained in this study will provide insights for designing lanthanide coordination polymers for developing temperature-sensing devices based on their luminescence. In future, the chameleon luminophore $[Tb_{0.99}Eu_{0.01}(hfa)_3(dpbp)]_n$ are expected to be promising candidates for temperature-sensitive dyes, which are used for temperature distribution measurements of material surfaces such as an aerospace plane in wind tunnel experiments. Such lanthanide coordination polymers with thermo-sensing properties have the potential to open up new fields in supramolecular chemistry, polymer science, and molecular engineering.

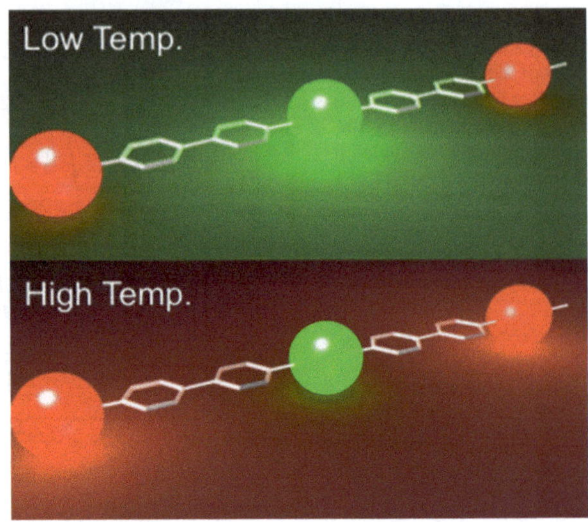

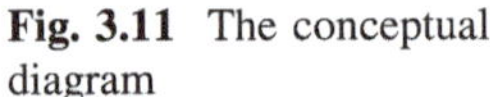
Fig. 3.11 The conceptual diagram

References

1. S. Shindo, *Low-Speed Wind Tunnel Testing Technique,* Chap. 1 (ed. By Corona publishing Co., Ltd., Japan, 1992)
2. *Proceedings of MOSAIC International Workshop*, November 10–11, 2003, Tokyo, Japan
3. N. Weibel, L.J. Charbonniere, M. Guardigli, A. Roda, R. Ziessel, J. Am. Chem. Soc. **126**, 4888 (2004)
4. S. Faulkner, B.P. Burton-Pye, pH dependent self-assembly of dimetallic lanthanide complexes. Chem. Commun. **49**, 259 (2005)
5. J. Yu, D. Parker, R. Pal, R.A. Poole, M.J. Cann, J. Am. Chem. Soc. **128**, 2294 (2006)
6. B. McMahon, P. Mauer, C.P. McCoy, T.C. Lee, T. Gunnlaugsson, J. Am. Chem. Soc. **131**, 17542 (2009)
7. K. Binnemans, Chem. Rev. **109**, 4283 (2009)
8. J.-C.G. Bünzli, Chem. Rev. **110**, 2729 (2010)
9. S.V. Eliseeva, J.-C.G. Bünzli, Chem. Soc. Rev. **39**, 189 (2010)
10. M. Tropiano, N.L. Kilah, M. Morten, H.R.J.J. Davis, P.D. Beer, S. Faulkner, J. Am. Chem. Soc. **133**, 11847 (2011)
11. L.D. Carlos, R.A.S. Ferreira, V. de Zea Bermudez, B. Julian-Lopez, P. Escribano, Chem. Soc. Rev. **40**, 536 (2011)
12. M. Mitsuishi, S. Kikuchi, T. Miyashita, Y. Amao, J. Mater. Chem. **13**, 2875 (2003)
13. G.E. Khalil, K. Lau, G.D. Phelan, B. Carlson, M. Gouterman, J.B. Callis, L.R. Dalton, Rev. Sci. Instrum. **75**, 192 (2004)
14. S. Katagiri, Y. Hasegawa, Y. Wada, S. Yanagida, Chem. Lett. **33**, 1438 (2004)
15. M.S. Tremblay, M. Halim, D. Sames, J. Am. Chem. Soc. **129**, 7570 (2007)
16. N. Kerbellec, D. Kustaryono, V. Haquin, M. Etienne, C. Daiguebonne, O. Guillou, Inorg. Chem. **48**, 2837 (2009)
17. Y. Xiao, Z. Ye, G. Wang, J. Yuan, Inorg. Chem. **51**, 2940 (2012)
18. S. Comby, S.A. Tuck, L.K. Truman, O. Kotova, T. Gunnlaugsson, Inorg. Chem. **51**, 10158 (2012)
19. K. Miyata, T. Ohba, A. Kobayashi, M. Kato, T. Nakanishi, K. Fushimi, Y. Hasegawa, ChemPlusChem **77**, 277 (2012)

20. D.J. Lewis, P.B. Glover, M.C. Solomons, Z. Pikramenou, J. Am. Chem. Soc. **133**, 1033 (2011)
21. S. Sato, M. Wada, Bull. Chem. Soc. Jpn. **43**, 1955 (1970)
22. G. Blasse, B.C. Grabmaier, *Luminescent materials* (Springer-Verlag, New York, 1994), p. 92
23. C. Piguet, J.-C.G. Bünzli, G. Bernardinelli, G. Hopfgartner, A.F. Williams, J. Am. Chem. Soc. **115**, 8197 (1993)

Chapter 4
Characteristic Structures and Photophysical Properties of Nona-Coordinated Eu(III) Complexes with Tridentate Phosphine Oxides

4.1 Introduction

Lanthanide complexes with characteristic narrow emission bands and long emission lifetimes have been regarded as attractive luminescent materials for use in electroluminescent (EL) devices [1–4], lasers [5], and luminescent biosensing applications [6–10]. The coordination number of lanthanide ions in solution varies from eight to twelve, depending on the nature of the ligating molecules [11]. Specific coordination structures result in lanthanide complexes with strong luminescence [12–14]. As described in Chap. 1, the coordination structure of the lanthanide complex dominates the characteristics of the two main parameters that describe the luminescence properties, namely, the radiative (k_r) and nonradiative rate constants (k_{nr}). Firstly, the radiative rate constants observed for lanthanide complexes greatly depend on the geometrical symmetry of the coordination structure. Richardson has estimated the transition intensity parameters of lanthanide complexes on the basis of the ligand field [15]. Binnemans has proposed evaluation of intensity enhancement versus symmetry and polarization by using Judd–Ofelt analysis [16]. Since these studies, it has been widely accepted that the radiative transition probability between the 4f orbitals is enhanced by reducing the geometrical symmetry of the coordination structures [17–19]. Secondly, the emission properties of lanthanide complexes also depend on their vibronic properties, which dominate the kinetics of nonradiative transitions. According to the energy gap theory, nonradiative transitions are promoted by ligands and solvents with high-frequency vibrational structures [20]. For these reasons, lanthanide complexes with large k_r and small k_{nr} constants are the most favorable for strongly luminescent materials. Additionally, efficient energy transfer from the various ligand states [singlet, triplet, and intraligand charge transfer (ILCT)] and the presence of charge transfer (CT) bands are also key factors for strongly luminescent Eu(III) complexes [21–23].

Recently, considerable attentions have been focused on lanthanide complexes with polydentate (tridentate, tetradentate, pentadentate and hexadetate) ligands and their characteristic luminescence properties to suppress the symmetry of the ligand field and vibronic quenching [24–31]. In most of these previous works, however, the

K. Miyata, *Highly Luminescent Lanthanide Complexes with Specific Coordination Structures*, Springer Theses,
DOI: 10.1007/978-4-431-54944-4_4, © Springer Japan 2014

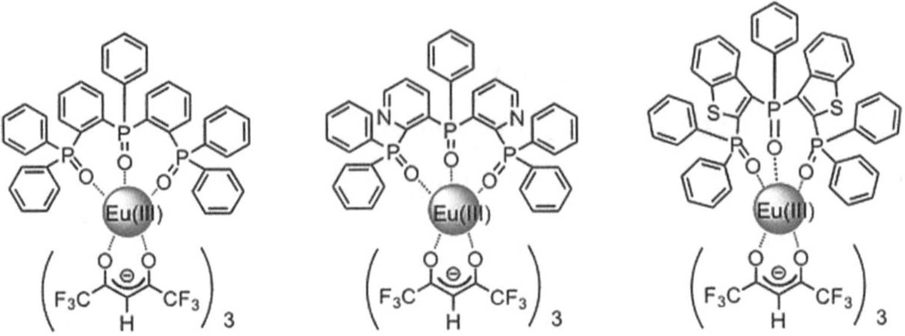

Fig. 4.1 Chemical structures of Eu(hfa)$_3$(dpppo), Eu(hfa)$_3$(dppypo), and Eu(hfa)$_3$(dpbtpo)

lanthanide complexes with polydentate ligands have C–H units close to the metal center. The author here attempted to prepare luminescent Eu(III) complexes with a tridentate phosphine oxide and three hfa as LVF ligands, which are anticipated to construct an asymmetrical nona-coordinated structure (mono-capped square antiprism: 9-SAP). Hasegawa et al. have also reported that the radiative rate constant of Sm(III) complex with 9-SAP structure is larger than those of symmetrical ten-coordinated Sm(III) complexes [32]. Polydentate ligands may also provide stable coordination bindings which show promise toward improved emission properties.

In this chapter, the author reports on luminescent properties of Eu(III) complexes with three kinds of novel tridentate phosphine oxide ligands: Eu(hfa)$_3$(dpppo) (dpppo: bis(*o*-diphenylphosphorylphenyl)phenylphosphine oxide), Eu(hfa)$_3$(dppypo) (dppypo: bis(*o*-diphenylphosphorylpyridyl)phenylphos phine oxide), and Eu(hfa)$_3$(dpbtpo) (dpbtpo: bis(*o*-diphenylphosphorylbenzo thienyl)phenylphosphine oxide) as shown in Fig. 4.1. The geometrical structures of the Eu(III) complexes were characterized using NMR and X-ray single crystal analyses. The author also compared the emission spectral shapes, the emission quantum yields, the emission lifetimes, the radiative and the non-radiative rate constants in nona-coordinated Eu(III) complexes with those of octa-coordinated Eu(III) complexes; Eu(hfa)$_3$(biphepo) and Eu(hfa)$_3$(tppo)$_2$. The photophysical properties of nona-coordinated Eu(III) complexes with tridentate phosphine oxide ligand are elucidated in terms of geometrical, vibrational and chemical structures.

4.2 Experimental Section

4.2.1 Materials

Acetone-d_6 (D, 99.9 %) and chloroform-d (D, 99.8 %) were obtained from Cambridge Isotope laboratories, Inc. Dry THF was prepared by distillation over benzophenone and Na metal. All other chemicals and solvents were reagent grade and were used without further purification.

4.2.2 Apparatus

Infrared spectra were recorded on JASCO FT/IR–420 spectrometer. ^{1}H (300 MHz), ^{19}F (500 MHz) and ^{31}P NMR (500 MHz) spectra were recorded on JEOL ECP–500. Chemical shifts are reported in d ppm, referenced to internal tetramethylsilane standard for ^{1}H NMR, external trifluoroacetic acid standard (δ–76.5 ppm) for ^{19}F NMR and internal 85 % H_3PO_4 standard for ^{31}P NMR. ESI–mass spectra were measured on JEOL JMS–700 M Station. Elemental analyses were performed by Perkin Elmer 2400II.

4.2.3 Syntheses

4.2.3.1 Preparation of (2-bromophenyl)diphenylphosphine

In Schlenk tube, 1-bromo-2-iodobenzene (2.5 g, 8.8 mmol), diphenylphosphine (1.4 mL, 8.0 mmol), triethylamine (1.5 mL, 11 mmol), and a catalytic amount of $Pd(PPh_3)_4$ (50 mg, 4.4×10^{-2} mmol) were dissolved in 2 mL of toluene under nitrogen atmosphere. The solution was heated at 80 °C under stirring for 16 h. The reaction mixture was washed with brine and extracted with diethyl ether three times. The organic layer was dried over anhydrous magnesium sulfate, and concentrated to dryness. Column chromatography on silica gel (hexane/dichloromethane = 4:1) afforded the titled compound as a white solid [33].

Yield: 2.1 g (79 %). ^{1}H NMR (500 MHz, $CDCl_3$, 25 °C) δ 7.59–7.61 (m, 1H, Ar), 7.34–7.38 (m, 6H, Ar), 7.26–7.30 (m, 4H, Ar), 7.19–7.21 (m, 2H, Ar), 6.74–6.76 (m, 1H, Ar) ppm.

4.2.3.2 Preparation of bis(o-diphenylphosphorylphenyl)phenylphosphine oxide (dpppo)

Bis(*o*-diphenylphosphorylphenyl)phenylphosphine (dpppo) was synthesized according to the published procedure. (2-bromophenyl)diphenylphosphine (3.0 g, 9.6 mmol) and dry THF (90 mL), were placed in a 200 mL, three necked flask equipped with a dropping funnel and a nitrogen balloon. The solution was cooled to –80 °C and then 1.6 M *n*-butyllithium hexane solution (6.0 mL, 9.64 mmol) was added dropwise via dropping funnel. After the mixture was stirred at –80 °C for 30 min, dichlorophenylphosphine (0.86 g, 4.8 mmol) was added. The reaction mixture was warmed to room temperature and stirred at room temperature for 3 h. The reaction was quenched by addition of water (ca. 30 mL), and then the resulting white solid was collected by filtration. The solid was dissolved in chloroform (ca. 100 mL) and washed with water three times. The organic layer was separated and dried over anhydrous magnesium sulfate, and concentrated to dryness. The obtained powder and dichloromethane (50 mL) were placed in a 100 mL flask. The solution was cooled to 0 °C and then 30 % H_2O_2 aqueous solution (8 mL) was added to it. The reaction mixture was stirred at −0 °C for 2 h. The organic layer was separated and washed with water three times, then dried over anhydrous magnesium sulfate and concentrated to dryness. Recrystallization from chloroform/hexane gave white solid of the titled compound [34].

Yield: 1.4 g (50 %). MALDI–TOF Mass (*m/z*) = 679.2 $[M+H]^+$. ^1H NMR (500 MHz, $CDCl_3$, 25 °C) δ 7.98 (m, 2H, Ar), 7.25–7.60 (m, 31H, Ar) ppm. ^{31}P NMR (200 MHz, $CDCl_3$, 25 °C) δ 36.78 (1P), 33.65 (2P) ppm. Anal. Calcd for $C_{42}H_{33}O_3P_3 \cdot 0.3CHCl_3$; C, 70.77; H, 4.68. Found: C, 70.88; H, 4.60.

4.2.3.3 Preparation of (*o*-bromopyridyl)diphenylphosphine

2,3-dibromopyridine (8.0 g, 34 mmol), diphenylphosphine (6.3 g, 34 mmol), triethylamine (6.9 g, 68 mmol), and a catalytic amount of $Pd(PPh_3)_4$ (390 mg, 0.34 mmol) were dissolved in 40 mL of toluene under nitrogen atmosphere. The solution was heated at 80 °C under stirring for 16 h. The reaction mixture was washed with brine and extracted with diethyl ether three times. The organic layer was dried over anhydrous magnesium sulfate, and concentrated to dryness. Recrystallization from ethyl acetate/hexane gave colorless block crystals of the titled compound.

Yield: 9.3 g (81 %). IR (ATR) 1559, 1478, 1436, 1382, 1011 cm^{-1}. MALDI–TOF Mass (*m/z*) = 342.0 $[M+H]^+$. ^1H NMR (300 MHz, $CDCl_3$, 25 °C) δ 8.58–8.56 (m, 1H, Ar), 7.78–7.83 (m, 1H, Ar), 7.35–7.40 (m, 10H, Ar), 7.06–7.10 (m, 1H, Ar) ppm. Anal. Calcd for $C_{17}H_{13}BrNP$: C, 59.67; H, 3.83; N 4.09. Found: C, 59.82; H, 3.71; N, 4.10.

4.2.3.4 Preparation of bis(*o*-diphenylphosphorylpyridyl)phenylphosphine oxide (dppypo)

(*o*-bromopyridyl)diphenylphosphine (1.7 g, 5.0 mmol) and dry THF (40 mL) were placed in a 200 mL, three necked flask equipped with a dropping funnel and a nitrogen balloon. The solution was cooled to −80 °C and then 1.6 M *n*-butyllithium hexane solution (3.3 mL, 5.3 mmol) was added dropwise via syringe. After the mixture was stirred at −80 °C for 1 h, dichlorophenylphosphine (0.45 g, 2.5 mmol) was added dropwise via dropping funnel. The reaction mixture was warmed to room temperature and stirred at room temperature for 3 h. The reaction was quenched by addition of water (ca. 5 mL) and extracted with chloroform three times. The organic layer was dried over anhydrous magnesium sulfate, and concentrated to dryness. The obtained powder was dissolved in dichloromethane (15 mL) in a 100 mL flask. The solution was cooled to 0 °C, and then 30 % H_2O_2 aqueous solution (3 mL) was added. The reaction mixture was stirred at 0 °C for 2 h, and then reaction mixture was extracted with dichloromethane three times.

The organic layer was dried over anhydrous magnesium sulfate, and concentrated to dryness. Recrystallization from acetone/hexane gave white powder of the titled compound.

Yield: 0.36 g (21 %). IR (ATR) 1548, 1434, 1390, 1197, 1119, 1107 cm^{-1}. MALDI–TOF Mass (*m/z*) = 681.5 $[M + H]^+$. ^1H NMR (300 MHz, $CDCl_3$, 25°C) δ 8.73 (s, 2H, Ar), 7.92–7.99 (m, 2H, Ar), 7.21–7.64 (m, 27H, Ar) ppm. ^{31}P NMR (200 MHz, $CDCl_3$, 25 °C) δ 29.71 (s,1P), 25.20 (br. 2P) ppm. Anal. Calcd for $C_{40}H_{31}N_2O_3P_3 \cdot H_2O$: C, 68.77; H, 4.76; N, 4.01. Found: C, 69.02; H, 4.66; N, 3.97.

4.2.3.5 Preparation of 2,3-dibromobenzothiophene

2,3-dibromobenzothiophene was synthesized according to the published procedure. Benzothiophene (19 g, 140 mmol) were dissolved in chloroform (200 mL). Then, Br_2 (16 mL, 310 mmol) was added dropwise to the solution. The reaction mixture was stirred at room temperature for 24 h. The resulting solution was washed with $Na_2S_2O_3$ aqueous solution, and extracted with ethyl acetate three times. The organic layer was dried over anhydrous magnesium sulfate, and concentrated to dryness. Column chromatography on silica gel (hexane/dichloromethane = 4:1) afforded the titled compound as a white solid [35].

Yield: 40 g (96 %). IR (ATR) 1419, 1298, 1245, 987, 893, 743, 716 cm^{-1}. MALDI–TOF Mass (*m/z*) = 290.8 $[M+H]^+$. ^1H NMR (300 MHz, $CDCl_3$, 25 °C) δ 7.69–7.76 (m, 2H, Ar), 7.35–7.45 (m, 2H, Ar) ppm.

4.2.3.6 Preparation of (*o*-bromobenzothienyl)diphenylphosphine

2,3-dibromobenzothiophene (20 g, 69 mmol) and dry THF (500 mL) were placed in a 1000 mL, three-necked flask equipped with a dropping funnel and a nitrogen balloon. The solution was cooled to −80 °C and then 1.6 M *n*-butyllithium hexane solution (43 mL, 69 mmol) was added dropwise via dropping funnel. After the mixture was stirred at −80 °C for 1 h, chlorodiphenylphosphine (15 g, 69 mmol) was added dropwise. The reaction mixture was warmed to room temperature and stirred at room temperature for 12 h. The reaction was quenched by addition of water, and extracted with diethyl ether 3 times. The organic layer was dried over anhydrous magnesium sulfate, and concentrated to dryness. Recrystallization from chloroform/hexane for a month gave white yellow crystals of the titled compound.

Yield: 15 g (54 %). IR (ATR) 1431, 1243, 999, 891, 743, 726, 692 cm^{-1}. MALDI–TOF Mass (*m/z*) = 397.3 $[M+H]^+$. ^1H NMR (300 MHz, $CDCl_3$, 25 °C) δ 7.83–7.86 (d, *J* = 9 Hz, 1H, Ar), 7.68–7.71 (d, *J* = 9 Hz, 1H, Ar), 7.37–7.47 (m, 12H, Ar) ppm. Anal. Calcd for $C_{20}H_{14}BrPS$: C, 60.47; H, 3.55. Found: C, 60.26; H, 3.36.

4.2.3.7 Preparation of bis(*o*-diphenylphosphorylbenzothienyl) phenylphos phine oxide (dpbtpo)

(*o*-bromobenzothienyl)diphenylphosphine (4.0 g, 10 mmol) and dry THF (100 mL) were placed in a 300 mL, three necked flask equipped with a dropping funnel and a nitrogen balloon. The solution was cooled to −80 °C and then 1.6 M *n*-butyllithium hexane solution (6.6 mL, 11 mmol) was added dropwise via syringe. After the mixture was stirred at −80 °C for 1 h, dichlorophenylphosphine (0.90 g, 5.0 mmol) was added dropwise via dropping funnel. The reaction mixture was warmed to room temperature and stirred at room temperature for 3 h. The reaction was quenched by addition of water (ca. 5 mL), and extracted with chloroform three times. The organic layer was dried over anhydrous magnesium sulfate, and concentrated to dryness. The obtained powder was dissolved in dichloromethane (30 mL) in a 100 mL flask. The solution was cooled to 0 °C, and then 30 % H_2O_2 aqueous solution (6 mL) was added. The reaction mixture was stirred at 0 °C for 2 h, and then reaction mixture was extracted with dichloromethane three times. The organic layer was dried over anhydrous magnesium sulfate, and concentrated to dryness. Recrystallization from acetone/hexane gave white powder of the titled compound.

Yield: 1.0 g (26 %). IR (ATR) 1457, 1437, 1388, 1203, 1116, 1102, 1010, 898, 801 cm^{-1}. MALDI-TOF Mass (*m/z*) = 791.5 $[M+H]^+$. ^1H NMR (300 MHz, $CDCl_3$, 25 °C) δ 8.34–8.42 (d, $J = 24$ Hz, 2H, Ar), 7.64–7.71 (m, 6H, Ar), 7.45–7.47 (m, 9H, Ar), 7.24–7.30 (m, 14H, Ar), 6.86 (br, 2H, Ar) ppm. ^{31}P NMR (200 MHz, $CDCl_3$, 25 °C) δ 21.80–21.83 (d, $J = 6$ Hz, 2P), 14.81–14.87 (t, $J = 6$ Hz, 1P) ppm. Anal. Calcd for $C_{46}H_{33}S_2O_3P_3{\cdot}0.5CH_2Cl_2$: C, 67.02; H, 4.11. Found: C, 67.33; H, 4.11.

4.2.3.8 General Procedure for the Preparation of Eu(III) Complexes

Phosphine oxide ligand (1 equiv) and $Eu(hfa)_3(H_2O)_2$ (1.2 equiv) were dissolved in methanol (20 mL). The solution was refluxed while stirring for 8 h, and the reaction mixture was concentrated to dryness. The residue was washed with chloroform several times. The insoluble material was removed by filtration, and the filtrate was concentrated. Recrystallization from acetone gave colorless crystals of the Eu(III) complexes.

[$Eu(hfa)_3$(dpppo)]: Yield: 90 mg (26 %). IR (ATR) 1657, 1537, 1439, 1252, 1176, 1136, 789, 746, 727, 692, 660 cm^{-1}. ^{19}F NMR (470 MHz, acetone–d_6, 25 °C) δ–79.36 ppm. ^{31}P NMR (200 MHz, acetone–d_6, 25 °C) δ 0.55 (1P), –29.03 (1P), –68.37 (1P) ppm. ESI–Mass (*m/z*) = 1245.052 $[M–(hfa)]^+$. Anal. Calcd for $C_{57}H_{36}EuF_{18}O_9P_3$: C, 47.16; H, 2.50. Found: C, 47.10; H 2.50.

[$Eu(hfa)_3$(dppypo)]: Yield: 380 mg (59 %). IR (ATR) 1653, 1550, 1525, 1507, 1440, 1252, 1191, 1138, 1124, 1097 cm^{-1}. ^{19}F NMR (470 MHz, acetone–d_6, 25 °C) δ –78.71 (s) ppm. ^{31}P NMR (200 MHz acetone–d_6, 25 °C) δ–12.08 (s, 1P), –15.81 (s, 1P), –64.75 (s, 1P) ppm. ESI–Mass (*m/z*) = 1247.052 $[M–(hfa)]^+$. Anal.

Calcd for $C_{55}H_{34}Eu\ F_{18}N_2O_9P_3 \cdot 1.5CHCl_3$: C, 41.59; H, 2.13; N, 1.72. Found: C, 41.71; H, 2.13; N, 1.70.

[Eu(hfa)$_3$(dpbtpo)]: Yield: 170 mg (28 %). IR (ATR) 1657, 1534, 1440, 1251, 1185, 1138, 1125, 1104 cm^{-1}. ^{19}F NMR (470 MHz, acetone–d_6, 25 °C) δ–78.55 (s) ppm. ^{31}P NMR (200 MHz, acetone–d_6, 25 °C) δ–47.78 (s, 1P), –53.29 (s, 1P), –78.18 (s, 1P) ppm. ESI–Mass (m/z) = 1357.001 [M–(hfa)]$^+$. Anal. Calcd for $C_{61}H_{36}EuF_{18}S_2O_9P_3$: C, 46.85; H, 2.32. Found: C, 46.68; H, 2.24.

4.2.4 Crystallography

Colorless single crystals of Eu(hfa)$_3$(dpppo) and Eu(hfa)$_3$(dpbtpo) obtained from acetone solution were mounted on a glass fiber using epoxy resin glue. All measurements were made on a Rigaku RAXIS RAPID imaging plate area detector with graphite monochromated MoKα radiation. The data were collected at a temperature range of -120 ± 1 °C to a maximum 2θ value of 48.8°. Corrections for decay and Lorentz-polarization effects were made with empirical absorption correction, solved by direct methods and expanded using Fourier techniques. The non-hydrogen atoms were refined anisotropically. Hydrogen atoms were refined using the riding model. The final cycle of full-matrix least-squares refinement was based on observed reflections and variable parameters. All calculations were performed using the crystal structure crystallographic software package. CCDC-737332 (for Eu(hfa)$_3$(dpppo)) and -737333 (for Eu(hfa)$_3$(dpbtpo)) contain the supplementary crystallographic data for this paper. CIF data confirmed by using checkCIF/PLATON service.

4.2.5 Optical Measurements

UV–Vis absorption spectra were recorded on a JASCO V–550 spectrometer. Emission spectra of the Eu(III) complexes were measured with a Hitachi F–4500 spectrometer and corrected for the response of the detector system. The emission quantum yields of Eu(hfa)$_3$(dpppo), Eu(hfa)$_3$(dppypo) and Eu(hfa)$_3$(dpbtpo) (5.0 mM in acetone-d_6) were obtained by comparison with the emission signal integration (570–640 nm) of Eu(hfa)$_3$(biphepo) as a reference ($\Phi_{Ln} = 0.60$: 50 mM in acetone-d_6) with excitation wavelength of 465 nm [36]. Emission lifetimes of Eu(hfa)$_3$(dpppo), Eu(hfa)$_3$(dppypo) and Eu(hfa)$_3$(dpbtpo) (1.0 mM in acetone-d_6) were measured with the third harmonics (355 nm) of a Q-swiched Nd:YAG laser (Spectra Physics, INDI-50, fwhm = 5 ns, $\lambda = 1064$ nm) and a photomultiplier (Hamamatsu photonics, R5108, response time $\leq$1.1 ns). The Nd:YAG laser response was monitored with a digital oscilloscope (Sony Tektronix, TDS3052, 500 MHz) synchronized to the single-pulse excitation. Emission lifetimes were determined from the slope of logarithmic plots of decay profiles. High-resolution emission spectra were measured with a SPEX-Fluorolog τ-3 (JOBIN YVON).

4.3 Results and Discussion

4.3.1 Syntheses and Coordination Structures

Eu(III) complexes with tridentate phosphine oxides, $Eu(hfa)_3$(dpppo), $Eu(hfa)_3$(dppypo) and $Eu(hfa)_3$(dpbtpo) were synthesized by complexation of phosphine oxide ligands with $Eu(hfa)_3(H_2O)_2$ in methanol for 8 h as illustrated in Scheme 4.1. The IR spectra and elemental analyses indicate that no water molecule existed in the coordination sites of Eu(III) complexes. The author also observed the signals of corresponding $[Eu(hfa)_2(dpppo)]^+$, $[Eu(hfa)_2(dppypo)]^+$ and $[Eu(hfa)_2(dpbtpo)]^+$ in ESI-mass spectra. The ^{31}P NMR spectra of the Eu(III) complexes in acetone-d_6 exhibited three specific signals. These results indicate that the magnetic environments of phosphine oxides were different from each other in the solution phase.

The single crystals of $Eu(hfa)_3$(dpppo) and $Eu(hfa)_3$(dpbtpo) were successfully prepared for X-ray single-crystal analyses by way of recrystallization from acetone solutions. The ORTEP views and crystal data from the X-ray single-crystal analyses are summarized in Fig. 4.2 and Table 4.1. The ORTEP views of both $Eu(hfa)_3$(dpppo) and $Eu(hfa)_3$(dpbtpo) show the nona-coordinated structures with one tridentate phosphine oxide (dpppo or dpbtpo) and three hfa ligands. The coordination geometries of $Eu(hfa)_3$(dpppo) and $Eu(hfa)_3$(dpbtpo) are categorized as Mono-capped square antiprisms (9-SAP) [37]. In $Eu(hfa)_3$(dpppo) and $Eu(hfa)_3$(dpbtpo), the Eu–O bond lengths that are attached with two terminal phosphine oxides in Table 4.2 are similar to those of the reported $Eu(hfa)_3$(biphepo), 2.32 and 2.34 Å [38]. The Eu–O bond length for the central phosphine oxide in $Eu(hfa)_3$(dpppo) (Eu–O2: 2.51 Å) and $Eu(hfa)_3$(dpbtpo) (Eu–O1: 2.48 Å) are considerably longer than those of the terminal phosphine oxides. Specific interactions between ligands, such as CH–F hydrogen bonds, CH–π, π–π interactions, have not been observed. The coordination structures of $Eu(hfa)_3$(dpppo) and $Eu(hfa)_3$(dpbtpo) are thus considerably distorted compared with the usual 9-SAP shapes and would be categorized according to their quasi-C_1 symmetry, consistent with octa-coordinated $Eu(hfa)_3$(biphepo) (Fig. 4.3).

Two terminal phosphine oxides of $Eu(hfa)_3$(dpppo) are located on the upper square of 9-SAP, while those of $Eu(hfa)_3$(dpbtpo) are attached to the upper and lower squares of 9-SAP. The author thus considers that the coordination structures of Eu(III) complexes with tridentate phosphine oxide ligands are markedly influenced by the moiety between the phosphine oxides.

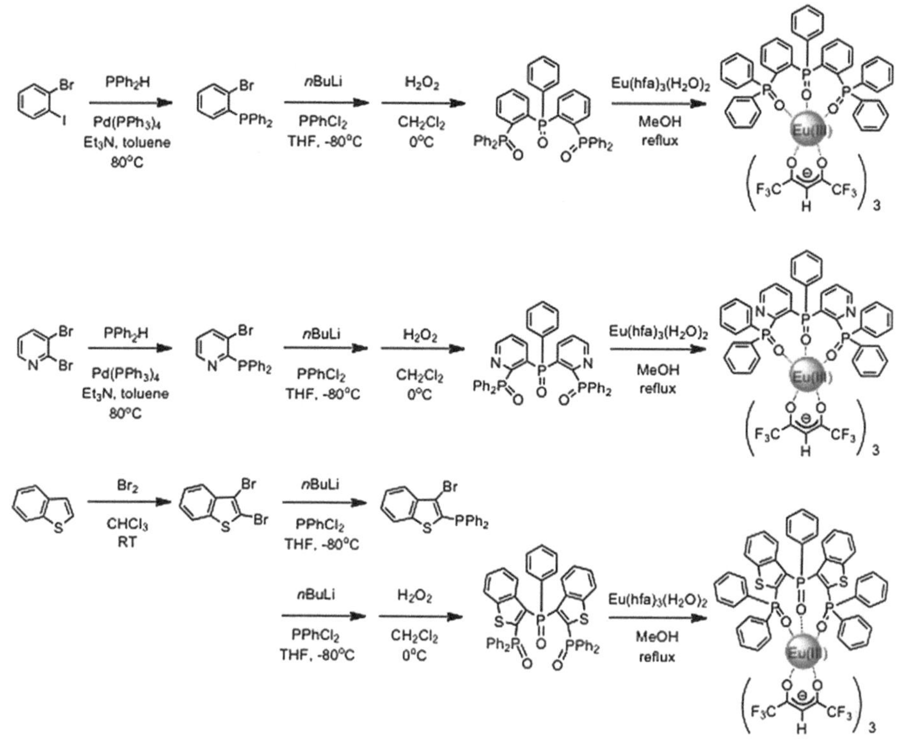

Scheme 4.1 Synthetic routes of nona-coordinated Eu(III) complexes

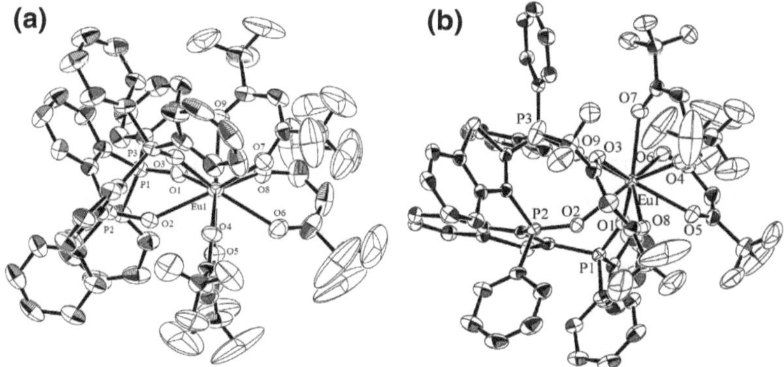

Fig. 4.2 ORTEP drawings of single crystals of **a** $Eu(hfa)_3(dpppo)$ from acetone/water solution and **b** $Eu(hfa)_3(dpbtpo)$ from acetone solution. Hydrogen atoms have been omitted for clarity and thermal ellipsoids are shown at the 50 % probability level

Based on the X-ray single crystal analyses, the author carried out calculations of charge densities of phosphorus atoms by using DFT calculation (6-31G(d)/B3LYP). According to the calculation, charge densities of phosphorus atoms were

Table 4.1 Crystal data of Eu(III) complexes with tridentate phosphine oxides

	Eu(hfa)$_3$(dpppo)	2Eu(hfa)$_3$(dpbtpo) + 0.5 acetone
Chemical formula	$C_{57}H_{36}EuF_{18}O_9P_3$	$C_{125}H_{78}Eu_2F_{36}O_{19}P_6S_4$
Formula weight	1451.76	3185.93
Crystal color, habit	Colorless, block	Colorless, block
Crystal system	Trigonal	Monoclinic
Space group	R3c	C2/c
a/Å	40.3572(7)	24.6695(14)
b/Å		12.3999(6)
c/Å	24.5364(5)	42.565(2)
β/°		99.3595(14)
V/Å^3	34608.5(10)	12847.4(11)
Z	18	4
d_{calc}/g cm^{-3}	1.254	1.645
T/ °C	−80 ± 1	−120 ± 1
μ (Mo Kα)/cm^{-1}	9.617	12.226
max 2θ/deg	50.6	48.8
No. of measured reflections	91569	44801
No. of unique reflections	13372	10505
R ($I > 2\sigma$(I))[a]	0.0344	0.0351
R_w ($I > 2\sigma$(I))[b]	0.1106	0.0870

[a] $R = \Sigma ||F_o| - |F_c||/\Sigma |F_o|$
[b] $R_w = [(\Sigma w (|F_o| - |F_c|)^2 / \Sigma w F_o^2)]^{1/2}$

Table 4.2 Selected bond lengths [Å] of Eu(III) complexes

Bond	Eu(hfa)$_3$(dpppo)	Eu(hfa)$_3$(dpbtpo)
Eu–O1 (P=O)	2.360 (terminal)	2.482 (central)
Eu–O2 (P=O)	2.511 (central)	2.403 (terminal)
Eu–O3 (P=O)	2.322 (terminal)	2.355 (terminal)

found to be P1: 0.22, P2: 0.14, P3 0.29 (dpppo), P1: 0.04, P2: 0.35, P3: 0.11 (dpbtpo), respectively (Fig. 4.4). The author thus considers that three peaks in ^{31}P NMR spectra of Eu(III) complexes might be caused by different electron densities of phosphorus atoms.

4.3.2 Photophysical Properties

The steady-state emission spectra of Eu(hfa)$_3$(dpppo), Eu(hfa)$_3$(dppypo) and Eu(hfa)$_3$(dpbtpo) in acetone-d_6 are shown in Fig. 4.5a. Emission bands were observed at around 578, 592, 614, 650, and 700 nm, and are attributed to the f–f transitions of $^5D_0$–^7F_J with J = 0, 1, 2, 3 and 4, respectively. The spectra were normalized with respect to the magnetic dipole transition intensity ($^5D_0$–$^7F_1$) at

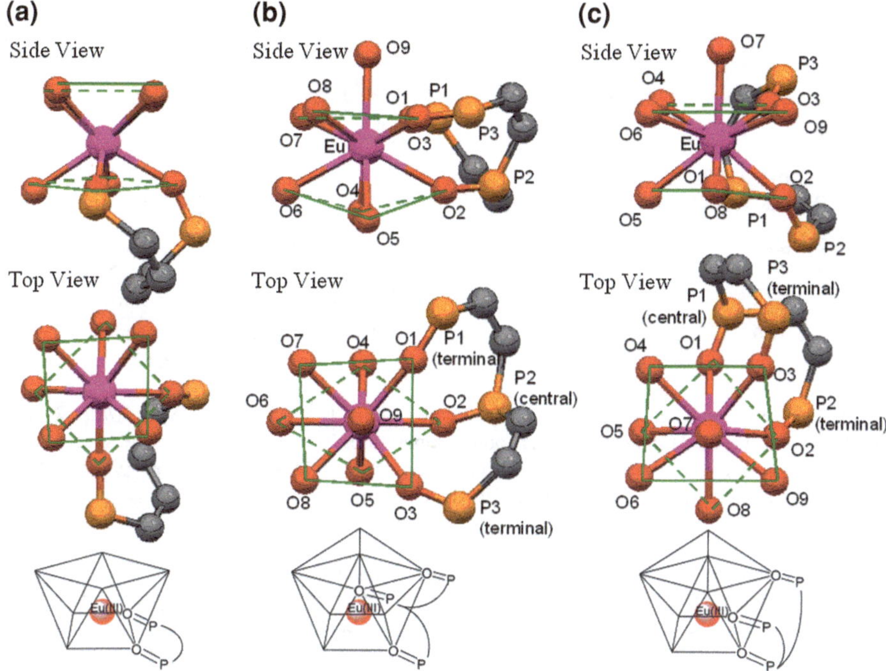

Fig. 4.3 Coordination polyhedrons of **a** Eu(hfa)$_3$(biphepo), **b** Eu(hfa)$_3$(dpppo) and **c** Eu(hfa)$_3$(dpbtpo)

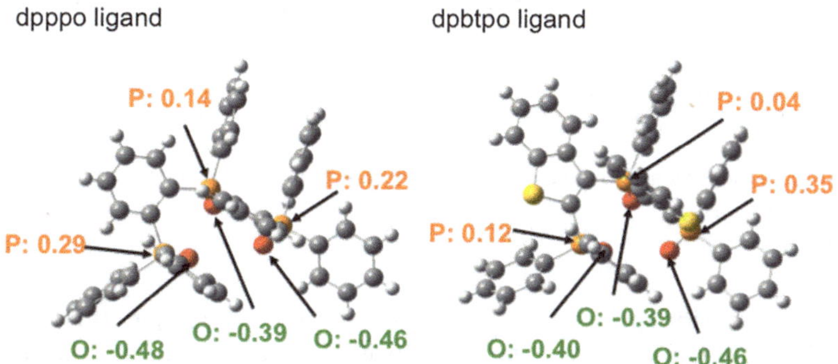

Fig. 4.4 Charge densities of phosphorus atoms in **a** dpppo and **b** dpbtpo ligand

592 nm which is known to be insensitive to the surrounding environment of the Eu(III) ion. The emission band at 614 nm ($^5D_0$–$^7F_2$) is due to the electric dipole transition for which intensity is greatly dependent on the chemical structures. In order to analyze the transition intensity, the author here estimated the relative emission intensity of $^5D_0$–$^7F_2$ transition with respect to that of $^5D_0$–$^7F_1$ as $I_{rel} = I_{614}/I_{592}$ in the

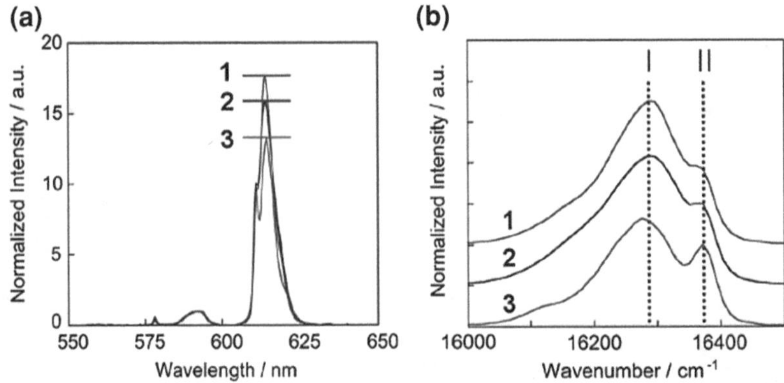

Fig. 4.5 a Emission spectra of Eu(hfa)$_3$(dpbtpo) (*line* 1), Eu(hfa)$_3$(dppypo) (*line* 2) and Eu(hfa)$_3$(dpppo) (*line* 3) in acetone-d_6 at room temperature. Excited at 465 nm. The spectra were normalized with respect to the magnetic dipole transition ($^5D_0$–$^7F_1$). **b** Stark splitting at the electric dipole transition ($^5D_0$–$^7F_2$) of the Eu(III) complexes

Table 4.3 Photophysical properties of Eu(III) complexes at room temperature[a]

Complex	Φ_{Ln}/%[b]	τ_{obs}/ms[c]	k_r/s^{-1}	k_{nr}/s^{-1}	I_{rel}[d]
Eu(hfa)$_3$(dpbtpo)	62	1.3	5.0 × 10^2	3.0 × 10^2	18
Eu(hfa)$_3$(dppypo)	61	1.2	5.1 × 10^2	3.2 × 10^2	16
Eu(hfa)$_3$(dpppo)	60	1.2	5.0 × 10^2	3.3 × 10^2	13
Eu(hfa)$_3$(biphepo)[e]	60	1.3	4.6 × 10^2	3.4 × 10^2	-
Eu(hfa)$_3$(tppo)$_2$[e]	65	1.2	5.4 × 10^2	3.0 × 10^2	-

[a] The emission spectra, quantum yield (Φ_{Ln}) and lifetime (τ_{obs}) were measured by excitation at 465 nm. Radiative rate constant $k_r = \Phi_{Ln}/\tau_{obs}$. Nonradiative rate constant $k_{nr} = 1/\tau_{obs} - k_r$

[b] Emission quantum yields were determined by comparing with the emission signal integration (570–640 nm) of Eu(hfa)$_3$(biphepo) as $\Phi_{Ln} = 0.60$

[c] Emission lifetimes (τ_{obs}) of the Eu(III) complexes were measured by excitation at 355 nm (Nd:YAG 3ω)

[d] Relative emission intensity of the electric dipole transition ($^5D_0$–$^7F_2$) to the magnetic dipole transition ($^5D_0$–$^7F_1$)

[e] [13]

normalized emission spectra. The I_{rel} values of Eu(hfa)$_3$(dpbtpo), Eu(hfa)$_3$(dppypo) and Eu(hfa)$_3$(dpppo) were found to be 18, 16 and 13, respectively, as summarized in Table 4.3. The emission spectra with Stark splittings at the electric dipole transition ($^5D_0$–$^7F_2$) of Eu(III) complexes in acetone-d_6 are also shown in Fig. 4.5b. Five-fold degenerated $^7F_2$ states of Eu(III) complexes are known to split into some Stark levels between one and five depending on symmetry of the coordination structure. The author found that the emission spectral profiles and the wavenumbers of transition bands I and II of Eu(hfa)$_3$(dpppo) and Eu(hfa)$_3$(dpbtpo) well agreed with those of Eu(hfa)$_3$(dppypo) (16280 cm^{-1}, 16370 cm^{-1}). The energy gaps between the transition bands I and II of these Eu(III) complexes were approximately 85 cm^{-1}.

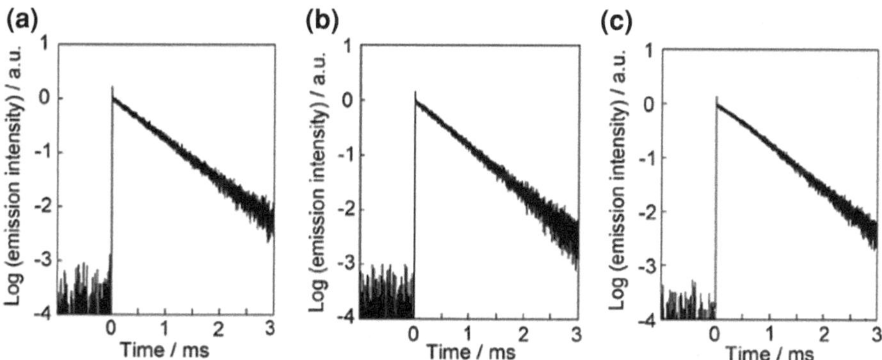

Fig. 4.6 The decay profiles of **a** $Eu(hfa)_3(dpppo)$, **b** $Eu(hfa)_3(dppypo)$, and **c** $Eu(hfa)_3$ (dpbtpo) in acetone-d_6

Since the energy gap between the Stark splitting levels reflects its geometrical symmetry [32], $Eu(hfa)_3(dppypo)$ might be similar to $Eu(hfa)_3(dpppo)$ and $Eu(hfa)_3(dpbtpo)$ in its structure in acetone-d_6. The emission quantum yields of $Eu(hfa)_3(dpppo)$, $Eu(hfa)_3(dppypo)$ and $Eu(hfa)_3(dpbtpo)$ in acetone-d_6 were found to be 0.60, 0.61 and 0.62, respectively (Table 4.3). These emission quantum yields are similar to those reported for $Eu(hfa)_3(biphepo)$ ($\Phi_{Ln} = 0.60$ in acetone-d_6) and $Eu(hfa)_3(tppo)_2$ ($\Phi_{Ln} = 0.65$ in acetone-d_6), both of which have been reported as highly luminescent Eu(III) complexes [38].

The time-resolved emission profiles of all Eu(III) complexes revealed single-exponential decays with lifetimes in the millisecond time scale as shown in Fig. 4.6. The emission lifetimes were determined from the slopes of logarithmic plots of the decay profiles. The radiative (k_r) and nonradiative (k_{nr}) rate constants estimated using the emission lifetimes and the emission quantum yields are summarized in Table 4.3. The author observed that the radiative rate constants of nona-coordinated $Eu(hfa)_3(dpppo)$ ($5.0 \times 10^2\ s^{-1}$), $Eu(hfa)_3(dppypo)$ ($5.1 \times 10^2\ s^{-1}$) and $Eu(hfa)_3(dpbtpo)$ ($5.0 \times 10^2\ s^{-1}$) were similar to those of the octa-coordinated $Eu(hfa)_3(biphepo)$ ($4.6 \times 10^2\ s^{-1}$) and $Eu(hfa)_3(tppo)_2$ ($5.4 \times 10^2\ s^{-1}$). Generally, reduction of the geometrical symmetry of coordination structure leads to a larger radiative rate constant. The radiative rate constant of Eu(III) complex with phosphine oxides seems to depend on the symmetry of the coordination sites, but not on the coordination number. The nonradiative rate constants of $Eu(hfa)_3(dpppo)$ ($3.3 \times 10^2\ s^{-1}$), $Eu(hfa)_3(dppypo)$ ($3.2 \times 10^2\ s^{-1}$) and $Eu(hfa)_3(dpbtpo)$ ($3.0 \times 10^2\ s^{-1}$) were also quite similar to that of $Eu(hfa)_3$(biphepo) ($3.4 \times 10^2\ s^{-1}$) and $Eu(hfa)_3(tppo)_2$ ($3.0 \times 10^2\ s^{-1}$). The nonradiative rate constant of Eu(III) might not be affected by vibrational, electric and steric structures of the chemical species attached with LVF phosphine oxides.

In this chapter, the author found that the emission intensities at the electric dipole transition of Eu(III) complexes depended on the organic linker species in phosphine oxide ligands, although the energy gap between Stark splitting levels, radiative and non-radiative rate constants were not affected by them. It seems that

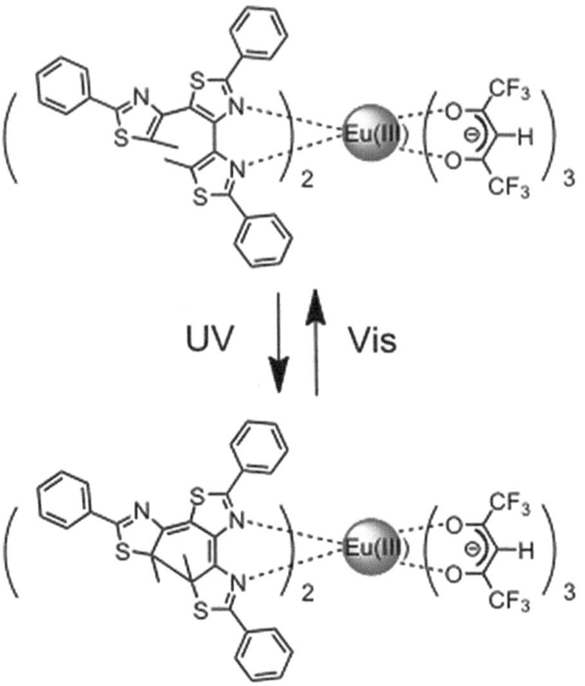

Fig. 4.7 Eu(III) complexes with photochromic terarylene ligands

the transition intensity at the electric dipole transition is affected by chemical structures of attached tridentate phosphine oxide ligands. Recently, Nakagawa has observed that the transition intensity of Eu(III) complex with photochromic terarylene ligands is influenced by the polarizability of attached photochromic units (Fig. 4.7) [39–41]. The effect of polarizability on lanthanide complexes has also been suggested [42, 43]. The author reasons that the emission spectral shapes of Eu(III) complexes with tridentate phosphine oxide ligands might be related not only to geometrical and vibrational structures of Eu(III) complexes, but also to the chemical structures of their ligands.

4.4 Conclusions

The author successfully synthesized novel Eu(III) complexes containing characteristic tridentate phosphine oxide ligands, dpppo, dppypo and dpbtpo with high emission quantum yields ($\Phi_{Ln} > 60$ %). These Eu(III) complexes showed characteristic emission properties depending on their moiety between phosphine oxides. In this chapter, the author has reported on nona-coordinated lanthanide complexes with characteristic photophysical properties. The photophysical properties of Eu(III) complexes with tridentate phosphine oxide ligands would be discussed on the basis of geometrical, vibrational and chemical structures of the ligands.

References

1. T. Justel, H. Nikol, C. Ronda, Angew. Chem. Int. Ed. **37**, 3084 (1998)
2. J. Kido, Y. Okamoto, Chem. Rev. **102**, 2357 (2002)
3. J. Yu, L. Zhou, H. Zhang, Y. Zheng, H. Li, R. Deng, Z. Peng, Z. Li, Inorg. Chem. **44**, 1611 (2005)
4. H. Xu, K. Yin, W. Huang, J. Phys. Chem. C **114**, 1674 (2010)
5. K. Kuriki, Y. Koike, Y. Okamoto, Chem. Rev. **102**, 2347 (2002)
6. N. Weibel, L.J. Charbonniere, M. Guardigli, A. Roda, R. Ziessel, J. Am. Chem. Soc. **126**, 4888 (2004)
7. J.-C.G. Bunzli, C. Piguet, Chem. Soc. Rev. **34**, 1048 (2005)
8. S. Faulkner, B.P. Burton-Pye, Chem. Commun. 259 (2005)
9. J. Yu, D. Parker, R. Pal, R.A. Poole, M.J. Cann, J. Am. Chem. Soc. **128**, 2294 (2006)
10. B. McMahon, P. Mauer, C.P. McCoy, T.C. Lee, T. Gunnlaugsson, J. Am. Chem. Soc. **131**, 17542 (2009)
11. V.S. Sastri, J.-C.G. Bünzli, V. Ramachandra Rao, G.V.S. Rayudu, J.R. Perumareddi, *Modern Aspects of Rare Earths and Their Complexes*, (Elsevier, Amsterdam, 2003)
12. K. Lunstroot, P. Nockemann, K.V. Hecke, L.V. Meervelt, C. Görller-Walrand, K. Binnemans, K. Driesen, Inorg. Chem. **48**, 3018 (2009)
13. S. Petoud, S.M. Cohen, J.-C.G. Bunzli, K.N. Raymond, J. Am. Chem. Soc. **125**, 13324 (2003)
14. Y. Hasegawa, Y. Wada, S. Yanagida, J. Photochem. Photobiol. C: Photochem. Rev. **5**, 183–202 (2004)
15. A.F. Kirby, F.S. Richardson, J. Phys. Chem. **87**, 2544 (1983)
16. K. Binnemans, R.V. Deun, C. Görller-Walrand, S.R. Collinson, F. Martin, D.W. Bruce, C. Wickleder, Phys. Chem. Chem. Phys. **2**, 3753 (2000)
17. S.F. Mason, J. Indian Chem. Soc. **63**, 73 (1986)
18. F. Gan, *Laser Materials* (World Scientific, Singapore, 1995)
19. A. Wada, M. Watanabe, Y. Yamanoi, T. Nankawa, K. Namiki, M. Yamasaki, M. Murata, H. Nishihara, Bull. Chem. Soc. Jpn **80**, 335 (2007)
20. G. Stein, E. Würzberg, J. Chem. Phys. **62**, 208 (1975)
21. S.V. Eliseeva, J.-C.G. Bünzli, Chem. Soc. Rev. **39**, 189 (2010)
22. S.V. Eliseeva, O.V. Kotova, F. Gumy, S.N. Semenov, V.G. Kessler, L.S. Lepnev, J.-C.G. Bünzli, N.P. Kuzmina, J. Phys. Chem. A **112**, 3614 (2008)
23. Q.-B. Bo, H.-Y. Wang, D.-Q. Wang, Z.-W. Zhang, J.-L. Miao, G.-X. Sun, Inorg. Chem. **50**, 10163 (2011)
24. L.J. Charbonnire, R. Ziessel, M. Montalti, L. Prodi, C. Boehme, G. Wipff, J. Am. Chem. Soc. **124**, 7779 (2002)
25. S. Faulkner, J.A. Pope, J. Am. Chem. Soc. **125**, 10526 (2003)
26. G.S. Kottas, M. Mehlstäubl, R. Fröhlich, L. De Cola, Eur. J. Inrog. Chem. **22**, 3465 (2007)
27. M. Seitz, E.G. Moore, A.J. Ingram, G. Muller, K.N. Raymond, J. Am. Chem. Soc. **129**, 15468 (2007)
28. S. Petoud, G. Muller, E.G. Moore, J. Xu, J. Sokolnicki, J.P. Riehl, U.N. Le, S.M. Cohen, K.N. Raymond, J. Am. Chem. Soc. **129**, 77 (2007)
29. E. Deiters, B. Song, A. Chauvin, C.D.B. Vandevyver, F. Gumy, J.-C.G. Bünzli, Chem. Eur. J. **15**, 885 (2009)
30. D. Imperio, G.B. Giovenzana, G.-L. Law, D. Parker, J.W. Walton, Dalton Trans. **39**, 9897 (2010)
31. J. Xu, T.M. Corneillie, E.G. Moore, G.-L. Law, N.G. Butlin, K.N. Raymond, J. Am. Chem. Soc. **133**, 19900 (2011)
32. Y. Hasegawa, S. Tsuruoka, T. Yoshida, H. Kawai, T. Kawai, J. Phys. Chem. A **112**, 803 (2008)
33. M.T. Whited, E. Rivard, J.C. Peters, Chem. Commun. 1613 (2006)
34. J.G. Hartley, L.M. Venanzi, D.C. Goodall, J. Chem. Soc. 3930 (1963)

35. A. Heynderickx, A. Samat, R. Guglielmetti, Synthesis **2**, 213 (2002)
36. K. Nakamura, Y. Hasegawa, H. Kawai, N. Yasuda, Y. Tsukahara, Y. Wada, Thin Solid Films **516**, 2376 (2008)
37. R.B. King, J. Am. Chem. Soc. **91**, 7211 (1969)
38. K. Nakamura, Y. Hasegawa, H. Kawai, N. Yasuda, N. Kanehisa, Y. Kai, T. Nagamura, S. Yanagida, Y. Wada, J. Phys. Chem. A **111**, 3029 (2007)
39. T. Nakagawa, Y. Hasegawa, T. Kawai, J. Phys. Chem. A **112**, 5096 (2008)
40. T. Nakagawa, K. Atsumi, T. Nakashima, Y. Hasegawa, T. Kawai, Chem. Lett. **36**, 372 (2007)
41. T. Nakagawa, Y. Hasegawa, T. Kawai, Chem. Commun. 5630 (2009)
42. S.F. Mason, J. Indian Chem. Soc. **63**, 73 (1986)
43. J.J. Dallara, M.F. Reid, F.S. Richardson, J. Phys. Chem. **88**, 3587 (1984)

Chapter 5
Photophysical Properties of Lanthanide Complexes with Asymmetric Dodecahedron Structures

5.1 Introduction

As described in Chap. 4, the geometrical symmetry of lanthanide complex is regarded as a significant factor influencing the transition probability. It has been widely accepted that the radiative transition probability between 4f orbitals is enhanced by reducing the coordination structure's geometrical symmetry [1–9]. In 2000, Raymond has proposed a symmetric factor, the shape measure *S*, to estimate the geometrical distortion of lanthanide complexes [10]. Harada has also found that distortion of the coordination structures of Eu(III) complexes leads to enhancement of the electric dipole transition probability [11, 12]. Here, the author focused on oxo-linked bidentate phosphine framework as a novel ligand of lanthanide complex for enhancement of geometrical distortion. The oxo-linked bidentate phosphine ligands have been known to provide a large bite angle between a metal ion and phosphorus atoms in metal complex [13]. The author considered that the introduction of oxo-linked bidentate phosphine framework into coordination sites of lanthanide complexes might lead to expansion of their bite angles related to geometrical distortion (Fig. 5.1a). Based on these structural and photophysical findings, a lanthanide complex with distorted coordination structure composed of oxo-linked bidentate phosphine oxide ligands is expected to give rise to an increase in the electric dipole transition probability.

In this chapter, the author reports on the characteristic photophysical properties of lanthanide complexes with novel asymmetric structures (trigonal dodecahedron structures) composed of three kinds of oxo-linked bidentate phosphine oxide ligands (4,5-bis(diphenylphosphoryl)-9,9-dimethylxanthene: xantpo, 4,5-bis(di-*tert*-butylphosphoryl)-9,9-dimethylxanthene: *t*Bu-xantpo, and bis[(2-diphenylphosphoryl)phenyl] ether: dpepo) as shown in Fig. 5.1b. Robertson has recently reported that Eu(III) complexes with hfa and dpepo ligands exhibited strong luminescence properties in polymer thin films, [14] however, the geometrical structure and radiative rate constants of lanthanide complexes with dpepo ligands have not been reported. The author characterized the geometrical structures of the Eu(III) and Sm(III) complexes using X-ray single crystal analyses and shape

K. Miyata, *Highly Luminescent Lanthanide Complexes with Specific Coordination Structures*, Springer Theses,
DOI: 10.1007/978-4-431-54944-4_5, © Springer Japan 2014

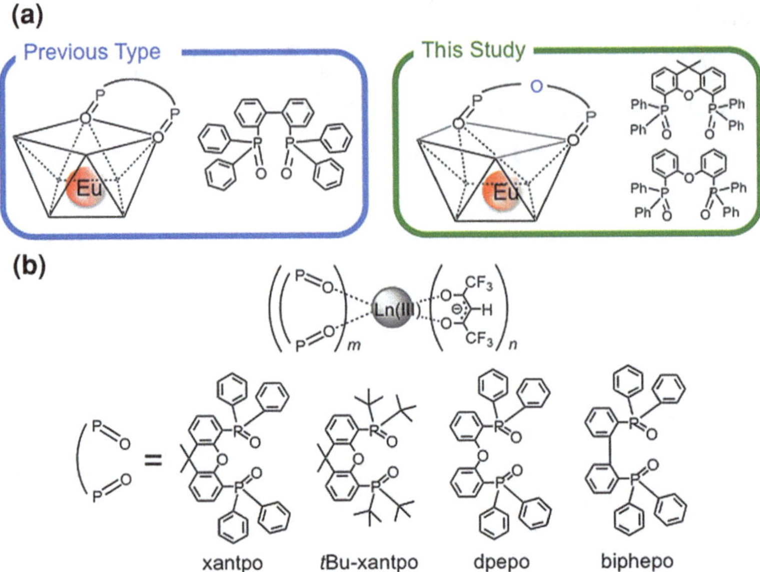

Fig. 5.1 **a** The research concept in this study and **b** chemical structures of lanthanide(III) complexes (Ln = Eu, Sm)

measure criteria. The luminescence properties were characterized by their emission quantum yields, emission lifetimes, and their radiative and nonradiative rate constants. Based on the structural and photophysical results, remarkable luminescence properties of lanthanide complexes with dodecahedron structures were demonstrated for the first time. The author also reports that the emission quantum yield of $Sm(hfa)_3(dpepo)$ with a trigonal dodecahedron structure is the highest in previous reported Sm(III) complexes. These remarkable luminescence properties will be elucidated in terms of their distorted coordination structures.

5.2 Experimental Section

5.2.1 Materials

Europium acetate monohydrate (99.9 %), samarium acetate tetrahydrate (99.9 %), acetone-d_6 (D, 99.9 %) and $CDCl_3$ (D, 99.8 %) were purchased from Wako Pure Chemical Industries Ltd. 1,1,1,5,5,5-hexafluoro-2,4-pentanedione, 4,5-bis(diphenyl-pho-sphino)-9,9-dimethylxanthene, 4,5-bis(di-*tert*-butylphosphino)-9,9-dimethylx-anthene and bis[(2-diphenylphosphino)phenyl]ether were obtained from Tokyo Kasei

Organic Chemicals and Aldrich Chemical Company Inc. All other chemicals and solvents were reagent grade and were used without further purification.

5.2.2 Apparatus

Infrared spectra were recorded on a JASCO FT/IR–420 spectrometer. ^{1}H (300 and 500 MHz) and ^{31}P NMR (200 MHz) spectra were recorded on a JEOL ECP–500. Chemical shifts are reported in δ ppm, referenced to an internal tetramethylsilane standard for ^{1}H NMR and an external 85 % H_3PO_4 standard for ^{31}P NMR. Mass spectra were measured using a JEOL JMS–700 M Station. Elemental analyses were performed using a Perkin Elmer 2400II.

5.2.3 Syntheses

5.2.3.1 Preparation of Tris(hexafluoroacetylacetonato)europium Dihydrates [Eu(hfa)$_3$(H$_2$O)$_2$]

Europium acetate monohydrate (8.0 g, 23 mmol) was dissolved in distilled water (100 mL) in a 300 mL flask. A solution of 1,1,1,5,5,5-hexafluoro-2,4-pentanedione (16 g, 77 mmol) was added dropwise to the solution. The reaction mixture produced a precipitation of white yellow powder after stirring for 2 h at room temperature. The reaction mixture was filtered, and the resulting powder was recrystallized from methanol/water to afford colorless needle crystals of the titled compound.

Yield: 15 g (79 %). IR (KBr): 1650 (st, C=O), 1145–1258 (st, C–F) cm^{-1}. Anal. Calcd for $C_{15}H_7EuF_{18}O_8$: C, 22.48; H, 0.88. Found: C, 22.12; H, 1.01.

5.2.3.2 Preparation of Tris(hexafluoroacetylacetonato)samarium Dihydrates [Sm(hfa)$_3$(H$_2$O)$_2$]

Samarium acetate tetrahydrate (5.0 g, 13 mmol) was dissolved in distilled water (60 mL) in a 100 mL flask. A solution of 1,1,1,5,5,5-hexafluoro-2,4-pentanedione (10 g, 48 mmol) was added dropwise to the solution. The reaction mixture produced a precipitation of white yellow powder after stirring for 2 h at room temperature. The reaction mixture was filtered, and the resulting powder was recrystallized from methanol to afford colorless needle crystals of the titled compound.

Yield: 8.3 g (82 %). IR (KBr): 1646 (st, C=O), 1094–1251 (st, C–F) cm^{-1}. Anal. Calcd for $C_{15}H_7SmF_{18}O_8$: C, 22.31; H, 0.87. Found: C, 21.92; H, 1.10.

5.2.3.3 Preparation of 4,5-bis(diphenylphosphoryl)-9,9-dimethylxanthene (xantpo)

4,5-bis(diphenylphosphino)-9,9-dimethylxanthene (1.0 g, 1.7 mmol) was dissolved in dichloromethane (20 mL) in a 100 mL flask. The solution was cooled to 0 °C, and then a 30 % H_2O_2 aqueous solution (4.0 mL) was added. The reaction mixture was stirred at 0 °C for 2 h, and was then washed with water and extracted three times with dichloromethane. The organic layer was dried over anhydrous magnesium sulfate, and concentrated to dryness. Reprecipitation from hexane gave a white powder of the titled compound.

Yield: 1.1 g (99 %). IR (ATR) 1190 (st, P=O), 1100–1229 (st, C–O–C) cm^{-1}. ^{1}H NMR (300 MHz, $CDCl_3$, 25 °C) δ 7.58–7.60 (d, J = 6 Hz, 2H, Ar), 7.30–7.47 (m, 20H, Ar), 6.94–7.00 (t, J = 6 Hz, 2H, Ar), 6.78–6.85 (m, 2H, Ar), 1.69 (s, 6H, 2Me) ppm. ^{31}P NMR (200 MHz, $CDCl_3$, 25 °C) δ 33.55 (1P), 30.32 (1P) ppm. FAB–Mass (*m/z*) = 611 $[M+H]^+$.

5.2.3.4 Preparation of 4,5-bis(di-*tert*-butylphosphoryl)-9,9-dimethylxanthene (*t*Bu-xantpo)

4,5-bis(di-*tert*-butylphosphino)-9,9-dimethylxanthene (1.0 g, 2.0 mmol) was dissolved in dichloromethane (20 mL) in a 100 mL flask. The solution was cooled to 0 °C, and then a 30 % H_2O_2 aqueous solution (4.5 mL, 40 mmol) was added. The reaction mixture was stirred at 0 °C for 2 h, and was then washed with water and extracted three times with dichloromethane. The organic layer was dried over anhydrous magnesium sulfate, and concentrated to dryness. Reprecipitation from hexane gave a white powder of the titled compound.

Yield: 0.97 g (91 %). IR (ATR) 1180 (st, P=O), 1103–1200 (st, C–O–C) cm^{-1}. ^{1}H NMR (300 MHz, $CDCl_3$, 25 °C) δ 7.82–7.84 (d, J = 6 Hz, 2H, Ar), 7.53–7.60 (m, 2H, Ar), 7.40–7.46 (m, 2H, Ar), 1.67 (s, 6H, 2Me), 1.35–1.46 (m, 36H, 4*t*Bu) ppm. ^{31}P NMR (200 MHz, $CDCl_3$, 25 °C) δ 69.53 (1P), 58.36 (1P) ppm. FAB–Mass (*m/z*) = 531 $[M+H]^+$.

5.2.3.5 Preparation of Bis[(2-diphenylphosphoryl)phenyl]ether (dpepo)

Bis[(2-diphenylphosphino)phenyl]ether (5.0 g, 9.3 mmol) was dissolved in dichloromethane (100 mL) in a 300 mL flask. The solution was cooled to 0 °C, and then a 30 % H_2O_2 aqueous solution (20 mL) was added. The reaction mixture was stirred at 0 °C for 2 h, and was then washed with water and extracted three times with dichloromethane. The organic layer was dried over anhydrous magnesium sulfate, and concentrated to dryness. Reprecipitation from hexane gave a white powder of the titled compound.

Yield: 5.0 g (94 %). IR (ATR) 1183 (st, P=O), 1070–1226 (st, C–O–C) cm^{-1}. ^{1}H NMR (300 MHz, $CDCl_3$, 25 °C) δ 7.06–7.71 (m, 26H, Ar), 6.02–6.07 (m, 2H, Ar) ppm. ^{31}P NMR (200 MHz, $CDCl_3$, 25 °C) δ 26.41 (2P) ppm. FAB–Mass (*m/z*) = 571 $[M\text{+H}]^+$.

5.2.3.6 General Procedure for the Preparation of Eu(III) and Sm(III) Complexes

Phosphine oxide ligand (1 equiv) and $Ln(hfa)_3(H_2O)_2$ (1.2 equiv) were dissolved in methanol (30 mL). The solution was refluxed while stirring for 8 h, and the reaction mixture was concentrated to dryness. The residue was washed with chloroform several times. The insoluble material was removed by filtration, and the filtrate was concentrated. The obtained powder was dissolved in hot methanol solution, and was then permitted to stand at room temperature. Recrystallization from methanol gave colorless block crystals of the lanthanide complexes.

[Eu(hfa)$_2$(xantpo)$_2$]: Yield: 0.32 g (49 %). IR (ATR) 1653 (st, C=O), 1137 (st, P=O), 1095–1251 (st, C–O–C and st, C–F) cm^{-1}. ^{1}H NMR (300 MHz, $CDCl_3$, 25 °C) δ 6.74–7.65 (m), 1.88 (s, Me) ppm. ^{31}P NMR (200 MHz, acetone-d_6, 25 °C) δ –92.12 (2P), –98.86 (2P) ppm. ESI–Mass (*m/z*) Calcd for $C_{88}H_{66}EuF_{12}O_{10}P_4$ $[M\text{–(hfa)}]^+$: 1787.264; Found: 1787.264. Anal. Calcd for $C_{93}H_{67}EuF_{18}O_{12}P_4$ $1.5CHCl_3$: C, 52.22; H, 3.18. Found: C, 52.11; H, 3.25.

[Sm(hfa)$_2$(xantpo)$_2$]: Yield: 0.79 g (46 %). IR (ATR) 1653 (st, C=O), 1138 (st, P=O), 1100–1252 (st, C–O–C and st, C–F) cm^{-1}. ^{1}H NMR (500 MHz, acetone-d_6, 25 °C) δ 5.81–8.09 (m), 1.67 (s, Me), 1.52 (s, Me) ppm. ^{31}P NMR (200 MHz, acetone-d_6, 25 °C) δ 33.03 (2P), 32.80 (2P) ppm. ESI–Mass (*m/z*) Calcd for $C_{88}H_{66}SmF_{12}O_{10}P_4$ $[M\text{–(hfa)}]^+$: 1786.257; Found: 1786.261. Anal. Calcd for $C_{93}H_{78}SmF_{15}O_{15}P_4$: C, 55.99; H, 3.94. Found: C, 55.54; H, 3.53.

[Eu(hfa)$_3$(*t*Bu-xantpo)]: Yield: 0.35 g (66 %). IR (ATR) 1653 (st, C=O), 1138 (st, P=O), 1098–1249 (st, C–O–C and st, C–F) cm^{-1}. ^{1}H NMR (300 MHz, $CDCl_3$, 25 °C) δ 7.41 (m, 2H, Ar), 7.07 (m, 2H, Ar), 6.83 (m, 2H, Ar), 5.92 (s, 3H, hfa-H), 2.97–3.02 (m, 6H, 2Me), 1.41–1.68 (m, 36H, 4*t*Bu) ppm. ^{31}P NMR (200 MHz, acetone-d_6, 25 °C) δ 68.41 (2P) ppm. ESI–Mass (*m/z*) Calcd for $C_{41}H_{50}EuF_{12}O_7P_2$ $[M\text{–(hfa)}]^+$: 1097.206; Found: 1097.206. Anal. Calcd for $C_{46}H_{51}EuF_{18}O_9P_2$: C, 42.38; H, 3.94. Found: C, 42.93; H, 4.00.

[Sm(hfa)$_3$(*t*Bu-xantpo)]: Yield: 0.39 g (53 %). IR (ATR) 1653 (st, C=O), 1137 (st, P=O), 1100–1252 (st, C–O–C and st, C–F) cm^{-1}. ^{1}H NMR (500 MHz, acetone-d_6, 25 °C) δ 8.06–8.08 (d, J = 7.5 Hz, 2H, Ar), 7.73–7.77 (m, 2H, Ar), 7.50–7.53 (t, J = 7.5 Hz, 2H, Ar), 6.73 (s, 3H, hfa-H), 1.83 (s, 6H, 2Me), 0.52–0.55 (d, J = 15 Hz, 36H, 4*t*Bu) ppm. ^{31}P NMR (200 MHz, acetone, 25 °C) δ 62.37 (2P) ppm. ESI–Mass (*m/z*) Calcd for $C_{41}H_{50}SmF_{12}O_7P_2$ $[M\text{–(hfa)}]^+$: 1096.204. Found: 1096.200. Anal. Calcd for $C_{46}H_{51}SmF_{18}O_9P_2$: C, 42.43; H, 3.95. Found: C, 42.68; H, 3.71.

[Eu(hfa)$_3$(dpepo)]: Yield: 0.62 g (74 %). IR (ATR) 1653 (st, C=O), 1135 (st, P=O), 1098–1251 (st, C–O–C and st, C–F) cm^{-1}. ^{1}H NMR (500 MHz, acetone,

25 °C) δ 7.32–7.64 (m, 22H, Ar), 7.10–7.13 (t, $J = 7.5$ Hz, 2H, Ar), 6.90–6.95 (dd, $J = 7.5$ Hz, 2H, Ar), 6.73 (s, 3H, hfa-H), 6.29–6.30 (m, 2H, Ar) ppm. ^{31}P NMR (200 MHz, acetone-d_6, 25 °C) δ −113.42 (2P) ppm. ESI–Mass (*m/z*) Calcd for $C_{46}H_{30}EuF_{12}O_7P_2$ [*M*–(hfa)]$^+$: 1137.049; Found: 1137.049. Anal. Calcd for $C_{51}H_{31}EuF_{18}O_9P_2$: C, 45.59; H, 2.33. Found: C, 45.76; H, 2.11.

[Sm(hfa)$_3$(dpepo)]: Yield: 0.66 g (79 %). IR (ATR) 1653 (st, C=O), 1134 (st, P=O), 1098–1250 (st, C–O–C and st, C–F) cm^{-1}. ^{1}H NMR (500 MHz, acetone-d_6, 25 °C) δ 7.32–7.64 (m, 22H, Ar), 7.10–7.13 (t, $J = 7.5$ Hz, 2H, Ar), 6.90–6.95 (dd, $J = 7.5$ Hz, 2H, Ar), 6.73 (s, 3H, hfa-H), 6.29–6.30 (m, 2H, Ar) ppm. ^{31}P NMR (200 MHz, acetone-d_6, 25 °C) δ 29.17 (2P) ppm. ESI–Mass (*m/z*) Calcd for $C_{46}H_{30}SmF_{12}O_7P_2$ [*M*–(hfa)]$^+$: 1136.048; Found: 1136.044. Anal. Calcd for $C_{51}H_{31}SmF_{18}O_9P_2$: C, 45.64; H, 2.33. Found: C, 45.54; H, 2.18.

5.2.4 Crystallography

Colorless single crystals of lanthanide complexes obtained from the methanol solution were mounted on a glass fiber using epoxy resin glue. All measurements were made on a Rigaku RAXIS RAPID imaging plate area detector with graphite monochromated MoKα radiation. Corrections for decay and Lorentz-polarization effects were made using empirical absorption correction, solved by direct methods and expanded using Fourier techniques. Non-hydrogen atoms were refined anisotropically. Hydrogen atoms were refined using the riding model. The final cycle of full-matrix least-squares refinement was based on observed reflections and variable parameters. All calculations were performed using the crystal structure crystallographic software package. The author confirmed the CIF data using the check CIF/PLATON service.

5.2.5 Optical Measurements

UV–Vis absorption spectra were recorded on a JASCO V–660 spectrometer. Emission spectra of the lanthanide complexes were measured with a Hitachi F–4500 spectrometer and corrected for the response of the detector system. The emission quantum yields of lanthanide complex solutions degassed with an argon (10 mM in acetone-d_6) were obtained by comparison with the integrated emission signal (550–750 nm) of Eu(hfa)$_3$(biphepo) as a reference ($\Phi_{Ln} = 0.60$: 50 mM in acetone-d_6) with an excitation wavelength of 465 nm (direct excitation of Eu(III) ions) for Eu(III) complexes [15] or Sm(hfa)$_3$(H$_2$O)$_2$ as a reference ($\Phi_{Ln} = 0.031$: 100 mM in DMSO-d_6) with an excitation wavelength of 481 nm (direct excitation of Sm(III) ions) for Sm(III) complexes. Emission lifetimes of lanthanide complexes (10 mM in acetone-d_6) were measured using the third harmonics (355 nm) of a Q-switched Nd:YAG laser (Spectra Physics, INDI-50, fwhm = 5 ns, $\lambda = 1064$ nm) and a photomultiplier (Hamamatsu photonics, R5108, response

Table 5.1 Crystal data of Eu(III) complexes

	$Eu(hfa)_2(xantpo)_2$	$Eu(hfa)_3$ (*t*Bu-xantpo)	$Eu(hfa)_3$(dpepo)	$Eu(hfa)_3$(biphepo)
Chemical formula	$C_{93}H_{67}F_{18}O_{12}P_4Eu$	$C_{46}H_{51}F_{18}O_9P_2Eu$	$C_{51}H_{31}F_{18}O_9P_2Eu$	$C_{51}H_{31}F_{18}O_8P_2Eu$
Formula weight	1994.37	1303.78	1343.68	1327.68
Crystal color, habit	Colorless, block	Colorless, block	Colorless, block	Colorless, block
Crystal system	Triclinic	Monoclinic	Triclinic	Monoclinic
Space group	*P*-1(#2)	$P2_1/c$(#14)	*P*-1(#2)	$P2_1/n$(#14)
a/Å	12.8434(2)	21.9134(5)	12.3802(5)	13.1819(2)
b/Å	17.9406(3)	20.1740(5)	13.3535(5)	31.5572(6)
c/Å	19.2651(4)	24.1073(6)	18.6388(9)	13.5087(3)
α/deg	84.3311(7)		82.4190(13)	
β/deg	82.6768(7)	149.9133(7)	77.2612(15)	111.4181(7)
γ/deg	80.9203(7)		62.3602(9)	
$V/Å^3$	4333.60(14)	5342.7(2)	2660.94(19)	5231.31(17)
Z	2	4	2	4
$d_{calc}/g\ cm^{-3}$	1.528	1.621	1.677	1.686
T/°C	-170 ± 1	-90 ± 1	-90 ± 1	-170 ± 1
μ (Mo Kα)/cm^{-1}	8.967	13.449	13.535	13.745
Max 2θ/deg	55.0	50.6	50.7	55.0
No. of measured reflections	43429	42405	21981	51911
No. of unique reflections	19805	9700	9698	11985
$R\ (I > 2\sigma(I))^a$	0.0453	0.0287	0.0385	0.0340
$R_w\ (I > 2\sigma(I))^b$	0.1366	0.0720	0.1062	0.0910

[a] $R = \Sigma\ ||F_o| - |F_c||/\Sigma\ |F_o|$
[b] $R_w = [(\Sigma\ w\ (|F_o| - |F_c|)^2/\Sigma\ w\ F_o^2)]^{1/2}$

time ≤1.1 ns). The Nd:YAG laser response was monitored with a digital oscilloscope (Sony Tektronix, TDS3052, 500 MHz) synchronized to the single-pulse excitation. Emission lifetimes were determined from the slope of logarithmic plots of the decay profiles. High-resolution spectra of the emission were measured with a HORIBA SPEX fluorolog.

5.3 Results and Discussion

5.3.1 Coordination Structures

Single crystals of the lanthanide complexes with oxo-linked bidentate phosphine oxides were successfully prepared for X-ray single-crystal analyses by recrystallization from methanol solutions. The resulting crystal data are summarized in Tables 5.1 and 5.2. The ORTEP views of all the lanthanide complexes showed octa-coordinated structures (Figs. 5.2 and 5.3). The coordination sites of $Eu(hfa)_3$(*t*Bu-xantpo), $Eu(hfa)_3$(dpepo), $Eu(hfa)_3$(biphepo), $Sm(hfa)_3$(*t*Bu-xantpo)

Table 5.2 Crystal data of Sm(III) complexes

	$Sm(hfa)_2(xantpo)_2$	$Sm(hfa)_3(t\text{Bu-xantpo})$	$Sm(hfa)_3(dpepo)$
Chemical formula	$C_{93}H_{78}F_{15}O_{15}P_4Sm$	$C_{46}H_{51}F_{18}O_9P_2Sm$	$C_{51}H_{31}F_{18}O_9P_2Sm$
Formula weight	1994.9	1302.22	1342.12
Crystal color, habit	colorless, block	colorless, block	colorless, block
Crystal system	Triclinic	Monoclinic	Triclinic
Space group	P-1(#2)	$P2_1/n$(#14)	P-1(#2)
a/Å	12.9433(4)	12.1153(5)	12.3661(5)
b/Å	17.7795(7)	20.1706(7)	13.4079(6)
c/Å	19.4357(7)	21.8687(8)	18.5261(7)
α/deg	85.0606(10)		82.2082(12)
β/deg	82.9841(10)	95.0228(12)	77.0833(10)
γ/deg	82.0480(11)		61.9059(11)
V/Å^3	4385.6(3)	5323.6(3)	2639.56(19)
Z	2	4	2
d_{calc}/g cm^{-3}	1.511	1.625	1.689
T/°C	-140 ± 1	-100 ± 1	-150 ± 1
μ (Mo Kα)/cm^{-1}	8.143	12.797	12.939
Max 2θ/deg	50.6	50.6	50.6
No. of measured reflections	35841	42569	21444
No. of unique reflections	15874	9688	9574
R ($I > 2\sigma$(I))[a]	0.0309	0.0287	0.0378
R_w ($I > 2\sigma$(I))[b]	0.0705	0.0984	0.1055

[a] $R = \Sigma \,||F_o| - |F_c||/\Sigma\, |F_o|$
[b] $R_w = [(\Sigma\, w\, (|F_o| - |F_c|)^2/\Sigma\, w\, F_o^2)]^{1/2}$

and $Sm(hfa)_3(dpepo)$ comprised three hexafluoroacetylacetonato (hfa) ligands and one bidentate phosphine oxide ligand. In contrast, the coordination sites of $Eu(hfa)_2(xantpo)_2$ and $Sm(hfa)_2(xantpo)_2$ comprised two hfa ligands and two bidentate phosphine oxide ligands with hexafluoroacetylacetonato anion (hfa$^-$: $CF_3COC^-HCOCF_3$) and trifluoroacetate anion (CF_3COO^-) counter ions, respectively. The author considers that the trifluoroacetate anion in Fig. 5.3 might be formed by the oxidation reaction of the hexafluoroacetylacetonato ligand [16].

Based on the crystal data, the author carried out the calculations of the shape factor S in order to estimate the degree of distortion of the coordination structure in first coordination sphere [10]. The S value is given by:

$$S = \min\sqrt{\left(\frac{1}{m}\right)\sum_{i=1}^{m}(\delta_i - \theta_i)^2} \tag{5.1}$$

where m, δ_i and θ_i are the number of possible edges ($m = 18$ in this study), the observed dihedral angle between planes along the ith edge and the dihedral angle for the ideal structure, respectively. The estimated S values of lanthanide complexes are summarized in Tables 5.3, 5.4, 5.5, 5.6, 5.7, 5.8, 5.9, 5.10, 5.11 and Figs. 5.4, 5.5, 5.6. For $Eu(hfa)_2(xantpo)_2$, the S value for the octa-coordinated

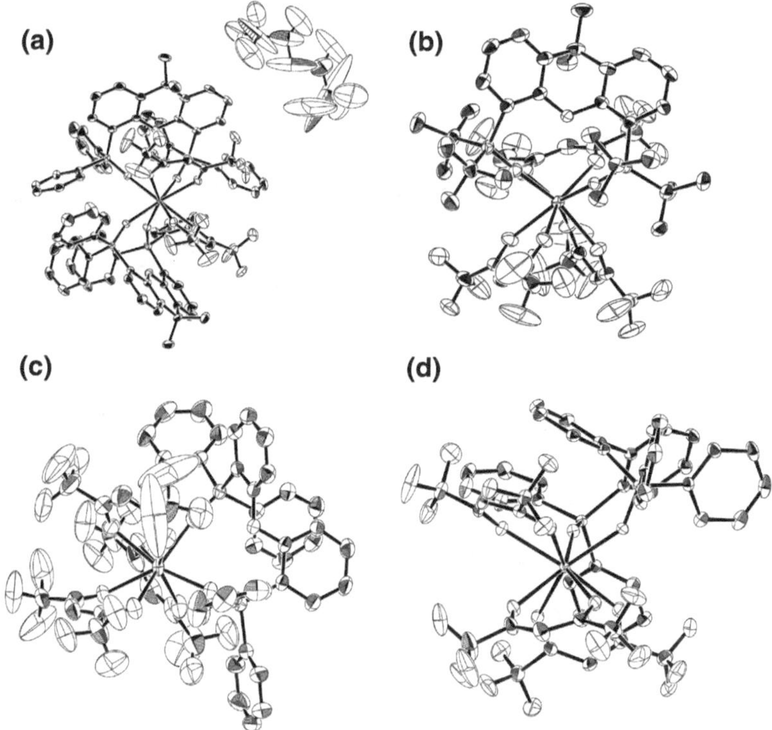

Fig. 5.2 ORTEP drawings of **a** $Eu(hfa)_2(xantpo)_2$, **b** $Eu(hfa)_3$(*t*Bu-xantpo), **c** $Eu(hfa)_3$(dpepo), and **d** $Eu(hfa)_3$(biphepo). Hydrogen atoms have been omitted for clarity and thermal ellipsoids are shown at the 50 % probability level

square antiprism structure (8-SAP, point group: D_{4d}, $S = 3.0°$) is smaller than that for the octa-coordinated trigonal dodecahedron structure (8-TDH, point group: D_{2d}, $S = 13°$), suggesting that the 8-SAP structure is less distorted than the 8-TDH structure. The author thus determined that the coordination geometry of $Eu(hfa)_2(xantpo)_2$ is 8-SAP. Based on the minimum value of *S*, the coordination geometries of lanthanide complexes with *t*Bu-xantpo and dpepo ligands are classified as 8-TDH, while those with xantpo and biphepo ligands are classified as 8-SAP. These estimations also suggest that the geometrical symmetry in first coordination sphere of lanthanide complexes with *t*Bu-xantpo and dpepo ligands (point group: D_{2d}) is lower than those of lanthanide complexes with xantpo and biphepo ligands (point group: D_{4d}).

The author also determined the symmetrical point groups of the lanthanide complexes concerned with the locations of phosphine oxide and β-diketonato linker species (Fig. 5.7). The symmetrical point groups of $Eu(hfa)_3$(*t*Bu-xantpo), $Eu(hfa)_3$(dpepo), $Sm(hfa)_3$(*t*Bu-xantpo) and $Sm(hfa)_3$(dpepo) were found to be quasi-C_1. On the other hand, those of $Eu(hfa)_2(xantpo)_2$ and $Sm(hfa)_2(xantpo)_2$ were categorized as quasi-C_2 because of their two coordinated xantpo ligands.

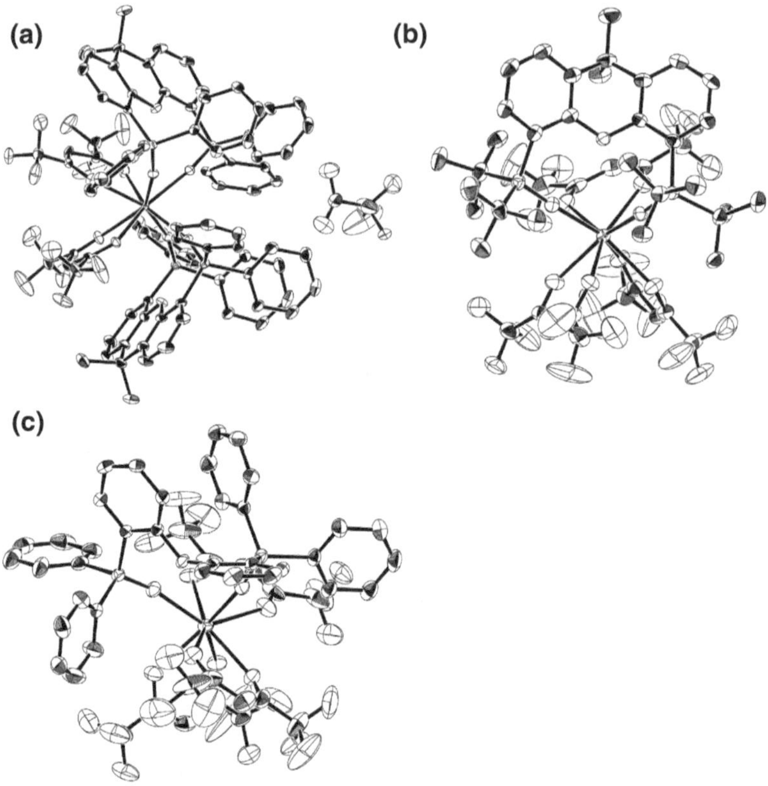

Fig. 5.3 ORTEP drawings of **a** $Sm(hfa)_2(xantpo)_2$, **b** $Sm(hfa)_3$(*t*Bu-xantpo), and **c** $Sm(hfa)_3$(dpepo). Hydrogen atoms have been omitted for clarity and thermal ellipsoids are shown at the 50 % probability level

Hasegawa has reported that the coordination structure of $Eu(hfa)_3$(biphepo) was classified as 8-SAP with quasi-C_1 symmetry [17]. These results indicate that lanthanide complexes with 8-TDH geometry (D_{2d}) and quasi-C_1 symmetry are expected to enhance the electric transition probability in the 4f orbitals related to change of odd parity. The characteristic structures of these lanthanide complexes are likely to significantly affect their photophysical properties.

5.3.2 Photophysical Properties

The steady-state emission spectra of lanthanide complexes in acetone-d_6 are shown in Fig. 5.8. Emission bands of Eu(III) complexes are observed at around 578, 592, 613, 650, and 698 nm, and are attributed to the f–f transitions of $^5D_0$–^7F_J with $J = 0$, 1, 2, 3 and 4 respectively. Emission bands of Sm(III) complexes are

Table 5.3 Observed dihedral angles (δ_i), dihedral angles of idealized square antiprism (θ_i) and measure shape criteria, $S(D_{4d})$ for $Eu(hfa)_2(xantpo)_2$ and $Eu(hfa)_3$(*t*Bu-xantpo)

	$Eu(hfa)_2(xantpo)_2$			$Eu(hfa)_3$(*t*Bu-xantpo)		
θ_i	Edge	δ_i	$\delta_i - \theta_i$	Edge	δ_i	$\delta_i - \theta_i$
0	O6–O7	4.35	4.35	O2–O4	24.70	24.70
0	O1–O10	5.68	5.68	O6–O9	24.48	24.48
77.1	O4–O6	76.58	−0.52	O2–O5	72.71	−4.39
77.1	O4–O7	78.74	1.64	O1–O2	61.45	−15.65
77.1	O7–O8	73.66	−3.44	O1–O4	81.23	4.13
77.1	O6–O8	75.02	−2.08	O4–O5	68.02	−9.08
77.1	O3–O10	77.25	0.15	O7–O9	57.50	−19.6
77.1	O1–O3	76.56	−0.54	O8–O9	69.73	−7.37
77.1	O1–O9	74.25	−2.85	O6–O8	67.37	−9.73
77.1	O9–O10	72.41	−4.69	O6–O7	75.26	−1.84
51.6	O3–O6	52.52	0.92	O5–O9	51.81	0.21
51.6	O3–O4	52.89	1.29	O2–O9	40.08	−11.52
51.6	O1–O4	52.66	1.06	O2–O8	58.73	7.13
51.6	O1–O7	47.36	−4.24	O1–O8	60.64	9.04
51.6	O7–O9	54.47	2.87	O1–O6	54.28	2.68
51.6	O8–O9	48.13	−3.47	O4–O6	32.52	−19.08
51.6	O8–O10	54.43	2.83	O4–O7	57.73	6.13
51.6	O6–O10	47.79	−3.81	O5–O7	57.23	5.63
	$S(D_{4d})$	**3.03**		**$S(D_{4d})$**	**12.51**	

Table 5.4 Observed dihedral angles (δ_i), dihedral angles of idealized square antiprism (θ_i) and measure shape criteria, $S(D_{4d})$ for $Eu(hfa)_3$(dpepo) and $Eu(hfa)_3$(biphepo)

	$Eu(hfa)_3$(dpepo)			$Eu(hfa)_3$(biphepo)		
θ_i	Edge	δ_i	$\delta_i - \theta_i$	Edge	δ_i	$\delta_i - \theta_i$
0	O2–O5	33.82	33.82	O1–O3	1.64	1.64
0	O1–O7	33.02	33.02	O6–O8	1.97	1.97
77.1	O4–O5	70.62	−6.48	O1–O4	81.95	4.85
77.1	O2–O4	61.01	−16.09	O4–O3	75.87	−1.23
77.1	O2–O9	69.51	−7.59	O3–O2	73.75	−3.35
77.1	O5–O9	54.52	−22.58	O2–O1	80.23	3.13
77.1	O6–O7	81.43	4.33	O5–O6	80.41	3.31
77.1	O1–O6	82.45	5.35	O6–O7	76.13	−0.97
77.1	O1–O8	79.32	2.22	O7–O8	73.96	−3.14
77.1	O7–O8	80.43	3.33	O8–O5	78.55	1.45
51.6	O5–O6	29.60	−22.00	O1–O5	45.97	−5.63
51.6	O4–O6	64.21	12.61	O5–O4	56.55	4.95
51.6	O1–O4	60.29	8.69	O4–O6	47.12	−4.48
51.6	O1–O2	55.42	3.82	O6–O3	55.47	3.87
51.6	O2–O8	32.51	−19.09	O3–O7	47.45	−4.15
51.6	O8–O9	65.07	13.47	O7–O2	55.08	3.48
51.6	O7–O9	57.46	5.86	O2–O8	46.22	−5.38
51.6	O5–O7	56.22	4.62	O8–O1	55.53	3.93
	$S(D_{4d})$	**15.81**		**$S(D_{4d})$**	**3.66**	

Table 5.5 Observed dihedral angles (δ_i), dihedral angles of idealized square antiprism (θ_i) and measure shape criteria, $S(D_{2d})$ for Eu(hfa)$_2$(xantpo)$_2$ and Eu(hfa)$_3$(tBu-xantpo)

Eu(hfa)$_2$(xantpo)$_2$				Eu(hfa)$_3$(tBu-xantpo)			
θ_i	Edge	δ_i	$\delta_i - \theta_i$	θ_i	Edge	δ_i	$\delta_i - \theta_i$
61.48	O1–O3	76.62	15.14	61.48	O1–O2	61.45	−0.030
61.48	O1–O4	52.63	−8.85	74.29	O1–O4	81.23	6.94
29.86	O1–O7	47.34	17.48	61.48	O1–O6	54.28	−7.20
74.29	O1–O9	74.84	0.55	53.12	O1–O8	60.64	7.52
29.86	O1–O10	5.56	−24.30	29.86	O2–O4	24.7	−5.16
53.12	O3–O4	52.85	−0.27	74.29	O2–O5	72.71	−1.58
61.48	O3–O6	52.57	−8.91	61.48	O2–O8	58.73	−2.75
74.29	O3–O10	77.24	2.95	29.86	O2–O9	40.08	10.22
61.48	O4–O6	76.53	15.05	61.48	O4–O5	68.02	6.54
74.29	O4–O7	78.72	4.43	29.86	O4–O6	32.52	2.66
29.86	O6–O7	4.42	−25.44	61.48	O4–O7	57.73	−3.75
74.29	O6–O8	75.00	0.71	53.12	O5–O7	57.23	4.11
29.86	O6–O10	47.76	17.90	61.48	O5–O9	51.81	−9.67
61.48	O7–O8	73.67	12.19	74.29	O6–O7	75.26	0.97
61.48	O7–O9	54.51	−6.97	61.48	O6–O8	67.37	5.89
53.12	O8–O9	48.09	−5.03	29.86	O6–O9	24.48	−5.38
61.48	O8–O10	54.48	−7.00	61.48	O7–O9	57.5	−3.98
61.48	O9–O10	72.47	10.99	74.29	O8–O9	69.73	−4.56
		S(D$_{2d}$)	**12.69**			**S(D$_{2d}$)**	**5.64**

Table 5.6 Observed dihedral angles (δ_i), dihedral angles of idealized square antiprism (θ_i) and measure shape criteria, $S(D_{2d})$ for Eu(hfa)$_3$(dpepo) and Eu(hfa)$_3$(biphepo)

Eu(hfa)$_3$(dpepo)				Eu(hfa)$_3$(biphepo)			
θ_i	Edge	δ_i	$\delta_i - \theta_i$	θ_i	Edge	δ_i	$\delta_i - \theta_i$
61.48	O1–O2	55.42	−6.06	74.29	O1–O2	76.53	2.24
53.12	O1–O3	60.29	7.17	61.48	O1–O3	56.15	−5.33
61.48	O1–O6	59.53	−1.95	53.12	O1–O5	47.74	−5.38
74.29	O1–O7	80.46	6.17	61.48	O1–O7	74.88	13.40
61.48	O1–O8	61.01	−0.47	29.86	O2–O3	48.01	18.15
29.86	O2–O3	33.82	3.96	61.48	O2–O4	55.83	−5.65
29.86	O2–O4	32.51	2.65	29.86	O2–O7	1.98	−27.88
74.29	O2–O6	69.51	−4.78	61.48	O2–O8	83.23	21.75
61.48	O3–O4	70.62	9.14	74.29	O3–O4	80.77	6.48
61.48	O3–O8	64.21	2.73	61.48	O3–O5	76.31	14.83
29.86	O4–O5	29.60	−0.26	29.86	O3–O6	2.99	−26.87
61.48	O4–O6	56.22	−5.26	61.48	O4–O6	79.24	17.76
61.48	O4–O8	54.52	−6.96	53.12	O4–O8	45.31	−7.81
74.29	O5–O6	77.45	3.16	74.29	O5–O6	74.56	0.27
29.86	O5–O7	28.77	−1.09	61.48	O5–O7	56.10	−5.38
61.48	O5–O8	57.73	−3.75	29.86	O6–O7	45.26	15.4
53.12	O6–O7	57.46	4.34	61.48	O6–O8	55.70	−5.78
61.48	O7–O8	65.07	3.59	74.29	O7–O8	81.29	7.00
		S(D$_{2d}$)	**4.71**			**S(D$_{2d}$)**	**13.49**

Table 5.7 Observed dihedral angles (δ_i), dihedral angles of idealized square antiprism (θ_i) and measure shape criteria, $S(D_{4d})$ for $Sm(hfa)_2(xantpo)_2$ and $Sm(hfa)_3(t\text{Bu-xantpo})$

	$Sm(hfa)_2(xantpo)_2$			$Sm(hfa)_3(t\text{Bu-xantpo})$		
θ_i	Edge	δ_i	$\delta_i - \theta_i$	Edge	δ_i	$\delta_i - \theta_i$
0	O1–O10	2.67	2.67	O3–O8	24.98	24.98
0	O4–O8	4.97	4.97	O5–O6	19.81	19.81
77.1	O1–O2	76.85	−0.25	O3–O9	72.55	−4.55
77.1	O2–O10	79.81	2.71	O1–O3	61.61	−15.49
77.1	O1–O9	75.63	−1.47	O1–O8	80.96	3.86
77.1	O9–O10	74.98	−2.12	O8–O9	68.13	−8.97
77.1	O4–O7	80.85	3.75	O5–O7	57.44	−19.66
77.1	O7–O8	76.66	−0.44	O6–O7	75.37	−1.73
77.1	O5–O8	77.12	0.02	O4–O6	67.13	−9.97
77.1	O4–O5	80.06	2.96	O4–O5	69.79	−7.31
51.6	O1–O4	53.63	2.03	O5–O9	51.97	0.37
51.6	O2–O4	52.65	1.05	O3–O5	40.10	−11.50
51.6	O2–O5	52.19	0.59	O3–O4	58.41	6.81
51.6	O5–O10	47.54	−4.06	O1–O4	60.55	8.95
51.6	O8–O10	54.94	3.34	O1–O6	54.54	2.94
51.6	O8–O9	47.47	−4.13	O6–O8	32.69	−18.91
51.6	O7–O9	53.71	2.11	O7–O8	57.46	5.86
51.6	O1–O7	48.41	−3.19	O7–O9	57.07	5.47
	$S(D_{4d})$	**2.73**		**$S(D_{4d})$**	**12.03**	

Table 5.8 Observed dihedral angles (δ_i), dihedral angles of idealized square antiprism (θ_i) and measure shape criteria, $S(D_{4d})$ for $Sm(hfa)_3(dpepo)$

$Sm(hfa)_3(dpepo)$			
θ_i	Edge	δ_i	$\delta_i - \theta_i$
0	O3–O4	31.67	31.67
0	O7–O8	22.72	22.72
77.1	O1–O4	80.42	3.32
77.1	O1–O3	56.18	−20.92
77.1	O4–O5	66.06	−11.04
77.1	O3–O5	68.98	−8.12
77.1	O6–O7	80.19	3.09
77.1	O6–O8	87.36	10.26
77.1	O8–O9	86.53	9.43
77.1	O7–O9	80.91	3.81
51.6	O4–O6	30.02	−21.58
51.6	O1–O6	59.33	7.73
51.6	O1–O8	59.94	8.34
51.6	O3–O8	59.83	8.23
51.6	O3–O9	35.21	−16.39
51.6	O5–O9	54.08	2.48
51.6	O5–O7	57.36	5.76
51.6	O4–O7	56.60	5.00
	$S(D_{4d})$	**13.68**	

Table 5.9 Observed dihedral angles (δ_i), dihedral angles of idealized square antiprism (θ_i) and measure shape criteria, $S(D_{2d})$ for $Sm(hfa)_2(xantpo)_2$ and $Sm(hfa)_3(t$Bu-xantpo)

$Sm(hfa)_2(xantpo)_2$				$Sm(hfa)_3(t$Bu-xantpo)			
θ_i	Edge	δ_i	$\delta_i - \theta_i$	θ_i	Edge	δ_i	$\delta_i - \theta_i$
61.48	O1–O2	76.85	15.37	61.48	O1–O3	61.61	0.13
61.48	O1–O4	53.63	−7.85	53.12	O1–O4	60.55	7.43
29.86	O1–O7	48.41	18.55	61.48	O1–O6	54.54	−6.94
74.29	O1–O9	75.63	1.34	74.29	O1–O8	80.96	6.67
29.86	O1–O10	2.67	−27.19	61.48	O3–O4	58.41	−3.07
53.12	O2–O4	52.65	−0.47	29.86	O3–O5	40.10	10.24
61.48	O2–O5	52.19	−9.29	29.86	O3–O8	24.98	−4.88
74.29	O2–O10	79.81	5.52	74.29	O3–O9	72.55	−1.74
61.48	O4–O5	76.60	15.12	74.29	O4–O5	69.79	−4.50
74.29	O4–O7	77.46	3.17	61.48	O4–O6	67.13	5.65
29.86	O5–O7	4.82	−25.04	29.86	O5–O6	24.57	−5.29
74.29	O5–O8	73.59	−0.70	61.48	O5–O7	57.44	−4.04
29.86	O5–O10	47.54	17.68	61.48	O5–O9	51.97	−9.51
61.48	O7–O8	73.19	11.71	74.29	O6–O7	75.37	1.08
61.48	O7–O9	53.71	−7.77	29.86	O6–O8	32.69	2.83
53.12	O8–O9	47.47	−5.65	61.48	O7–O8	57.46	−4.02
61.48	O8–O10	54.94	−6.54	53.12	O7–O9	57.07	3.95
61.48	O9–O10	74.98	13.50	61.48	O8–O9	68.13	6.65
		S(D$_{2d}$)	**13.19**			**S(D$_{2d}$)**	**5.58**

Table 5.10 Observed dihedral angles (δ_i), dihedral angles of idealized square antiprism (θ_i) and measure shape criteria, $S(D_{2d})$ for $Sm(hfa)_3(dpepo)$

$Sm(hfa)_3(dpepo)$			
θ_i	Edge	δ_i	$\delta_i - \theta_i$
61.48	O1–O3	56.18	−5.30
74.29	O1–O4	80.42	6.13
61.48	O1–O6	59.33	−2.15
53.12	O1–O8	59.94	6.82
29.86	O3–O4	31.67	1.81
74.29	O3–O5	68.98	−5.31
61.48	O3–O8	59.83	−1.65
29.86	O3–O9	35.21	5.35
61.48	O4–O5	66.06	4.58
29.86	O4–O6	30.02	0.16
61.48	O4–O7	56.60	−4.88
53.12	O5–O7	57.36	4.24
61.48	O5–O9	54.08	−7.40
74.29	O6–O7	77.53	3.24
61.48	O6–O8	64.43	2.95
29.86	O6–O9	28.77	−1.09
61.48	O7–O9	56.92	−4.56
74.29	O8–O9	70.74	−3.55
		S(D$_{2d}$)	**4.42**

Table 5.11 Summary of shape-measure calculations of the Eu(III) and Sm(III) complexes

Complex	S value for 8-TDH[a]: $S(D_{2d})$	S value for 8-SAP[b]: $S(D_{4d})$	Determined coordination geometry
$Eu(hfa)_2(xantpo)_2$	13°	3.0°	8-SAP
$Eu(hfa)_3$(*t*Bu-xantpo)	5.6°	13°	8-TDH
$Eu(hfa)_3$(dpepo)	4.7°	16°	8-TDH
$Eu(hfa)_3$(biphepo)	13°	3.7°	8-SAP
$Sm(hfa)_2(xantpo)_2$	13°	2.7°	8-SAP
$Sm(hfa)_3$(*t*Bu-xantpo)	5.6°	12°	8-TDH
$Sm(hfa)_3$(dpepo)	4.4°	14°	8-TDH

[a] 8-TDH: octa-coordinated trigonal dodecahedron
[b] 8-SAP: octa-coordinated square antiprism

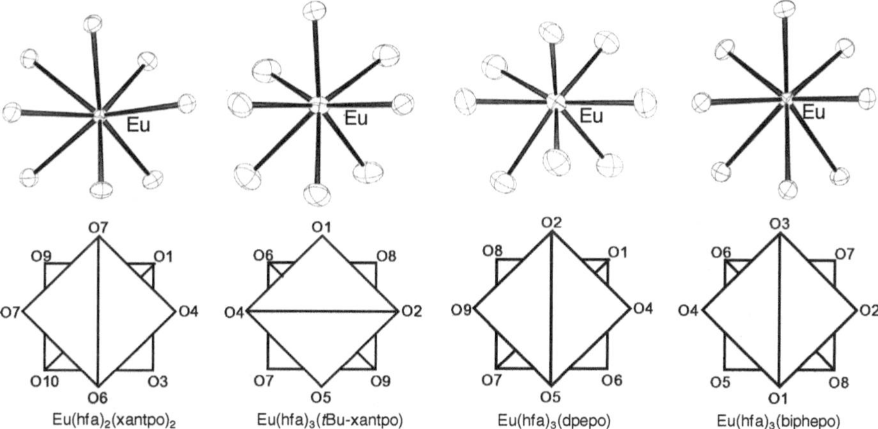

Fig. 5.4 Coordination environments around Eu(III) ion for the calculation of $S(D_{4d})$ values

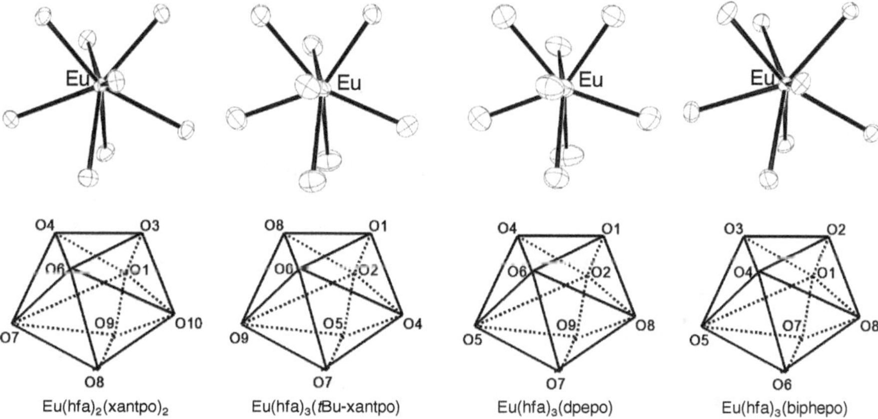

Fig. 5.5 Coordination environments around Eu(III) ion for the calculation of $S(D_{2d})$ values

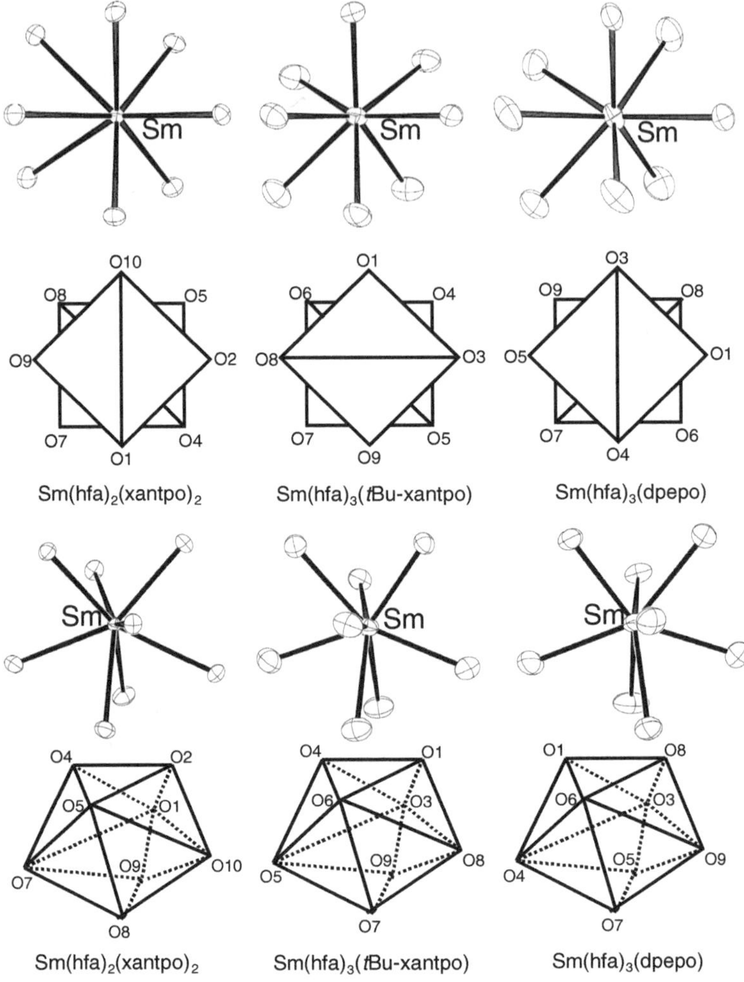

Fig. 5.6 Coordination environments around Sm(III) ion for the calculation of $S(D_{4d})$ and $S(D_{2d})$ values

also observed at around 562, 598, 642 and 704 nm, and are attributed to the f–f transitions of $^4G_{5/2}$–^6H_J with J = 5/2, 7/2, 9/2 and 11/2, respectively. The spectra are normalized with respect to the magnetic dipole transition intensities at 592 nm (Eu: $^5D_0$–$^7F_1$) and at 598 nm (Sm: $^4G_{5/2}$–$^6H_{7/2}$) which are known to be insensitive to the surrounding environment of the lanthanide ions [18, 19]. The emission bands at 613 nm (Eu: $^5D_0$–$^7F_2$) and 642 nm (Sm: $^4G_{5/2}$–$^6H_{9/2}$) are due to electric dipole transitions, which are strongly dependent on their coordination geometry. The emission quantum yields for Eu(III) and Sm(III) complexes in acetone-d_6 excited at their 4f orbitals (direct excitation) were found to be 55–72 % and 2.4–5.0 %, respectively (Table 5.12). These values are similar to those reported

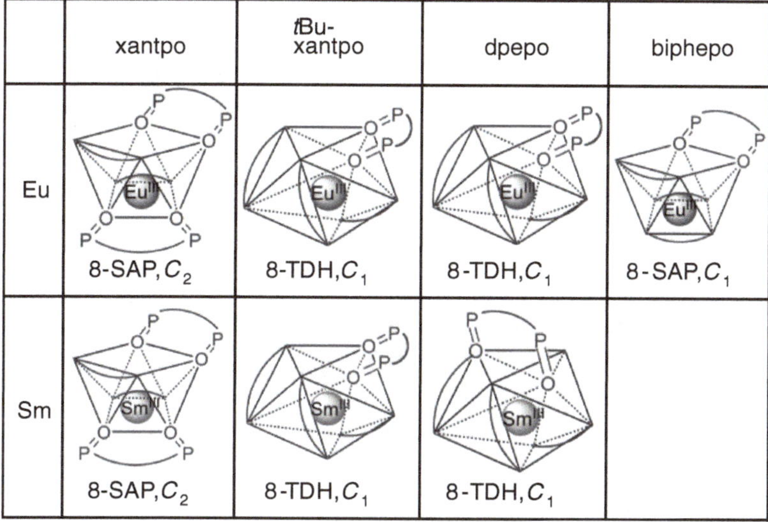

Fig. 5.7 Geometrical coordination structure images of Eu(III) and Sm(III) complexes. Coordinated oxygen atoms are shown as corner polyhedra. *Curved lines* between oxygen atoms represent phosphine oxide and hfa ligands

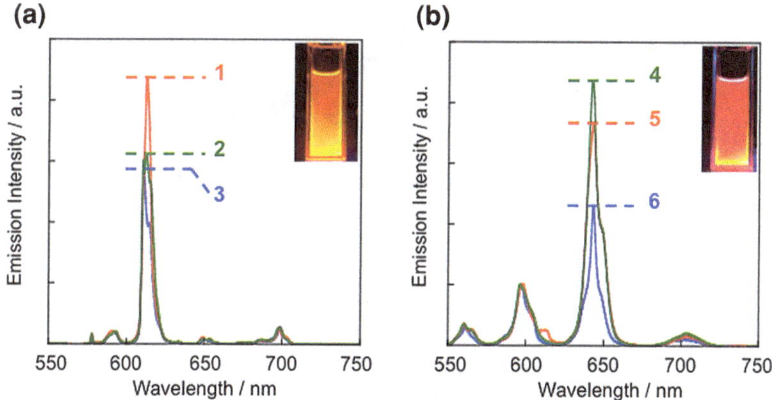

Fig. 5.8 **a** Emission spectra of Eu(hfa)$_3$(tBu-xantpo) (*red line* 1), Eu(hfa)$_3$(dpepo) (*green line* 2) and Eu(hfa)$_2$(xantpo)$_2$ (*blue line* 3) in acetone-d_6 at room temperature. Excited at 465 nm. The spectra were normalized with respect to the magnetic dipole transition ($^5D_0$–$^7F_1$). **b** Emission spectra of Sm(hfa)$_3$(dpepo) (*green line* 4), Sm(hfa)$_3$(tBu-xantpo) (*red line* 5) and Sm(hfa)$_2$(xantpo)$_2$ (*blue line* 6) in acetone d_6 at room temperature. Excited at 402 nm. The spectra were normalized with respect to the magnetic dipole transition ($^4G_{5/2}$–$^6H_{7/2}$)

for Eu(hfa)$_3$(tppo)$_2$ (Φ_{Ln} = 65 %) and Sm(hfa)$_3$(tppo)$_2$ (Φ_{Ln} = 4.1 %) in acetone-d_6, which have been reported as highly luminescent lanthanide complexes [20]. To the best of our knowledge, the emission quantum yield for Sm(hfa)$_3$(dpepo) (Φ_{Ln} = 5.0 %) is the highest in previous reported Sm(III) complexes.

Table 5.12 Photophysical properties of Eu(III) and Sm(III) complexes in acetone-d_6

Complex	Φ^{a}_{Ln}/%	τ^{b}_{obs}/ms	k^{c}_{r}/s^{-1}	k^{d}_{nr}/s^{-1}
Eu(hfa)$_2$(xantpo)$_2$	55	1.3	4.4×10^2	3.6×10^2
Eu(hfa)$_3$(*t*Bu-xantpo)	67	1.2	5.5×10^2	2.7×10^2
Eu(hfa)$_3$(dpepo)	72	1.5	4.7×10^2	1.8×10^2
Eu(hfa)$_3$(biphepo[e]	60	1.3	4.6×10^2	3.4×10^2
Sm(hfa)$_2$(xantpo)$_2$	3.8	0.35	1.1×10^2	2.8×10^3
Sm(hfa)$_3$(*t*Bu-xantpo)	2.4	0.15	1.6×10^2	6.5×10^3
Sm(hfa)$_3$(dpepo)	5.0	0.28	1.8×10^2	3.3×10^3

[a] Emission quantum yields for Eu(III) complexes were determined by comparing with the integrated emission signal (550–750 nm) of Eu(hfa)$_3$(biphepo) as $\Phi_{Ln} = 0.60$. Excitation at 465 nm. Emission quantum yields for Sm(III) complexes were determined by comparing with the integrated emission signal (550–750 nm) of Sm(hfa)$_3$(H$_2$O)$_2$ as $\Phi_{Ln} = 0.031$. Excitation at 481 nm

[b] Emission lifetime (τ_{obs}) of the lanthanide complexes were measured by excitation at 355 nm (Nd:YAG 3ω)

[c] Radiative rate constants $k_r = \Phi_{Ln}/\tau_{obs}$

[d] Nonradiative rate constants $k_{nr} = 1/\tau_{obs} - k_r$

[e] see ref. 20

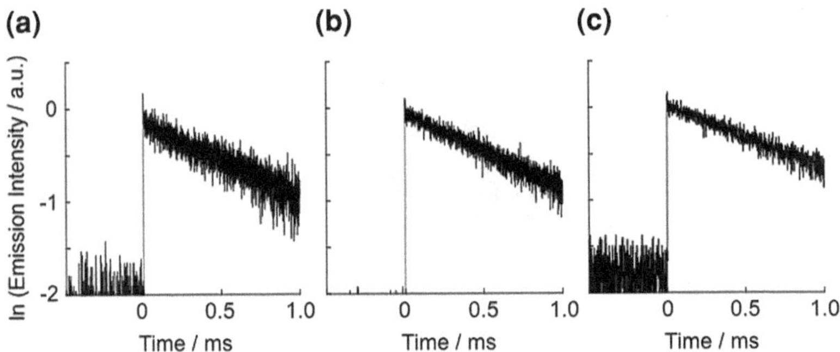

Fig. 5.9 The decay profiles of **a** Eu(hfa)$_2$(xantpo)$_2$, **b** Eu(hfa)$_3$(*t*Bu-xantpo), and **c** Eu(hfa)$_3$(dpepo) in acetone-d_6

The time-resolved emission profiles of lanthanide complexes revealed single-exponential decays with lifetimes in the millisecond timescale as shown in Fig. 5.9. The emission lifetimes were determined from the slopes of logarithmic plots of the decay profiles. The radiative (k_r) and nonradiative (k_{nr}) rate constants estimated using the emission lifetimes and the emission quantum yields are summarized in Table 5.12. The k_r values for Eu(hfa)$_3$(*t*Bu-xantpo) and Eu(hfa)$_3$(dpepo) were found to be 5.5×10^2 s^{-1} and 4.7×10^2 s^{-1}, respectively. These k_r values are larger than that for Eu(hfa)$_2$(xantpo)$_2$ (4.4×10^2 s^{-1}). The author also observed that the k_r values for Sm(hfa)$_3$(*t*Bu-xantpo) and Sm(hfa)$_3$(dpepo) are larger than that for Sm(hfa)$_2$(xantpo)$_2$. In general, reduction of the geometrical symmetry of the coordination structure leads to a larger k_r value. The author suggests that k_r for lanthanide

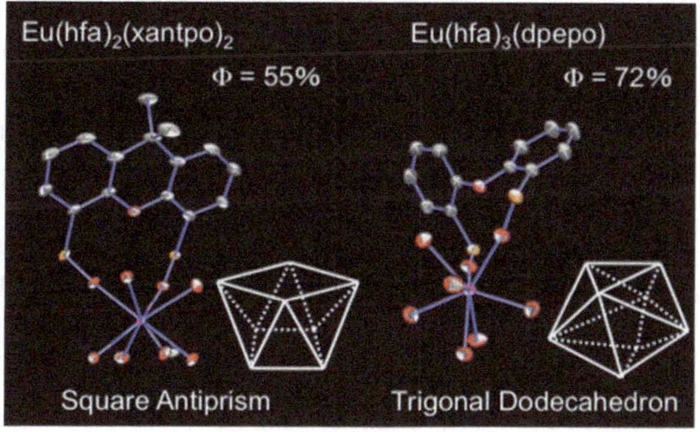

Fig. 5.10 The conceptual diagram in Chapter 5

complexes with phosphine oxides depend on the symmetry of the coordination sites. The characteristic 8-TDH structure of Eu(III) and Sm(III) complexes might be maintained even in acetone-d_6, because the emission spectrum in acetone-d_6 is similar to those of in the solid state. It can also be seen that k_{nr} for Sm(hfa)$_3$(*t*Bu-xantpo) is much larger than the values for the other lanthanide complexes as listed in Table 5.12. Note that the energy gap of the Sm(III) ion is smaller than that of the Eu(III) ion (Sm(III): 7500 cm^{-1}, Eu(III): 12500 cm^{-1}). The excited state of the Sm(III) ion with a smaller energy gap is effectively quenched by vibrational relaxation of the high-vibrational frequency C–H bonds. The author considered that the larger k_{nr} values for Sm(hfa)$_3$(*t*Bu-xantpo) might be attributed to effective vibrational relaxation caused by the 36 C–H bonds in the 4 *tert*-butyl groups attached to the phosphorus atoms.

From these results, the author considers that Eu(III) and Sm(III) complexes with 8-TDH structures lead to high emission quantum yields and large radiative rate constants, which result in strong luminescence.

5.4 Conclusions

The author successfully synthesized novel lanthanide complexes containing the oxo-linked bidentate phosphine oxide ligands xantpo, *t*Bu-xantpo and dpepo with high emission quantum yields (Fig. 5.10). They exhibit characteristic luminescence properties that depend on their coordination structures. The symmetry of coordination structures of lanthanide complexes is correlated with their photophysical properties. In order to produce strong luminescence in luminescent lanthanide complexes, the following intrinsic requirements should be fulfilled:

(1) low-vibrational structure
(2) asymmetrical point group
(3) asymmetrical coordination structure (quasi-C_1).

In this chapter, characteristic photophysical properties for octa-coordinated lanthanide complexes with 8-TDH structures have been demonstrated. Research on the driving force for the formation of the 8-TDH structure is expected to provide molecular strategies for obtaining strongly luminescent Eu(III) complexes directly. Ohkubo et al. have reported that the driving force for the construction of an asymmetric coordination geometry can be estimated from the dipole moments of the ligands [21]. Strongly luminescent Eu(III) complexes with 8-TDH structures may lead to the development of new fields in photophysical, coordination, and material chemistry.

References

1. S.F. Mason, R.D. Peacock, B. Stewart, Chem. Phys. Lett. **29**, 149 (1974)
2. S.F. Mason, J. Indian Chem. Soc. **63**, 73 (1986)
3. A.F. Kirby, F.S. Richardson, J. Phys. Chem. **87**, 2544 (1983)
4. M. Montalti, L. Prodi, N. Zaccheroni, L. Charbonnière, L. Douce, R. Ziessel, J. Am. Chem. Soc. **123**, 12694 (2001)
5. K. Driesen, P. Lenaerts, K. Binnemans, C. Görller-Walrand, Phys. Chem. Chem. Phys. **4**, 552 (2002)
6. W. Liu, T. Jiao, Y. Li, Q. Liu, M. Tan, H. Wang, L. Wang, J. Am. Chem. Soc. **126**, 2280 (2004)
7. J.P. Cross, M. Lauz, P.D. Badger, S. Petoud, J. Am. Chem. Soc. **126**, 16278 (2004)
8. P. Nockemann, B. Thijs, N. Postelmans, K.V. Hecke, L.V. Meervelt, K. Binnemans, J. Am. Chem. Soc. **128**, 13658 (2006)
9. A. Wada, M. Watanabe, Y. Yamanoi, T. Nankawa, K. Namiki, M. Yamasaki, M. Murata, H. Nishihara, Bull. Chem. Soc. Jpn **80**, 335 (2007)
10. J. Xu, E. Radkov, M. Ziegler, K.N. Raymond, Inorg. Chem. **39**, 4156 (2000)
11. T. Harada, Y. Nakano, M. Fujiki, M. Naito, T. Kawai, Y. Hasegawa, Inorg. Chem. **48**, 11242 (2009)
12. T. Harada, H. Tsumatori, K. Nishiyama, J. Yuasa, Y. Hasegawa, T. Kawai, Inorg. Chem. **51**, 6476 (2012)
13. P.W.N.M. van Leeuwen, P.C.J. Kamer, J.N.H. Reek, P. Dierkes, Chem. Rev. **100**, 2741 (2000)
14. O. Moudam, B.C. Rowan, M. Alamiry, P. Richardson, B.S. Richards, A.C. Jones, N. Robertson, *Chem. Commun.* 6649 (2009)
15. K. Nakamura, Y. Hasegawa, H. Kawai, N. Yasuda, Y. Tsukahara, Y. Wada, Thin Solid Films **516**, 2376 (2008)
16. D.A. Johnson, A.B. Waugh, T.W. Hambley, J.C. Taylor, J. Fluorine Chem. **27**, 371 (1985)
17. K. Nakamura, Y. Hasegawa, H. Kawai, N. Yasuda, N. Kanehisa, Y. Kai, T. Nagamura, S. Yanagida, Y. Wada, J. Phys. Chem. A **111**, 3029 (2007)
18. C. Görller-Walrand, L. Fluyt, A. Ceulemans, W.T. Carnall, J. Chem. Phys. **95**, 3099 (1991)
19. M.H.V. Werts, R.T.F. Jukes, J.W. Verhoeven, Phys. Chem. Chem. Phys. **4**, 1542 (2002)
20. H. Kawai, C. Zhao, S. Tsuruoka, T. Yoshida, Y. Hasegawa, T. Kawai, J. Alloys Compd. **488**, 612 (2009)
21. Y. Hasegawa, T. Ohkubo, T. Nakanishi, A. Kobayashi, M. Kato, T. Seki, H. Ito, K. Fushimi, Eur. J. Inorg. Chem. **2013**, 5911 (2013)

Chapter 6
Solvent-Dependent Luminescence of Octa-Coordinated Eu(III) Complexes

6.1 Introduction

As described in Chap. 5, in order to prepare an intensely luminescent lanthanide complex, a large radiative rate constant based on reducing the geometrical symmetry and a small nonradiative rate constant by introducing low-vibrational frequency organic ligands should be required [1, 2]. The author has reported on two asymmetric Eu(III) complexes with hfa and bidentate phosphine oxide ligands, $Eu(hfa)_2(xantpo)_2$ and $Eu(hfa)_3(t$Bu-xantpo) as shown in Fig. 6.1. Their coordination geometries were categorized as 8-SAP and 8-TDH structures, respectively. Their coordination structures composed of the low-vibrational frequency phosphine oxide and hfa provide Eu(III) complexes with high emission quantum yields and relatively large radiative rate constants in acetone-d_6. In the photophysical analyses, the author also found that the nonradiative rate constant of $Eu(hfa)_3(t$Bu-xantpo) ($k_{nr} = 2.7 \times 10^2\ s^{-1}$) is smaller than that for $Eu(hfa)_2(xantpo)_2$ ($k_{nr} = 3.6 \times 10^2\ s^{-1}$). The author here consider that smaller k_{nr} of $Eu(hfa)_3(t$Bu-xantpo) might be achieved by specific coordination structure, 8-TDH.

In general, the nonradiative process of Eu(III) complex is also affected by their coordination structures in liquid media [3]. The coordination structure in organic solvent could be directly linked to performance of coordination ability of the solvent molecule, which is related to dielectric constant of solvent [4]. In this chapter, the author focus on solvent-dependent luminescence of two types of octa-coordinatd Eu(III) complexes, $Eu(hfa)_2(xantpo)_2$ and $Eu(hfa)_3(t$Bu-xantpo). Their photophysical properties are estimated using the emission quantum yield, emission lifetime, and radiative and nonradiative rate constant in acetone, acetone-d_6, toluene, chloroform, and DMF. The relationship between photophysical properties and coordination structures of octa-coordinated Eu(III) complexes will be discussed.

K. Miyata, *Highly Luminescent Lanthanide Complexes with Specific Coordination Structures*, Springer Theses,
DOI: 10.1007/978-4-431-54944-4_6, © Springer Japan 2014

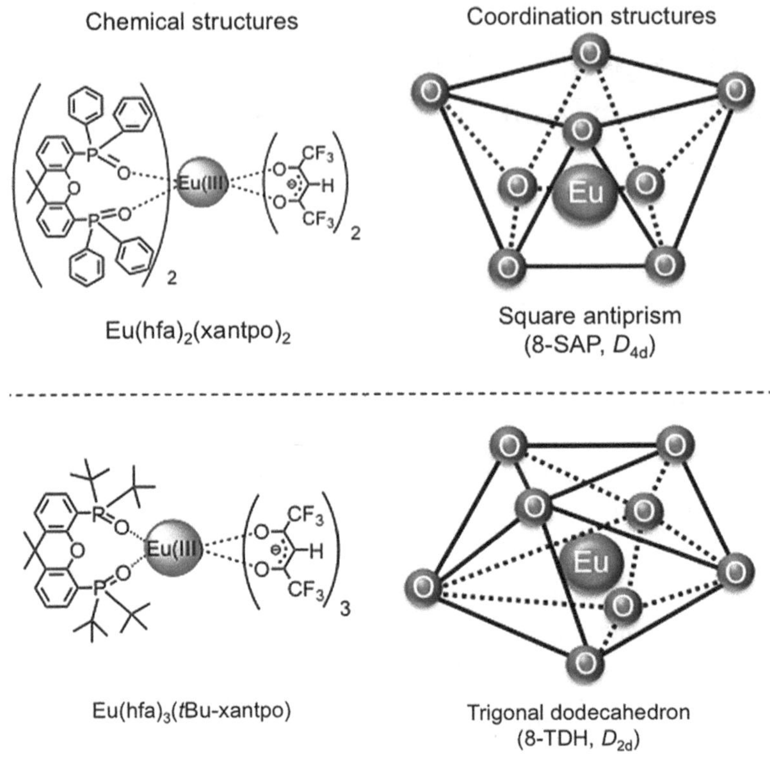

Fig. 6.1 Chemical and coordination structures of $Eu(hfa)_2(xantpo)_2$ and Eu(hfa)$_3$(*t*Bu-xantpo)

6.2 Experimental Section

6.2.1 *Sample Preparation*

$Eu(hfa)_2(xantpo)_2$ and Eu(hfa)$_3$(*t*Bu-xantpo) were prepared according to the procedure described in Chap. 5.

6.2.2 *Optical Measurements*

UV–Vis absorption spectra were recorded on a JASCO V–550 spectrometer. Emission spectra of the lanthanide complexes were measured with a JASCO F-6300-H spectrometer and corrected for the response of the detector system. The intrinsic emission quantum yields (Φ_{Ln}) of lanthanide complex solutions degassed with an argon (10 mM in toluene, chloroform, acetone, *N*,*N*-dimethylformamide; DMF) were obtained by comparison with the integrated emission signal

(550–750 nm) of Eu(hfa)$_3$(biphepo) as a reference ($\Phi_{Ln} = 0.60$: 50 mM in acetone-d_6) with an excitation wavelength of 465 nm. The emission quantum yields excited at 380 nm (ligand excitation: (Φ_{tot}) were estimated using JASCO F-6300-H spectrometer attached with JASCO ILF-533 integrating sphere unit ($\phi = 100\,\text{mm}$). The wavelength dependences of the detector response and the beam intensity of Xe light source for each spectrum were calibrated using a standard light source. Emission lifetimes of lanthanide complexes (1.0 mM in toluene, chloroform, acetone, DMF) were measured using the third harmonics (355 nm) of a Q-switched Nd:YAG laser (Spectra Physics, INDI-50, fwhm = 5 ns, $\lambda = 1064$ nm) and a photomultiplier (Hamamatsu photonics, R5108, response time ≤ 1.1 ns).

6.3 Results and Discussion

6.3.1 Effects of Deuterated Solvent

The steady-state emission spectra of Eu(hfa)$_2$(xantpo)$_2$ and Eu(hfa)$_3$(tBu-xantpo) in organic solvent are shown in Fig. 6.2. Emission bands are observed at around 578, 591, 613, 650, and 698 nm, and are attributed to the f–f transitions of $^5D_0$–^7F_J with $J = 0, 1, 2, 3$, and 4, respectively. The spectra are normalized with respect to the magnetic dipole transition intensity ($^5D_0$–$^7F_1$) at 591 nm which is known to be insensitive to the surrounding environment of the Eu(III) ion [5, 6]. The emission band at 613 nm ($^5D_0$–$^7F_2$) is due to electric dipole transition, which is strongly dependent on their coordination geometry. We also estimated the relative emission intensity of $^5D_0$–$^7F_2$ transition with respect to that of $^5D_0$–$^7F_1$ as $I_{rel} = I_{613}/I_{591}$ in the normalized emission spectra. The I_{rel} values of Eu(hfa)$_2$(xantpo)$_2$ and Eu(hfa)$_3$(tBu-xantpo) are summarized in Table 6.2. The intrinsic emission quantum yield (Φ_{Ln}) for Eu(hfa)$_3$(tBu-xantpo) in acetone excited at 4f orbitals (excited at 465 nm) was found to be 64 % (Table 6.1). This value is slightly smaller than that for reported Eu(hfa)$_3$(tBu-xantpo) in acetone-d_6 ($\Phi_{Ln} = 67$ %).

The radiative (k_r) and nonradiative (k_{nr}) rate constants estimated using the emission lifetimes (τ_{obs}) and the intrinsic emission quantum yields (Φ_{Ln}) are summarized in Table 6.1. The radiative rate constant for Eu(hfa)$_3$(tBu-xantpo) in acetone was estimated to be 5.4×10^2 s^{-1}. This k_r value is much similar to that for Eu(hfa)$_3$(tBu-xantpo) in acetone-d_6 (5.5×10^2 s^{-1}). The nonradiative rate constant for Eu(hfa)$_3$(tBu-xantpo) in acetone-d_6 (2.7×10^2 s^{-1}) is smaller than that for Eu(hfa)$_3$(tBu-xantpo) in acetone (3.0×10^2 s^{-1}). The relatively smaller k_{nr} for Eu(hfa)$_3$(tBu-xantpo) in acetone-d_6 is attributed to the suppression of vibrational relaxation surroundings of the Eu(III) complex. The nonradiative transitions of lanthanide complexes are affected by the high-vibrational frequency of C–H and O–H bonds of solvent. The author consider that introduction of deuterated solvent is effective for enhancement of emission quantum yield of octa-coordinated Eu(III) complexes.

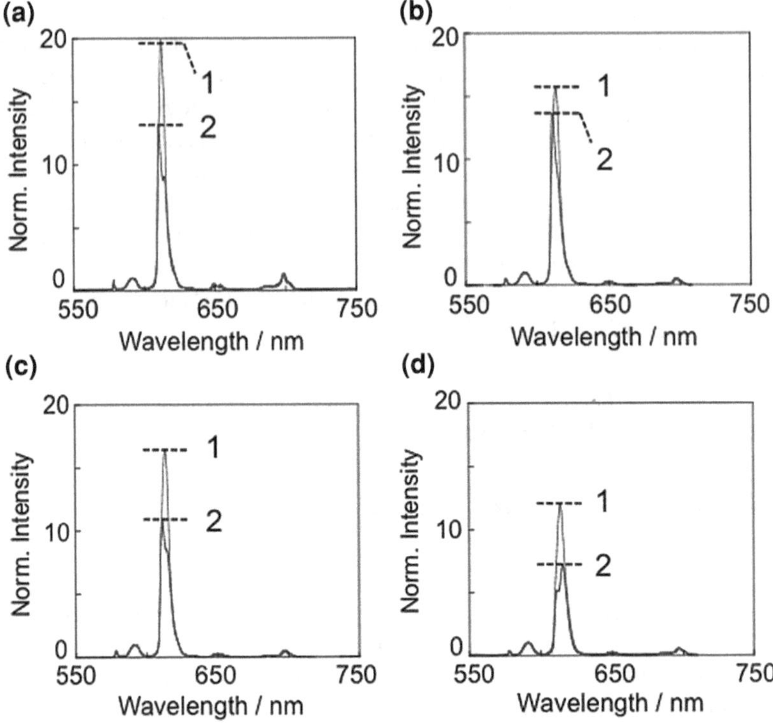

Fig. 6.2 Emission spectra of Eu(hfa)$_3$(*t*Bu-xantpo) (line 1) and Eu(hfa)$_2$(xantpo)$_2$ (line 2) in **a** acetone, **b** toluene, **c** chloroform, and **d** DMF at room temperature. Excitation wavelength is 465 nm. The spectra are normalized with respect to the magnetic dipole transition ($^5D_0$–$^7F_1$)

Table 6.1 Photophysical properties of Eu(hfa)$_3$(*t*Bu-xantpo) at room temperature

Solvent	τ_{obs} (ms)[a]	Φ_{Ln} (%)	k_r (s^{-1})	k_{nr} (s^{-1})
Acetone	1.2	64	5.4×10^2	3.0×10^2
Acetone-d_6	1.2	67	5.5×10^2	2.7×10^2

6.3.2 Photophysical Properties in Various Organic Solvent

The radiative (k_r) and nonradiative (k_{nr}) rate constants, the emission lifetimes (τ_{obs}), and the intrinsic emission quantum yields (Φ_{Ln}) of Eu(hfa)$_2$(xantpo)$_2$ and Eu(hfa)$_3$(*t*Bu-xantpo) are summarized in Table 6.2. The radiative rate constants of Eu(hfa)$_3$(*t*Bu-xantpo) ($k_r = 3.6$–5.6×10^2 s^{-1}) were larger than that for Eu(hfa)$_2$(xantpo)$_2$ ($k_r = 3.0$–4.9×10^2 s^{-1}). Generally, the radiative rate constants of lanthanide complexes are directly linked to their geometrical structures. The symmetrical point groups of Eu(hfa)$_2$(xantpo)$_2$ and Eu(hfa)$_3$(*t*Bu-xantpo) are D_{4d} and D_{2d}, respectively [7]. The larger radiative rate constants for

Table 6.2 Photophysical properties of Eu(III) complexes in various solvent at room temperature

Complex	Solvent	τ_{obs} (ms)	Φ_{Ln} (%)	Φ_{tot} (%)[a]	η_{sens} (%)[b]	I_{rel}[c]	k_r (10^2 s^{-1})	k_{nr} (10^2 s^{-1})
$Eu(hfa)_3(xantpo)_2$	Toluene	1.3	64	24	38	14	4.9	2.8
	Chloroform	1.3	58	22	38	11	4.3	3.1
	Acetone	1.3	50	20	40	13	3.9	3.8
	DMF	1.7	51	12	23	7.1	3.0	2.9
$Eu(hfa)_3$(*t*Bu-xantpo)	Toluene	1.3	67	29	43	16	5.3	2.6
	Chloroform	1.2	67	24	36	16	5.6	2.7
	Acetone	1.2	64	22	34	20	5.4	3.0
	DMF	1.6	58	12	21	12	3.6	2.6

[a] Total emission quantum yield (excitation at 380 nm)
[b] Photosensitized energy transfer efficiency $\eta_{sens} = \Phi_{tot}/\Phi_{Ln}$. Estimated relative errors: τ_{obs}, ±4 %; Φ_{Ln}, ±3 %; Φ_{tot}, ±9 %; η_{sens}, ±12 %
[c] Relative emission intensity of the electric dipole transition ($^5D_0$–$^7F_2$) to the magnetic dipole transition ($^5D_0$–$^7F_1$)

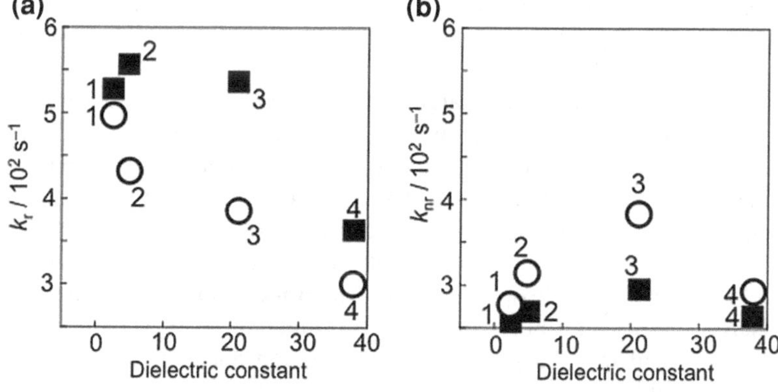

Fig. 6.3 The relationship between rate constants of Eu(III) complexes and dielectric constants of organic solvent. **a** The radiative (k_r) and **b** nonradiative constant (k_{nr}). (■: $Eu(hfa)_3$(*t*Bu-xantpo); 8-TDH, ○: $Eu(hfa)_2(xantpo)_2$; 8-SAP, *1* toluene, *2* chloroform, *3* acetone, *4* DMF)

$Eu(hfa)_3$(*t*Bu-xantpo) should come from asymmetrical D_{2d} structure related to change of odd parity in organic solvent.

The ability of solvent molecules for the coordination to metal centers can be estimated from the dielectric constant or the donor number. The relationship between rate constants of Eu(III) complexes and dielectric constants of organic solvent are shown in Fig. 6.3. In $Eu(hfa)_2(xantpo)_2$, k_r and k_{nr} are much dependent of the dielectric constant of organic solvent. These results indicate that coordination geometry and vibrational structure of $Eu(hfa)_2(xantpo)_2$(8-SAP) is affected by organic solvent with large dielectric constant, solvation to Eu(III) ion

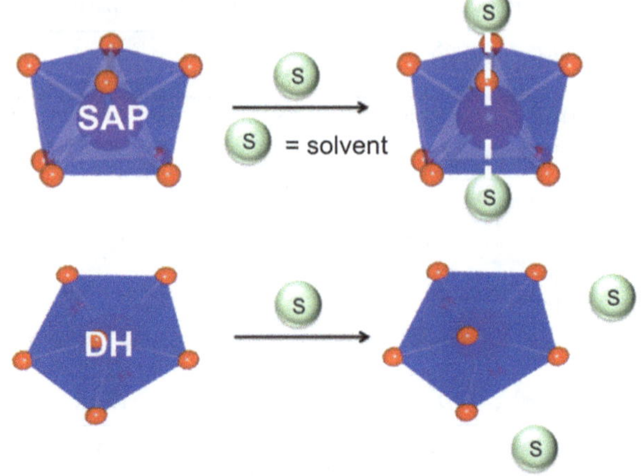

Fig. 6.4 Schematic illustrations of the difference between coordination ability of solvent molecules to metal center in **a** 8-SAP and **b** 8-TDH structures. The author considers that the solvent molecules might be difficult to coordinate to metal center in 8-TDH structures owing to their structural stability

(Fig. 6.4a). In contrast, we found that k_r and k_{nr} of Eu(hfa)$_3$(*t*Bu-xantpo) are maintained constant in acetone, toluene, and chloroform. The structure of 8-TDH in Eu(hfa)$_3$(*t*Bu-xantpo) might not be affected by solvent (Fig. 6.4b), although k_r and k_{nr} of 8-SAP are dependent of solvent. The author here considers that small nonradiative rate of Eu(hfa)$_3$(*t*Bu-xantpo) is caused by structural stability of 8-TDH in organic media.

The author also carried out charge density calculations of xantpo and *t*Bu-xantpo ligand by DFT calculation (6-31G(d)/B3LYP) based on the X-ray single crystal analyses. According to the calculation, charge densities oxygen atoms in phosphine oxides were found to be –0.51 and –0.48 (xantpo), –0.51 and –0.47 (*t*Bu-xantpo), respectively (Fig. 6.5). However, we found that charge densities of phosphorus atoms in *t*Bu-xantpo (–0.11 and 0.12) were higher than those in xantpo (0.22 and 0.23), because phosphorus atoms in *t*Bu-xantpo were affected by electron donating ability of *t*-butyl groups. The effective electron donating ability would be also lead to enhancement of coordination and structural stabilities of Eu(hfa)$_3$(*t*Bu-xantpo).

According to the emission process in DMF, k_r of Eu(hfa)$_3$(*t*Bu-xantpo) is similar to that for Eu(hfa)$_2$(xantpo)$_2$. Eu(hfa)$_2$(xantpo)$_2$ and Eu(hfa)$_3$(*t*Bu-xantpo) might be decomposed in DMF due to coordination to Eu(III) ion of DMF molecules with large dielectric constant. Decomposition of coordination structure by addition of solvent molecule with large dielectric constant has been previously reported [3]. The photosensitized energy transfer efficiency (η_{sens}) based on Φ_{Ln} and the total emission quantum yield (Φ_{tot}: excitation at 380 nm) of Eu(hfa)$_2$(xantpo)$_2$ and Eu(hfa)$_3$(*t*Bu- xantpo) are summarized in Table 6.2. The photosensitized energy transfer efficiencies in DMF (η_{sens} = 21–23 %) are smaller than those in toluene, chloroform, and acetone (34–43 %). The author considers that hfa ligands of the complexes might move away from Eu(III) ion in DMF.

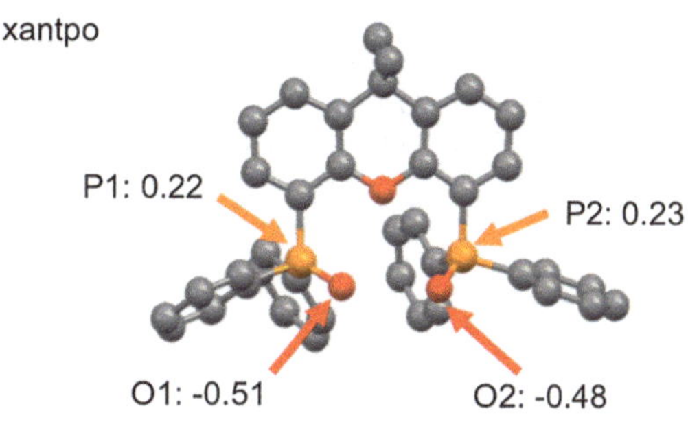

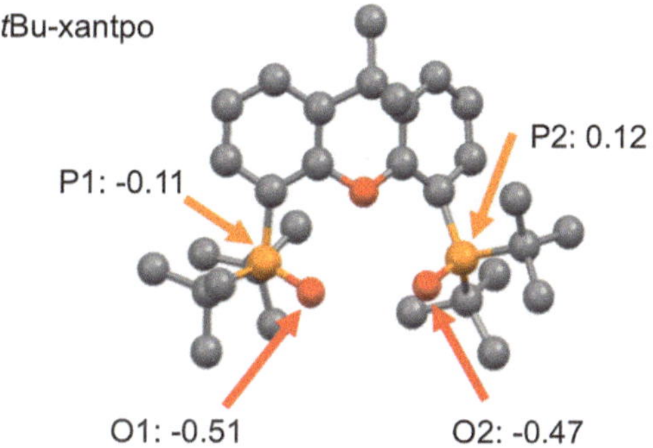

Fig. 6.5 Charge densities of phosphorus and oxygen atoms in xantpo and *t*Bu-xantpo ligand

6.4 Conclusion

Solvent-dependent luminescence of $Eu(hfa)_2(xantpo)_2$ and $Eu(hfa)_3$(*t*Bu-xantpo) are demonstrated. The author found that nonradiative rate process of Eu(III) complex is directly linked to characteristics of surrounding environment, organic solvent. Smaller k_{nr} of $Eu(hfa)_3$(*t*Bu-xantpo) is achieved by characteristic coordination structure, 8-TDH because of their structural stability in organic media. Solvent-dependent luminescence of lanthanide complex with characteristic octa-coordination structure is expected to provide a novel aspect in the field of lanthanide photochemistry.

References

1. Y. Hasegawa, M. Yamamuro, Y. Wada, N. Kanehisa, Y. Kai, S. Yanagida, J. Phys. Chem. A **107**, 1697 (2003)
2. K. Miyata, Y. Hasegawa, Y. Kuramochi, T. Nakagawa, T. Yokoo, T. Kawai, Eur. J. Inorg. Chem. **32**, 4777 (2009)
3. Y. Hasegawa, Y. Kimura, K. Murakoshi, Y. Wada, J. Kim, N. Nakashima, T. Yamanaka, S. Yanagida, J. Phys. Chem. **100**, 10201 (1996)
4. B.M. Kurdziel, K. Solymosi, J. Kruk, B. Böddi, K. Strzałka, Eur. Biophys. J. **37**, 1185 (2007)

5. C. Görller-Walrand, L. Fluyt, A. Ceulemans, W.T. Carnall, J. Chem. Phys. **95**, 3099 (1991)
6. M.H.V. Werts, R.T.F. Jukes, J.W. Verhoeven, Phys. Chem. Chem. Phys. **4**, 1542 (2002)
7. K. Miyata, T. Nakagawa, R. Kawakami, Y. Kita, K. Sugimoto, T. Nakashima, T. Harada, T. Kawai, Y. Hasegawa, Chem. Eur. J. **17**, 521 (2011)

Chapter 7
Summary

The luminescence and thermal properties of lanthanide(III) complexes can be manipulated and enhanced by designing organic ligands which coordinate to lanthanide(III) ions. In this thesis, molecular designs for the enhancement of the luminescence properties and functionalization (thermostability and temperature-sensing ability) of lanthanide complexes with LVF phosphine oxide ligands have been described. The results obtained in this study were briefly summarized as follows.

In Chap. 1, the scientific backgrounds about lanthanide(III) luminescence were introduced. Then, issues of former researchers were pointed out in terms of structures of lanthanide complexes. The strategies of molecular designs of lanthanide complexes with the strong-luminescence properties and high thermal stability were also mentioned.

In Chap. 2, novel thermostable luminophores comprised of Eu(III) coordination polymers; $[Eu(hfa)_3(dpb)]_n$, $[Eu(hfa)_3(dpbp)]_n$, and $[Eu(hfa)_3(dppcz)]_n$ were successfully synthesized. In particular, $[Eu(hfa)_3(dppcz)]_n$ exhibited both high emission quantum yields ($\Phi_{Ln} = 83$ %) and remarkable thermal stability (decomposition point = 300 °C) due to a tight-binding structure composed of Eu(III) ions and low-vibrational phosphine oxide, although many types of luminescent organic dyes are generally decomposed at temperatures under 200 °C. The emission quantum yields of these coordination polymers are similar to those of strong-luminescent coordination polymers in former chapters. These coordination polymers are expected to employ in optics applications such as luminescent plastics, displays, and opto-electronic devices.

In Chap. 3, a novel thermosensing dye composed of color-changing luminescent coordination polymers containing Eu(III) and Tb(III) ions was successfully synthesized. This coordination polymer exhibited a high emission quantum yield Φ_{tot} of 40 % at room temperature and an effective temperature-sensing ability over a wide range of 200–500 K. Temperature-sensitive dyes with thermostable structures and dual sensing units were reported for the first time. The results obtained in this chapter will provide insights for designing lanthanide coordination polymers for development of temperature-sensing devices based on their luminescence.

K. Miyata, *Highly Luminescent Lanthanide Complexes with Specific Coordination Structures*, Springer Theses,
DOI: 10.1007/978-4-431-54944-4_7, © Springer Japan 2014

In Chap. 4, novel nona-coordinated Eu(III) complexes containing characteristic tridentate phosphine oxide ligands, dpppo, dppypo, and dpbtpo were successfully synthesized. These Eu(III) complexes exhibited strong-luminescence with high emission quantum yields ($\Phi_{Ln} > 60$ %). The author found that the emission intensities at the electric dipole transition of Eu(III) complexes depended on the organic linker species in phosphine oxide ligands, although the energy gap between Stark splitting levels, radiative and non-radiative rate constants were not affected by them. This result revealed that the transition intensity at the electric dipole transition is affected by chemical structures of attached tridentate phosphine oxide ligands.

In Chap. 5, Eu(III) and Sm(III) complexes with novel asymmetric structures composed of oxo-linked bidentate phosphine oxide ligands (xantpo, *t*Bu-xantpo, and dpepo) have been reported. The author carried out the calculations of the shape factor *S* in order to estimate the degree of distortion of the coordination structure in first coordination sphere. Based on the structural and photophysical findings, remarkably strong luminescence properties of Eu(III) and Sm(III) complexes with characteristic trigonal dodecahedron structures (8-TDH) were demonstrated for the first time. The author also showed that the emission quantum yield of $Sm(hfa)_3(dpepo)$ with a trigonal dodecahedron structure is the highest value ($\Phi_{Ln} = 5.0$ %) in previous reported Sm(III) complexes.

In Chap. 6, solvent-dependent luminescence of $Eu(hfa)_2(xantpo)_2$ (8-SAP) and $Eu(hfa)_3$(*t*Bu-xantpo) (8-TDH) has been investigated. Their emission quantum yields, emission lifetimes, and radiative and nonradiative rate constants were characterized in acetone, acetone-d_6, toluene, chloroform, and DMF. The author found that nonradiative rate process of Eu(III) complex is directly linked to characteristics of surrounding environment, organic solvent. In particular, smaller nonradiative rate constant of $Eu(hfa)_3$(*t*Bu-xantpo) was achieved by characteristic coordination structure, 8-TDH because of their structural stability in organic media.

The luminescence properties of lanthanide complexes can be enhanced by design of chemical structures. However, the correlation between coordination structures and luminescence properties of lanthanide complexes has been scarcely investigated. In this thesis, the correlation between coordination structures and photophysical properties of lanthanide complexes with phosphine oxide ligands were demonstrated. Additionally, functionalization of lanthanide compounds was also described in terms of lanthanide coordination polymers. This thesis gives the first systematic studies on molecular photo-science between lanthanide(III) coordination chemistry and photo-functional materials science.

The author wishes the strategies for molecular design of luminescent lanthanide complexes described in this thesis would have contribution to development of applications in novel organic lanthanide devices, such as organic liquid lasers, luminescent plastics, optical fibers, EL devices, and temperature-sensing devices. The photochemistry of luminescent lanthanide complexes is expected to open up frontier fields between lanthanide coordination chemistry and photo-functional materials science.

Curriculum Vitae

Kohei Miyata, Ph.D.
Former affiliation address:
Graduate School of Chemical Sciences and Engineering, Hokkaido University
Sapporo, Hokkaido, Japan
Web: http://www.eng.hokudai.ac.jp/labo/amc

Career

- ADEKA Corporation (April 2013–)
- The Japan Society for the Promotion of Science (JSPS) Research Fellow (DC2, 2011–2013)

Education

- Doctor of Engineering, Hokkaido University, 2010–2013 (Supervisor: Prof. Yasuchika Hasegawa)
- Master of Engineering, Nara Institute of Science and Technology, 2008–2010 (Supervisor: Prof. Tsuyoshi Kawai)
- Bachelor of Engineering, The University of Shiga Prefecture, 2004–2008 (Supervisor: Prof. Tsutomu Kumagai)

Award

- The Japan Photochemistry Association, Student Presentation Award

K. Miyata, *Highly Luminescent Lanthanide Complexes with Specific Coordination Structures*, Springer Theses,
DOI: 10.1007/978-4-431-54944-4, © Springer Japan 2014

Zeitfracht Medien GmbH
Ferdinand-Jühlke-Straße 7
99095 Erfurt, Deutschland
produktsicherheit@kolibri360.de

Lineare Strukturgleichungsmodelle

SOZIALWISSENSCHAFTLICHE FORSCHUNGSMETHODEN

Band 9

Holger Steinmetz

Lineare Strukturgleichungsmodelle

Eine Einführung mit R

Rainer Hampp Verlag München und Mering 2015

Bibliografische Information der Deutschen Nationalbibliothek
Die Deutsche Nationalbibliothek verzeichnet diese Publikation in der Deutschen Nationalbibliografie; detaillierte bibliografische Daten sind im Internet über http://dnb.d-nb.de abrufbar.

ISBN 978-3-95710-049-8 (print)
ISBN 978-3-95710-149-5 (e-book)
SOZIALWISSENSCHAFTLICHE FORSCHUNGSMETHODEN: ISSN 1869-7151
ISBN-A/DOI 10.978.395710/1495
1. Auflage, 2014
2. Auflage, 2015

 Rainer Hampp Verlag München und Mering
Marktplatz 5 D – 86415 Mering
www.Hampp-Verlag.de

∞ *Dieses Buch ist auf säurefreiem und chlorfrei gebleichtem Papier gedruckt.*

Liebe Leserinnen und Leser!
Wir wollen Ihnen ein gutes Buch liefern. Wenn Sie aus irgendwelchen Gründen nicht zufrieden sind, wenden Sie sich bitte an uns.

Inhaltsverzeichnis

1 Einführung

Lineare Strukturgleichungsmodelle sind aus der verhaltenswissenschaftlichen Forschung nicht mehr wegzudenken. Sie sind ein sehr nützliches Werkzeug, um Hypothesen über Beziehungen zwischen Variablen zu prüfen und – mehr noch – Implikationen kausaler Strukturen zu testen. Während in den ersten Jahrzehnten der Nutzung von „SEM" (structural equation modeling)[1] der Enthusiasmus groß war und Parameterschätzungen vorschnell als wirkliche, kausale Effekte interpretiert wurden, hat sich in der letzten Zeit genau das Gegenteil eingestellt – nämlich, dass kausales Denken quasi aus SEM verbannt und die Parameter lediglich in Termini von „Beziehungen" interpretiert werden. Dies stellt allerdings einen Rückschritt dar. SEM sind, wenn sie richtig angewendet werden, eine mächtige Methode, um bei Fehlen von randomisierten Experimenten (d.h. bei Feldstudien) kausale Hypothesen zu testen, ohne gleichermaßen in die Falle der naiven kausalen Interpretation der Koeffizienten in Modellen zu tappen.

Dieses Buch bietet eine Einführung in die Hintergründe, Prinzipien, Möglichkeiten und Grenzen von SEM. Unter Verwendung eines realen Modells wird eine Schritt-für-Schritt-Anleitung gegeben, um Einsteigern einen schnellen und praktikablen Start zu ermöglichen. Wegen der Kürze können die Themen nicht so detailliert behandelt werden, wie dies in gängigen Standardwerken der Fall ist. Daher sei eine weitere Vertiefung angeraten. Vieles in diesem Buch wurde beeinflusst von Hayduk (1987), Kline (2011), Shipley (2000) und Bollen (1989). Die Illustration erfolgt mit der freien Statistiksoftware R, die in den letzten Jahren eine rasante Verbreitung gefunden hat. Dabei sollen nicht nur die spezielle Modellierung mit dem R-Paket „lavaan" behandelt werden (siehe dazu auch Beaujean, 2014), sondern auch wichtige Vorgehensweisen zur Vor- und Nachbereitung. Dies sind zum Beispiel das Einlesen von Daten aus anderen Programmen, oder das Überprüfen und Veranschaulichen der Daten im Rahmen der

[1] In diesem Buch soll die Abkürzung SEM sowohl für Strukturgleichungsmodelle als auch für den gesamten Ansatz der Modellierung benutzt werden, da diese Abkürzung auch von nicht-englischen Forschern und Forscherinnen üblicherweise verwendet wird.

Überprüfung zentraler Annahmen, die ein Modell beinhaltet. Vorkenntnisse von R werden nicht vorausgesetzt. Allerdings sind Vorkenntnisse in Varianzen und Kovarianzen von Variablen, der Regressionsanalyse und deskriptiver und Inferenzstatistik hilfreich. Der Fokus liegt dabei nicht auf methodischen und statistischen Aspekten von SEM (dies ist in den o.g. Büchern ausführlicher und auch fundierter behandelt), sondern vielmehr in der Verknüpfung von Theorie und Modell. Dabei werden diejenigen Aufgaben und Probleme diskutiert, die bei der Modellierung in der Regel auftauchen. Diese bestehen u.a. in der adäquaten Übersetzung von theoretischen Konstrukten in ein Modell, der Präzisierung der eigenen Vorstellungen, der Übersetzung dieser Vorstellungen in ein Modell und Reflektion der kausalen Implikationen dieser Vorstellungen. Schließlich soll der produktive Umgang mit „nicht-fittenden" Modellen behandelt werden.

1.1 Was ist ein Kausalmodell?

Das Ziel der Anwendung von SEM besteht darin, ein Kausalmodell zu spezifizieren, das die theoretisch erwarteten kausalen Effekte von **Variablen** repräsentiert. Eine Variable ist eine Dimension, auf der sich Fälle (z.B. Personen) graduell unterscheiden. Dies können beispielsweise Unterschiede in der Häufigkeit eines Verhaltens, im Ausmaß der Zustimmung zu bestimmten Aussagen, in der Anzahl von zählbaren Ereignissen oder in der Evaluation eines Zustandes (z.B. Zufriedenheit) sein. Die Variablen des Modells repräsentieren theoretische oder **hypothetische Konstrukte,** wie sie in wissenschaftlichen Theorien enthalten sind, oder – genauer – diejenigen **empirischen Phänomene**, die diese Konstrukte zu beschreiben versuchen (vgl. Kap. 3). Das Modell beschreibt daher einen Ausschnitt der Realität, da in ihm lediglich eine Auswahl von Variablen enthalten ist. Ob und wann dies problematisch ist, soll später erörtert werden.

Weiterhin beschreibt ein Modell eine mehr oder weniger ausgeprägte *Struktur von Beziehungen* zwischen den Variablen, die sich sehr gut in **Pfaddiagrammen** (siehe nächster Abschnitt) als Struktur von Pfeilen illustrieren lässt. Auch wenn in solchen Diagrammen meist die expliziten Beziehungen ins Auge fallen, sei hier bereits angemerkt, dass die Struktur erst durch die Nicht-Beziehungen („Restriktionen") innerhalb des Modells vollständig definiert ist. So hat in einem Modell nicht jede Variable auf jede andere einen Effekt. Es sind diese nicht-sichtbaren sowie impliziten Restriktionen, die sowohl einen **Test** des Modells ermöglichen, als auch den Grund für einen **misfit** (d.h. eine inakzeptable Vorhersage der Daten) ausmachen. Das Modell lässt sich schließlich als ein System von linearen **Strukturgleichungen** interpretieren, in dem abhängige Variablen als die Folge des gewichteten Einflusses anderer Variablen beschrieben werden .

1.2 Pfaddiagramme als Illustrationen von Modellen

Wie oben erwähnt, sind **Pfaddiagramme** eine sehr hilfreiche Methode, um ein Modell darzustellen. Ein Pfaddiagramm enthält alle postulierten Effekte der Variablen, sowie bivariate Beziehungen (Kovarianzen), für die keine spezifische kausale Hypothese besteht. Insbesondere in der frühen Phase der Konzeptualisierung eines Forschungsprozesses ist der Detailgrad eines postulierten Forschungsmodells oft sehr gering und hypothetisierte Beziehungen zwischen theoretischen Konstrukten eher abstrakt und vage. Ein Pfaddiagramm erfordert einen höheren Präzisionsgrad in der Konzeptualisierung der Variablen und ihrer Effekte, der über das theoretisch oft vage „Konstrukt A ist assoziiert mit Konstrukt B" hinaus geht. Dies ist v.a. der Fall wenn Konstrukte multidimensional sind. Ein Pfaddiagramm ist daher enorm hilfreich, um eine eher abstrakte Theorie in ein spezifisches Modell zu übersetzen.

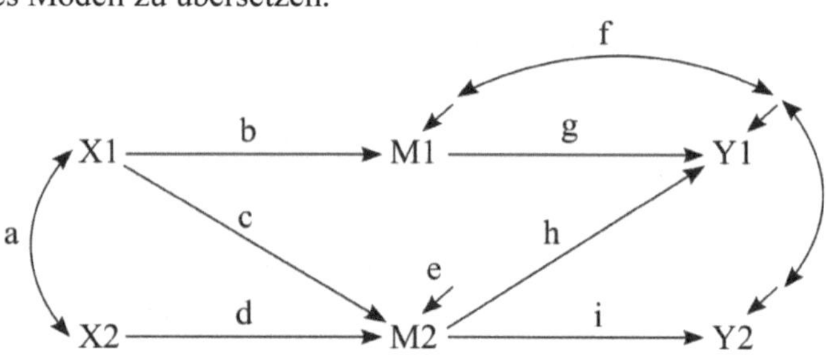

Abb. 1: Pfaddiagramm

Abb. 1 enthält ein exemplarisches Pfaddiagramm. In dem entsprechenden Modell gibt es sechs Variablen (X1, X2, M1, M2, Y1, Y2). Die Variablen X1 und X2 nennt man **exogene Variablen**; sie entsprechen den unabhängigen Variablen in herkömmlichen Forschungsdesigns, da von ihnen nur Effekte ausgehen, aber keine empfangen werden. Wir werden den Buchstaben „X" in der Regel für solche exogenen Variablen verwenden. Die Variablen M1 – Y2 sind **endogene Variablen**, weil sie Effekte empfangen; sie entsprechen den abhängigen Variablen in herkömmlichen Designs.

Weiterhin enthält das Pfaddiagramm vier verschiedene Arten von Pfeilen:

1) unidirektionale Pfeile zwischen zwei Variablen (z.B. b),

2) unidirektionale Pfeile ohne Ursprung (z.B. e),
3) bidirektionale Pfeile zwischen Variablen (z.B. a) und
4) bidirektionale Pfeile zwischen Pfeilen (z.B. f).

Die unidirektionalen Pfeile zwischen zwei Variablen symbolisieren die **direkten Effekte** einer Variablen auf die andere. Zum Beispiel sieht das Modell vor, dass X1 direkt M1 und M2 beeinflusst. Hintereinander geschaltete Pfeile, die zwei Variablen verbinden, definieren einen **Pfad** – unabhängig davon, ob die Pfeile in dieselbe Richtung weisen. So ist cd ein Pfad, der X1 und X2 verbindet. Besteht ein Pfad aus Pfeilen, die in dieselbe Richtung weisen, spricht man von einem **indirekten Effekt**. So beeinflusst X1 die endogene Variable Y1 indirekt oder *vermittelt* über M1. Damit übernimmt M1 die Rolle eines **Mediators** (MacKinnon, Fairchild, & Fritz, 2007). Der indirekte Effekt von X1 (über M1) auf Y1 ergibt sich aus dem Produkt der direkten Effekte b und g. Schließlich gibt der **totale Effekt** einer Variablen auf eine andere den Gesamteffekt an, den diese Variable über alle direkten und indirekten Pfade hat. Zum Beispiel ist der totale Effekt von X1 auf Y1 die Summe der beiden indirekten Effekte bg und ch.

Die unidirektionalen Pfeile ohne Ursprung kennzeichnen den **Fehlerterm** der Variable. Er repräsentiert alle nicht im Modell befindlichen weiteren Einflussvariablen der Variable sowie Messfehler. Die bidirektionalen Pfeile zwischen Variablen kennzeichnen **Kovarianzen** *zwischen exogenen Variablen*, deren Grund nicht im Modell spezifiziert ist. So könnte der Grund für die Kovarianz zwischen X1 und X2 ein Einfluss jeweils einer dieser Variablen auf die andere sein (z.B. X1 → X2), oder aber die Konsequenz einer oder mehrerer gemeinsamer Ursachen (vgl. 2.1). Kovarianzen zwischen endogenen Variablen gibt es in einem Modell nicht, da sie eine Funktion der Varianzen und Effekte vorangehender Variablen sind. So ist z.B. die Kovarianz zwischen Y1 und Y2 kein schätzbarer Modellparameter, sondern eine direkte Folge der Varianz von M2 und der beiden Effekte h und i. Stattdessen sind die bivariaten Pfeile zwischen den Fehlertermen **Kovarianzen zwischen den Fehlertermen** dieser Variablen und nicht zwischen den Variablen selbst. Sie repräsentieren *weitere*, nicht im Modell adressierte, gemeinsame Ursachen der betreffenden Variablen. Wie wir später noch sehen werden, macht es Sinn, diese Kovarianzen zu schätzen, außer man vertritt die Hypothese, dass alle gemeinsamen Ursachen im Modell enthalten sind. Auf diese Implikationen eines Modells für Varianzen und Kovarianzen von Variablen werden wir im nächsten Kapitel genauer eingehen.

2

Theorie und Realität – Das Grundprinzip hinter SEM

2.1 Korrelation und Kausalität

Jede/r Studierende kennt das Mantra, dass aus Korrelation keine Kausalität folgt. Dies ist in der Tat korrekt und bedeutet, dass zwei korrelierende Variablen zwar in einem Zusammenhang stehen, daraus aber nicht gefolgert werden kann, dass die eine Variable die andere beeinflusst. Anders ist es bei der Umkehrung dieser Aussage – denn *aus Kausalität folgt Korrelation*. Dies bedeutet, dass, wenn die Variable X eine Ursache von Variable Y ist, diese beiden Variablen folglich korrelieren müssen (siehe dazu eine ausführlichere Einführung in Hayduk & Pazderka-Robinson, 2007).

Sehen wir uns dazu ein einfaches Beispiel an. Grundlage des Modells ist die Annahme, dass X einen kausalen Einfluss auf Y hat. Ein Pfaddiagramm dieses kleinstmöglichen Modells sähe folgendermaßen aus (s. Abb. 2):

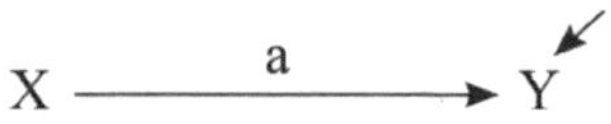

$$X \xrightarrow{\quad a \quad} Y \swarrow$$

Abb. 2: Ein einfaches Modell

Wenn diese Hypothese korrekt ist, *müssen* X und Y folglich korrelieren. Und ebenso, wie aus Kausalität Korrelationen folgen, folgen aus ihr *Kovarianzen* (da Korrelationen lediglich die standardisierte Version von Kovarianzen sind). Betrachtet man das Modell in Abb. 2, ergibt sich die Kovarianz zwischen X und Y aus dem Produkt des Effekts von X auf Y („a“) und der Varianz von X („Var(X)“):

$$\text{Kov}(X,Y) = a\text{Var}(X) \quad (1)$$

Wenn also X einen Effekt auf Y hat, folgt aus diesem Effekt unabdingbar eine bestimmte Kovarianz beider Variablen.

Das Modell impliziert aber nicht nur eine Kovarianz zwischen X und Y, sondern auch, dass die Varianz von Y eine Funktion der Varianz von X und dem Effekt b ist:

$$Var(Y) = a^2 Var(X) + Var(e) \quad (2)$$

Dies zeigt, dass sich die Varianz von Y zusammensetzt aus einer Komponente, die auf seine Ursache X zurückgeht, und der Fehlervarianz („Var(e)"), welche Messfehler und alle weiteren Ursachen von Y beinhaltet. In Abb. 2 ist die Fehlervarianz symbolisiert durch den Pfeil ohne Ursprung, der auf Y gerichtet ist.

Nach diesem Grundprinzip lassen sich die Korrelationen bzw. Kovarianzen (und Varianzen) aller Variablen eines komplexeren Modell ableiten. Das Grundprinzip hinter der Schätzung von Effekten und Möglichkeiten der Evaluation des Modells (vgl. Abschnitt 2.4) basiert auf dieser unabdingbaren Konsequenz der postulierten Modellstruktur für die Daten (d.h. Varianzen und Kovarianzen). Es ist daher essentiell, zu verstehen, welche Kovarianzen auch aus komplexeren Strukturen folgen. Um die die grundlegenden Prinzipien dabei zu verstehen, macht es Sinn, zunächst die zentralen Teil-Strukturen eines jeden Modells zu betrachten – dies sind a) *Common-Effect-Strukturen*, b) *Mediatorstrukturen* und c) *Common-Cause-Strukturen*. Um einen Wiedererkennungseffekt zu erreichen, sollen diese Teil-Strukturen als Elemente des Modells in Abb. 1 diskutiert werden.

Das Common-Effect-Modell beinhaltet mehrere (meist korrelierende) exogenen Variablen (s. Abb. 3). Dieses Modell impliziert die Annahme, dass die exogenen Variablen spezifische Effekte auf die endogene Variable (hier: M2) auch bei Kontrolle der jeweils anderen Variablen haben. Der Einfachheit halber wollen wir nur zwei exogene Variablen betrachten:

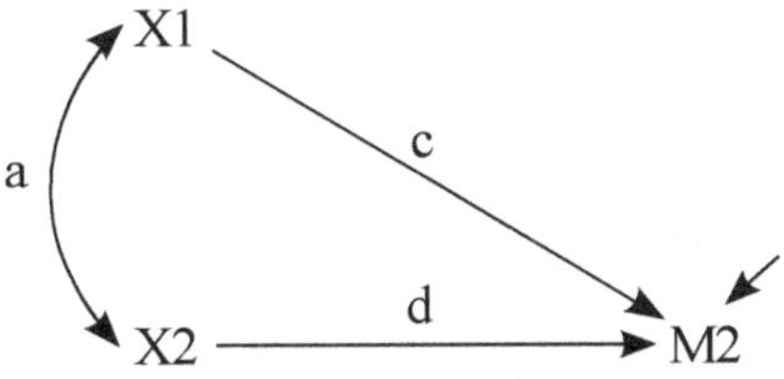

Abb. 3: Common-Effect-Struktur

Die Kovarianz von X1 und M2 ergibt sich nun sowohl aus dem direkten Effekt c als auch einem sog. „zusammengesetzten" Pfad (**compound path**) ad:

$$\begin{aligned} Kov(X1,M2) &= c + ad \qquad \text{bzw.} \\ &= c + Kov(X1,X2)d \end{aligned} \tag{3}$$

Dies hat interessante und wichtige Implikationen: Es zeigt zum einen, dass der Effekt c verzerrt geschätzt würde, wenn man X2 fälschlicherweise nicht in das Modell mit einbezieht, da der Pfad „ad" nicht mehr separat geschätzt und so die Kovarianz zwischen X1 und M2 nicht mehr in die Komponenten „c" und „ad" **dekomponiert** würde. Stattdessen würde ein verzerrter Effekt $c^* = c+ad$ geschätzt. Dieses „omitted variable"-Szenario stellt eines der zentralen Probleme in jeder Regressionsanalyse dar (Cohen, Cohen, West, & Aiken, 2003) und wird ausführlicher noch einmal in Kapitel 7 diskutiert. Zum anderen erklären sich so scheinbar paradoxe Widersprüche zwischen Kovarianzen oder Korrelationen zweier Variablen und den in einem Modell geschätzten Effekten. Zum Beispiel kann es zu folgenden Szenarien kommen:

a) Die Kovarianz (oder Korrelation) zwischen einer exogenen (z.B. X1) und einer endogenen Variable (z.B. M2) ist nicht-signifikant. Wird sie in ein Modell zusammen mit einer oder mehreren Kovariaten (z.B. X2) eingefügt, ist ihr Effekt aber signifikant.
b) Die Kovarianz zwischen X1 und M2 ist positiv – aber ihr Effekt ist nicht-signifikant.
c) Die Kovarianz zwischen X1 und M2 ist positiv, aber ihr Effekt ist negativ.
d) Die Kovarianz zwischen X1 und M2 ist negativ, aber ihr Effekt aber positiv.

All diese Szenarien haben den simplen Grund, dass verschiedene Kombinationen von c und ad zu entsprechenden Kovarianzen führen. Wenn z.B. „c" und „ad" unterschiedliche Vorzeichen haben, führt dies dazu, dass beide sich nivellieren und so in einer reduzierten oder völlig aufgehobenen Kovarianz zwischen der betreffenden exogenen und endogenen Variable äußern, oder aber die Kovarianz sogar einen zum Effekt gegenläufiges Vorzeichen hat. Beispielsweise würde ein Modell mit den Parametern $a = -.61$, $d = .33$ und $c = .20$ eine nicht-signifikante Kovarianz zwischen X1 und M2 ergeben, obwohl $c = .20$ ein substantieller Effekt ist. Grund ist, dass der compound path „ad" negativ und substantiell ist ($-.61 \times .33 = -.20$) und dieselbe Höhe hat wie der Effekt c. Analog kann eine Kovarianz zwischen X1 und M2 lediglich dadurch bedingt (und signifikant) sein, dass ad positiv ist, X1 selbst aber keinerlei Effekt hat (d.h. $c=0$).

Dies zeigt, dass das „**eye-balling**" von bivariaten Korrelationen wenig hilfreich ist, wenn es um kausale Hypothesen geht, weil einer Kovarianz ein komplexes Netz an Effekten und Kovarianzen der beteiligten exogenen Variablen zugrunde liegen kann, die sich je nach Höhe und Vorzeichen zu der betreffenden Kovarianz addieren. Aus diesem Grund folgt aus Korrelation nicht Kausalität.

Gleichung 3 zeigt auch, warum die statistische Kontrolle anderer exogener Variablen, die mit der Variable von Interesse korrelieren bzw. kovariieren, wichtig ist, um den Effekt c **unverzerrt** und **konsistent** zu schätzen. Die Kontrolle anderer Variablen ist nur unerheblich, wenn die Kovarianz zwischen beiden exogenen Variablen null beträgt, oder diese anderen Variablen selbst keinen Effekt haben. Aus beidem folgt, dass ad = 0 ist.

Das nächste Teilmodell ist eine Mediatorstruktur (MacKinnon et al., 2007), bei der eine exogene Variable auf eine endogene Variable wirkt – und diese Wirkung über eine dazwischen liegende Mediatorvariable *vermittelt* (d.h. mediiert) wird (s. Abb. 4). Wie schon unter Punkt 1.2 diskutiert, liegt hier ein indirekter Effekt vor.

Abb. 4: Ein Mediatormodell

Weil dieses Modell zwei endogene Variablen enthält (M1 und Y1) wollen wir Y1 als Outcome-Variable bezeichnen. Aus einer solchen Struktur folgt eine Kovarianz zwischen X1 und Y1, die sich als Produkt der beiden direkten Effekte b und g ergibt:

$$\text{Kov}(X1,Y1) = \text{Var}(X1)bg \quad (4)$$

Während Abb. 4 eine **vollständige Mediation** zeigt, bei der die exogene Variable nur über den Mediator mit der Outcome-Variable verbunden ist, liegt eine **partielle Mediation** dann vor, wenn es einen zusätzlichen direkten Effekt gibt. Gleichung 4 würde sich in diesem Fall dahingehend erweitern, dass dieser Effekt „c" hinzukommt:

$$\text{Kov}(X1,Y1) = \text{Var}(X1)bg + c \quad (5)$$

In solchen Fällen kann es zu den scheinbar paradoxen Situationen kommen, wie sie bereits bei dem Modell in Abb. 3 diskutiert wurden – d.h. die Kovarianz zwischen exogener und Outcome-Variable kann nicht-signifikant sein, oder ein anderes Vorzeichen haben, wenn der direkte und indirekte Effekt gegenläufige Vorzeichen haben. Dies ist auch der Grund, warum eine Mediationsanalyse keinen bivariaten Zusammenhang zwischen exogener und Outcome-Variable voraussetzt, wie dies lange Zeit gefordert wurde (Hayes, 2009).

Als letzte Struktur soll nun die sog. Common-Cause-Struktur besprochen werden, wie sie u.a. Faktormodellen zugrunde liegt (vgl. Abschnitt 3.4). Wie der Name sagt, betrifft diese Struktur die Situation, dass zwei oder mehrere Variablen eine gemeinsame Ursache haben.

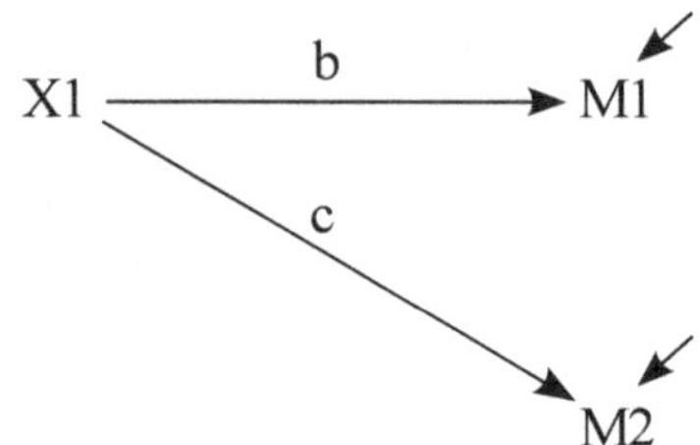

Abb. 5: Eine Common-Cause-Struktur

Wie Abb. 5 zeigt, werden M1 und M2 von ein und derselben Variable X1 beeinflusst. Aus einer solchen Struktur folgt eine Kovarianz zwischen M1 und M2. Diese ergibt sich aus

$$\text{Kov}(M1,M2) = \text{Var}(X1)bc \tag{6}$$

Dies zeigt, dass zwei Variablen einen Zusammenhang aufweisen können, obwohl sie überhaupt keine kausalen Verknüpfungen haben. Daher stellt das Common-Cause-Modell eine alternative Erklärung für Modelle dar, in denen der Einfluss einer Variablen auf eine andere postuliert wird (auch „Spurious-Association-Modell"). In der Realität werden häufig Mischformen auftreten – d.h. es gibt einen kausalen Effekt einer Variablen auf eine andere und zusätzlich werden beide von einer oder mehrerer „Drittvariablen" beeinflusst. Ignoriert man diese Drittvariable(n), wird der geschätzte Effekt genauso verzerrt, wie im omitted-variable-Szenario („**confounding**"). Genaugenommen ist dies eine Unterform des Omitted-Variable-Szenarios.

Zusammenfassend verfolgte die vorangehende Diskussion zwei Ziele: Erstens sollten die drei Modelltypen als konstituierende Teilmodelle aller denkbaren komplexeren Modelle vorgestellt werden, um so die Sicht auf komplexere Modelle etwas zu strukturieren. Zweitens sollte insbesondere durch die Erörterung der Strukturgleichungen, die diese Modelltypen implizieren, verdeutlicht werden, welche Kovarianzen der beteiligten Variablen aus ihr folgen. *D.h. auch wenn aus einer Kovarianz nicht ein bestimmtes Kausalmodell folgt, so folgt aus dem Kausalmodell ein bestimmtes und damit testbares Muster von Kovarianzen und Varianzen der Variablen.*

2.2 Die Pfadregeln von Sewell Wright

Die bisherigen Ausführungen beinhalteten die *impliziten Kovarianzen* einer bestimmten Kausalstruktur und waren auf sehr simple Teilstrukturen bezogen. Ein komplexeres Modell ist immer aus mehreren dieser Teilstrukturen zusammengesetzt, was es in der Praxis erst mal sehr schwer macht, die Implikationen zu verstehen. Es ist aber für das Verständnis der Arbeitsweise eines Modells und v.a. für das Verständnis eventueller Probleme bei der Testung des Modells sehr hilfreich und für eine erfolgreiche **Respezifikation** eines Modells unumgänglich, diese Implikationen des Modells für die Daten nachvollziehen zu können.

Ein sehr wirksames Hilfsmittel sind zu diesem Zweck die Pfadregeln des Erfinders der Pfadanalyse, Sewell Wright (1921). Mithilfe dreier einfacher Regeln lassen sich die impliziten *Korrelationen* jedes Modells aus dem Modell herauslesen. Wie oben ausführlich diskutiert, impliziert das Modell *Kovarianzen*, aber für ein Verständnis des Zusammenhangs zwischen einem Modell und den Daten ist die Ableitung der impliziten *Korrelationen* meist ausreichend. Der Grund ist, dass Korrelationen standardisierte Kovarianzen sind, und damit alle Varianzen der Variablen gleich 1 sind. Dies hat den Vorteil, dass die Varianz der jeweiligen exogenen Variable in den Strukturgleichungen ignoriert werden kann. Zum Beispiel zeigte Gleichung 6, dass in einer Common-Cause-Struktur die Kovarianz zwischen den beiden Variablen M1 und M2 aus dem Produkt der Effekte der gemeinsamen Ursache X1 und der Varianz von X1 bestand (d.h. Var(X1)bc). Betrachtet man stattdessen die implizite Korrelation zwischen M1 und M2, besteht sie nur noch aus dem Produkt der (standardisierten) Effekte. Hat man die Pfadregeln verstanden, lässt sich die Erweiterung zu den impliziten Kovarianzen durch Einbeziehung der exogenen Variablen leicht bewerkstelligen.

Die Grundidee der Pfadregeln ist, dass sich die Korrelation zwischen den betreffenden Variablen zusammensetzt aus allen Pfaden, die man entlang schreiten kann, um von der einen zur anderen Variable zu gehen. Eine Metapher könnte die eines Kanalsystems sein, in dem zwei Orte über Rohre verbunden sind. Die Frage lautet hier also, durch welche Rohre Wasser fließen kann, um von einem Ort zum anderen zu gelangen.

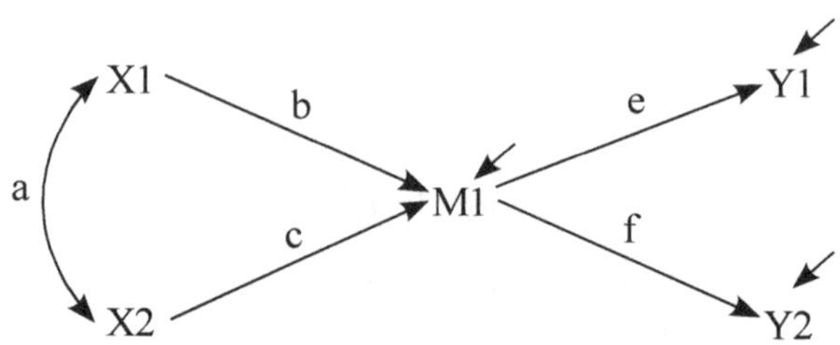

Abb. 6: Ein Mediatormodell

Nehmen wir als Beispiel die Outcome-Variable Y1 in Abb. 6. Sie hat eine proximale (M1) und zwei distale Ursachen (X1 und X2), die sie über M1 beeinflussen. Um herauszufinden, woraus sich z.B. die Korrelation zwischen Y1 und X1 ergibt, gehen wir die Pfade entlang, die beide Variablen verbindet und setzen diese additiv zusammen: Dies ist z.B. der indirekte (standardisierte) Effekt be, der beide Variablen verbindet sowie der Pfad über X2 (ace). Grund für die Einbeziehung von X2 ist, dass X2 mit X1 korreliert (hier zeigt sich also eine Kombination der Teilstrukturen „Mediatormodell“ (vgl. Abb. 4) und dem „Common-Effect-Modell“ (vgl. Abb. 3). Will man nun wissen, aus welchen Pfaden sich die implizite Korrelation zwischen X1 und Y1, zusammensetzt, addiert man die Teile zusammen:

$$\mathrm{Kor}(X1,Y1) = be + ace \tag{7}$$

Die Wright'schen Pfadregeln lassen sich nun als Verkehrsregeln verstehen, nach denen dieses Fortschreiten über die Pfade geschehen muss. Denn auch bei Modellen gibt es Einbahnstraßen und das Befahren ist nicht für alle Pfade und Richtungen erlaubt. Die drei Pfadregeln sollen anhand von Abb. 7 erläutert werden.

- **Regel 1 („Einmal vorwärts, immer vorwärts“):** In einem Pfad ist es erlaubt, mehrfach „rückwärts“ (d.h. entgegen der Pfeilrichtung) fortzuschreiten. Betritt man allerdings einen Pfad in der Vorwärtsrichtung, darf daraufhin kein Pfeil mehr in Rückwärtsrichtung beschritten werden. Zum Beispiel ist zur Herleitung der Korrelation zwischen M1 und M2 der Pfad bc erlaubt, weil hier erst rückwärts über c und anschließend vorwärts über b vorgegangen wird. Im Gegensatz dazu ist der Pfad ef nicht erlaubt. Dies entspricht der Logik, dass zwei

Variablen nicht allein deshalb korrelieren, nur weil sie einen Effekt auf dieselbe Outcome-Variable haben.

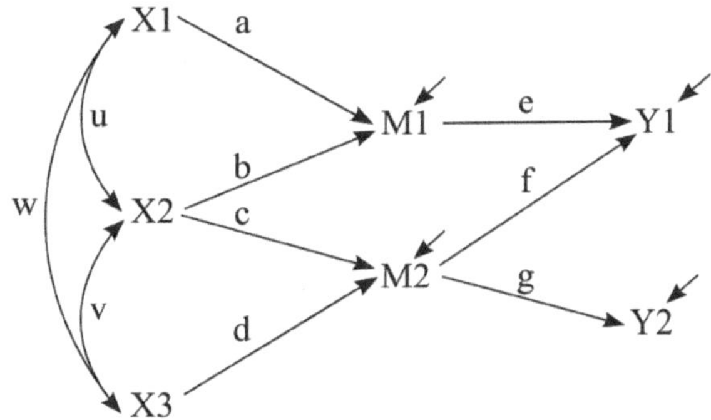

Abb. 7: Modell zur Veranschaulichung der Pfadregeln

- **Regel 2 („Nur ein bidirektionaler Pfeil pro Pfad"):** Innerhalb eines Pfades ist nur ein bidirektionaler Pfeil erlaubt. So sind zur Bestimmung der impliziten Korrelation zwischen M1 und M2 die Pfade bvd, auc als auch awd erlaubt, weil all diese Pfade nur über jeweils einen bidirektionalen Pfeil laufen – nicht aber dvua, denn bei diese würde man u und v passieren müssen.
- **Regel 3 („Kein mehrfaches Durchkreuzen"):** Eine Variable darf *innerhalb eines Pfades* nicht mehrfach passiert werden. So ist z.B. zur Herleitung der impliziten Korrelation zwischen Y1 und Y2 nicht der Weg über den Pfad fcvdg erlaubt – weil dies bedeuten würde, dass M2 innerhalb eines einzigen Pfades zweimal passiert wird.

Auf den ersten Blick erscheint die Anwendung dieser Regeln mühsam; allerdings ist man mit ihrer Hilfe und etwas Übung schnell in der Lage, die impliziten Korrelationen, die aus einer spezifizierten Kausalstruktur folgen – und damit die Implikationen, die das Modell für die Daten hat – abzuleiten. Außerdem helfen die Regeln zu verstehen, welch komplexes Muster an spezifischen Effekten und Zusammenhängen hinter einer simplen Korrelation stecken kann. So ist beispielsweise die Korrelation zwischen M1 und M2 bedingt durch all die besprochenen Pfade – nämlich bc + bvd + awd + auc. Die implizite Korrelation zwischen Y1 und Y2 folgt aus fg + ebcg + eaucg + eawdg. Die Kenntnis der Pfadregeln verhindert, beim Auftreten scheinbar paradoxer Widersprüche zwischen der Korrelation und dem Struktur- oder Regressionseffekt, wie dies im vorherigen Abschnitt diskutiert wurde, mit Überraschung oder gar Sorge zu reagieren.

2.3 Bedingte Unabhängigkeiten und d-separation

Der vergangene Abschnitt hat illustriert, welche Daten – sprich Kovarianzen und Korrelationen der empirischen Variablen – aus einer Kausalstruktur folgen. Der Test des Modells, der in Abschnitt 2.4 genauer erläutert werden wird, bezieht sich auf den Vergleich zwischen diesen impliziten Kovarianzen und den tatsächlichen, empirischen Kovarianzen. Ist das Modell korrekt, sollten diese sehr ähnlich sein. Bevor dieser Test erläutert werden soll, lohnt es sich aber, eine weitere Implikation von Modellen zu erörtern, die traditionell nicht in der SEM Literatur diskutiert wurden, sondern durch die Arbeiten von Pearl (2001) bekannt wurden: **bedingte Unabhängigkeiten** und das Konzept der **„d-separation“** („directional separation“). Zentraler Aspekt dieser Implikationen ist, dass aus einem Modell nicht nur folgt, welche Variablen miteinander korrelieren und wie hoch diese Korrelationen sind, sondern auch, welche Variablen *dann nicht mehr korrelieren sollten, wenn man andere konstant hält.* Neben dem Vergleich von impliziten und tatsächlichen Kovarianzen ist damit eine weitere, überprüfbare Implikation von Modellen vorhanden. Außerdem helfen bedingte Unabhängigkeiten, Anwender/innen von SEM zu verdeutlichen, welche (impliziten) Annahmen sie in ihrem Modell machen. Denn oft sind es diese impliziten Annahmen, deren Test in der Praxis häufig fehlschlägt, ohne dass der/die Anwender/in sich dessen bewusst ist.

Schließlich können d-separation-Implikationen genutzt werden, um bei der Planung des Forschungsprojekts Variablen zu erkennen, die in die Studie und damit in das Modell aufgenommen werden sollten, um eine Verzerrung der wesentlichen Modellstrukturen von Interesse zu vermeiden. Wir werden später das Programm „DAGitty“ kennenlernen, das dabei helfen kann.

Nehmen wir als einfaches Beispiel ein Mediatormodell mit einer vollständigen Mediation (also X → M → Y). Aufgrund der Pfadregeln ist intuitiv ersichtlich, dass X mit Y korrelieren muss und diese Korrelation aus dem Produkt der beiden direkten Effekte folgt. Eine weitere Implikation ist aber, dass diese Korrelation *dann* nicht-signifikant wird, *wenn* wir M konstant halten. Dies bedeutet: Wenn wir nur diejenigen Fälle aus der Population mit einem bestimmten Wert in M betrachten würden, dann wäre für diese Personen Kov(X,Y) = 0. Dies gilt für alle möglichen Werte von M. Die Notation dafür ist: Kov(X,Y | M) = 0 („Die Kovarianz zwischen X und Y *gegeben* M ist 0“). Ist diese bedingte Kovarianz nicht 0, enthält das Modell der vollständigen Mediation Fehler.

Aus der Perspektive der Pfadregeln führt ein solcher Fall dazu, dass die Kovarianz zwischen X und Y sich nicht nur aus dem direkten Effekt zusammensetzt und somit die implizite Korrelation von der tatsächlichen, d.h. empirischen abweicht. Es gibt in solch einem Fall daher noch weitere Prozesse, die eine Abhängigkeit von X und Y bewirken (z.B. einen direkten Effekt von X auf Y oder eine ausgeschlossene Drittvariable, die X und Y beeinflusst). Pfadregeln und d-separation sind demnach zwei verschiedene Blickwinkel auf dasselbe Problem.

Als ein weiteres, etwas komplexeres Beispiel greifen wir nochmals das Modell in Abb. 7 auf. Wie zu sehen, gibt es zwischen M1 und M2 keinerlei kausale Verbindungen. Dennoch müssen aufgrund der Modellstruktur beide Variablen miteinander korrelieren und mithilfe der Pfadregeln lässt sich illustrieren, warum dies so ist und woraus sich die Korrelation beider Variablen ergibt. Eine weitere Implikation dieses Modells ist aber auch, dass die empirische Korrelation beider Variablen *dann* 0 ist, wenn X1 und X2 oder X2 und X3 konstant gehalten werden (und folglich auch, wenn alle drei Ursachen konstant gehalten werden). Mit anderen Worten: die Korrelation zwischen M1 und M2 *gegeben* X1 und X2 (Kov[M1,M2 | X1,X2]) ist 0, d.h. sie sind „bedingt unabhängig". Ähnlich zum oben diskutierten Beispiel wird die Kovarianz von M1 und M2 dann 0, wenn wir sie nur bei Personen mit bestimmten Werte-Paaren für X1 und X2 betrachten würden.

Wenn wir die Metapher des Kanalsystems wieder aufgreifen, würde das bedeuten, dass von M1 zu M2 *dann* kein Wasser mehr fließen kann, wenn die Schleusen X1 und X2 oder X2 und X3 geschlossen werden. Durch die Kontrolle dieser Variablen werden die entsprechenden Pfade **blockiert**. Hält man nur eine der beiden Variablen konstant, besteht immer noch eine Verbindung zwischen M1 und M2 – Kontrolle beider ist also nötig, um alle Verbindungen zu kappen. Der Begriff „d-separation" bezeichnet nun den Zustand, dass zwei Variablen, die aufgrund der zugrundeliegenden Kausalstruktur auf natürliche Weise (d.h. „unbedingt") korrelieren, durch die Konstanthaltung einer oder mehrerer anderer Variablen separiert (also unabhängig) werden.

Für jedes Modell lassen sich nun eine ganze Reihe „d-separation"-Aussagen für alle möglichen Variablenpaare ableiten, die zwei einfachen Regeln (ähnlich den Pfadregeln) folgen und die im Falle der korrekten Spezifikation des Modells zutreffen müssen. Zur Erläuterung der Regeln benötigen wir zunächst die Definition einiger Begriffe (*collider, fork, chain, conditioning set* und *basis set*). Diese Begriffe als auch die Regeln sollen anhand der Abb. 8 illustriert werden.

Ein **collider** ist eine Variable, die mindestens zwei im Modell befindliche Ursachen hat. Beispielsweise ist in dem Modell in Abb. 8 Y2 ein collider, weil in ihr die Effekte von X1 und Y1 kollidieren. Ein collider tritt somit in einer Common-Effect-Struktur auf. Eine **fork** ist eine Teilstruktur innerhalb eines Modells, in der eine Variable Effekte auf mehrere andere Variablen aussendet. In der Abbildung ist Z1 eine fork. Wir haben dies anfangs als Common-Cause-Struktur bezeichnet. Eine **chain** liegt schließlich dann vor, wenn die Variablen in dem Pfad in einer kausalen Sequenz liegen. Wir haben diese im Rahmen der Mediatorstruktur kennengelernt.

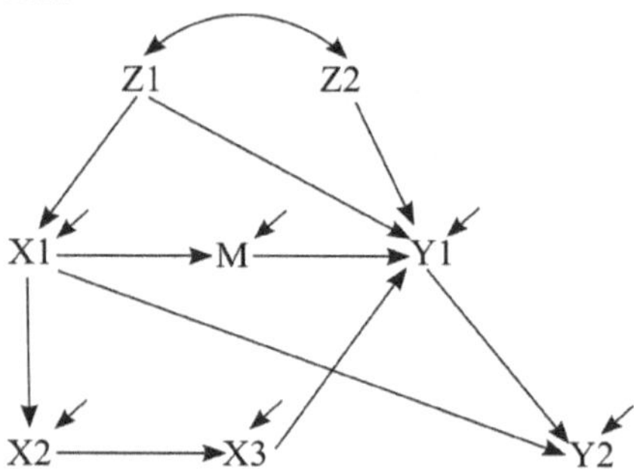

Abb. 8: Pfadmodell zur Illustration des d-separation-Prinzips

Schließlich ist der Begriff des **conditioning set** und des **basis set** bedeutsam. Ersterer umfasst eine Menge von Variablen, die jeweils kontrolliert werden, um einen Pfad zu blockieren – d.h. es ist u.U. nicht nur eine, sondern mehrere Variablen. Das basis set ist die minimale Menge von Variablen, die kontrolliert werden *müssen*, damit der Pfad blockiert und beide Variablen d-separiert sind.
Die Frage ist nun: Welche Variablen müssten statistisch kontrolliert werden, um zwei Variablen zu separieren? Mithilfe der **d-separation-Regeln** lässt sich dies beantworten. Die Bedingungen für eine Separierung lauten demnach:

(1) Ein Pfad ist d-separiert oder blockiert, wenn er die Form einer chain oder fork hat und die Variable(n) in der Mitte des Pfades sich im conditioning set befinden (also kontrolliert werden). Beispielsweise würde der Pfad, der X1 und Y1 über M verbindet, durch Kontrolle von M blockiert, weil M innerhalb einer chain liegt. Analog würde der Pfad über Z1 blockiert, wenn Z1 kontrolliert wird, weil Z1 innerhalb einer fork liegt.

(2) Ein Pfad ist d-separiert, wenn er einen collider enthält, und dieser collider *nicht* im conditioning set ist. So ist der Pfad zwischen X1 und Y1 über Y2 von vorne herein schon blockiert (dies folgt bereits aus Pfadregel Nr. 1, nach der ein Pfad nicht erst vorwärts und dann rückwärts beschritten werden darf). Dieser Pfad ist also blockiert, weil Y2 *nicht* kontrolliert wird. Wird hingegen Y2 kontrolliert, *öffnet dieses den Pfad.* Eine interessante und praktische Implikation davon ist z.B., dass zwei Variablen, die eigentlich nicht korrelieren, weil sie lediglich eine gemeinsame abhängige Variable haben, sonst aber keinerlei Verbindungen oder gemeinsame Ursachen haben, durch die statistische Kontrolle dieser abhängigen Variable plötzlich negativ korrelieren, was im Gesamtmodell zu verzerrten Effekten führen kann. Ein Fehler in der Regressionsanalyse ist daher, eine gemeinsame abhängige Variable des Prädiktors und Kriterium als Kontrollvariable einzufügen.

Wenden wir die Regeln auf das Modell in Abb. 8 an: Wie schon im Beispiel zu Regel 1 beschrieben, muss zunächst M kontrolliert werden. Dies blockiert den indirekten Effekt über M. Weiterhin gibt es die mit X1 korrelierende Variable X2, die Y1 über X3 beeinflusst. Hier muss entweder X2 oder X3 kontrolliert werden, um den Pfad zu unterbrechen. Beide zu kontrollieren ist unnötig, weil bereits die Kontrolle einer dieser Variablen den Pfad blockiert. Des Weiteren sind X1 und Y1 über Z1 und Z2 verbunden. Z2 zu kontrollieren reicht nicht aus, weil Z1 eine fork bzw. ein common cause ist, der X1 und Y1 verbindet. Kontrolliert man allerdings Z1, ist die Kontrolle von Z2 nicht mehr nötig. Schließlich darf man Y2 nicht kontrollieren, weil dies den Pfad zwischen beiden öffnen würde. Es gibt also folgende basis sets: {M, X2, Z1} und {M, X3, Z1}. Ist die postulierte Struktur korrekt, muss die Kovarianz oder Korrelation zwischen X1 und Y1, die durch die Modellstruktur erklärt werden soll, durch Kontrolle eines dieser beiden basis sets verschwinden.

Einen interessanten und nicht so intuitiven Fall finden wir bei der Existenz von ausgeschlossenen Drittvariablen: Angenommen, wir hypothetisieren ein simples vollständiges Mediatormodell, in dem X eine Wirkung auf Y über M haben soll (s. Abb. 9). Was wir nicht ahnen ist, dass M und Y eine gemeinsame Ursache Z haben. Aufgrund unseres simplen Modells müsste die Korrelation zwischen X und Y bei Konstanthaltung von M Null betragen. In diesem Fall aber schlägt dieser Test fehl. Stattdessen müssten wir M *und* Z konstant halten, um die Korrelation zu eliminieren. Warum? Die Lösung ist Regel Nr. 2: Im tatsächlichen Modell ist M ein collider (für X und Z). Durch die Konstanthaltung von M wird nun aufgrund von Regel Nr. 2 der Pfad zwischen X und Z geöffnet und eine Beziehung hergestellt. Dadurch wird dann auch der Pfad zwischen X und Y geöffnet (es herrscht nun eine Verbindung zwischen beiden über Z). Deshalb müsste Z ebenfalls kontrolliert werden (Regel Nr. 1), damit X und Y nicht mehr korrelieren. Da wir aber von Z nichts ahnen (und es daher kein Teil des Modells ist), führt die Konstanthaltung von M zu einer immer noch vorhandenen Korrelation von X und Y – was korrekterweise Zweifel am Modell aufkommen lässt. Nur sollte diese noch vorhandene Korrelation nicht vorschnell und automatisch im Sinne eines zusätzlichen direkten Effekts interpretiert werden. Insbesondere negative bedingte Korrelationen können ein Hinweis auf eine ausgeschlossene Drittvariable sein, da das Konstanthalten von collidern in der Regel zu negativen bedingten Korrelationen führt. Dieses Beispiel illustriert, wie wichtig Variablen wie X sind, um auf Probleme im „hinteren" Teils eines Modells hinzuweisen. Kap. 7 wird dies eingehender behandeln.

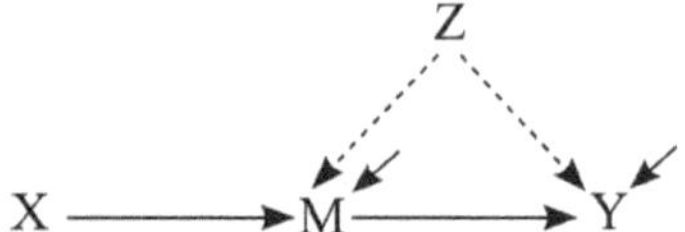

Abb. 9: D-separation bei ausgeschlossenen Drittvariablen

Zusammenfassend lässt sich damit feststellen, dass d-separation neben den Pfadregeln einen alternativen und ergänzenden Blick auf die Implikationen von Kausalmodellen und die impliziten Restriktionen bietet, die aus einem Modell folgen. D-separation eignet sich zur Auswahl wichtiger Modellvariablen (auch Kontrollvariablen) und kann helfen, abzuschätzen, welche Variablen essentiell für die Validität eines Modells sein dürften. Neben der Einbeziehung und Kontrolle aller essentiellen Variablen werden wir in Kapitel 7 einen Ansatz kennenlernen, den negativen Folgen der Auslassung von Variablen (und noch anderer Gefahren für kausale Interpretationen) zu begegnen. Das Beispiel in Abb. 9 wird uns dabei wieder begegnen.

Schließlich ist ein Nutzen von d-separation, dass ein Modell bereits mit simplen Mitteln (Partialkorrelation) getestet werden kann – und dies, ohne irgendwelche Modellparameter schätzen zu müssen. Ein hilfreiches Werkzeug ist hierbei die Software DAGitty (s. Abb. 10), die entweder als Browser-internes Programm (www.dagitty.net) genutzt oder heruntergeladen werden kann. Damit können Pfaddiagramme einfach gezeichnet werden; als Ergebnis werden die basis sets

(„adjustment for direct effect“) und eine Liste aller bedingten Unabhängigkeiten („Testable implications“) ausgegeben, die aus dem hypothetisierten Modell folgen.

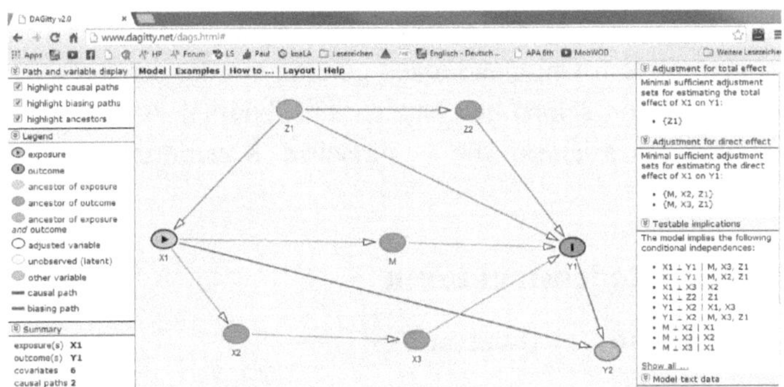

Abb. 10: Oberfläche der software DAGitty, mit der man sich alle d-separation-Relationen eines Modells ausgeben lassen kann.

DAGitty kann dabei behilflich sein, in der Planungsphase eines Forschungsprojekts diejenigen Variablen zu identifizieren, die miterhoben und entweder kontrolliert oder in das Modell einbezogen werden sollten. So könnte man den derzeitigen Stand des Wissens bzw. der Theorie im Rahmen eines Pfadmodells ausdrücken und diejenigen Variablen ausfindig machen, die für die Identifikation derjenigen kausalen Effekten von Interesse zentral sind. Wie oben erläutert, müssen das nicht viele Variablen sein, sondern die „richtigen“, d.h. diejenigen an den zentralen Stellen. Natürlich hängt die Sinnhaftigkeit dieses Vorgehens von der Korrektheit des angenommenen Modells ab, ist aber theoretisch begründeter als die standardmäßige Einbeziehung der „üblichen Verdächtigen“, wie Alter, Geschlecht oder Bildung.

Aus den o.g. Ausführungen dürfte auch klar geworden sein, dass mit der Anzahl von **Restriktionen** eines Modells die Anzahl von d-separation-Aussagen und damit die testbar Implikationen eines Modells zusammenhängen. Beispielsweise ist ein vollständiges Mediatormodell dahingehend **restriktiv**, dass es *keinen* direkten Effekt der unabhängigen Variablen X auf die abhängige Variable Y vorsieht. In einem Modell würde dieser Effekt daher auf null fixiert statt geschätzt. Ein solches Modell impliziert die bedingte Unabhängigkeit von X und Y bei Konstanthaltung von M. Ein partielles Mediatormodell dagegen ist **saturiert**; es gibt keine d-separation-Aussage. Dieses Prinzip lässt sich auf komplexere

Modelle übertragen. Es sind die Restriktionen, die das Modell testbar machen: Je saturierter die Struktur, umso weniger d-separation-Aussagen macht das Modell und umso weniger leicht fallen falsch spezifizierte Modelle durch eine Nicht-Passung von modell-impliziter und empirischer Kovarianz auf. Der folgende Abschnitt wird nun zeigen, wie in einem SEM die Parameter geschätzt werden und wie das Ergebnis mit dem Chi-Quadrattest getestet werden kann. Dieser testet, inwieweit das Modell mit seinen geschätzten und fixierten Parametern und den dadurch impliziten Kovarianzen die empirischen Kovarianzen reproduzieren kann.

2.4 Der Test von Modellstrukturen

2.4.1 Grundprinzip und Chi-Quadrat-Test

Das Grundprinzip, welches durch die Erläuterung der Pfadregeln und ihrer Anwendung klar geworden sein sollte, ist, dass aus einem postulierten Kausalmodell ein bestimmtes Muster von Kovarianzen folgen *muss.* Es ist somit zentral für die Beurteilung der Korrektheit des Modells, ob die modell-implizite Kovarianzmatrix mit der empirischen Kovarianzmatrix übereinstimmt. Da das Modell allerdings unter Umständen Dutzende von Variablen enthält, besteht nun das Problem, wie man eine Unterschiedlichkeit beider Matrizen feststellen soll. Immerhin hat eine Matrix viele Zellen, die sich in unterschiedlichem Maße über beide Matrizen hinweg unterscheiden werden. Außerdem stellt sich die Frage, was dasjenige Ausmaß von „Unähnlichkeit" ist, ab dem Zweifel an der Spezifikation des Modells aufkommen.

Die Lösung dieses Problem besteht darin, die Abweichung beider Matrizen mit einem statistischen Test auf einen signifikanten Unterschied zu testen. Auch wenn, wie bei allen Tests, das Kriterium für die Zurückweisung der Nullhypothese willkürlich ($p < .05$) ist (siehe die Kritik von Gigerenzer, 2004), so stellt ein Test ein objektives und transparentes Kriterium dar: Ist der Test signifikant, bedeutet dies, dass es eine überzufällige Abweichung beider Matrizen gibt. Der p-Wert des Tests gibt dabei an, wie wahrscheinlich eine Diskrepanz – basierend auf der Kovarianzmatrix der aktuellen Stichprobe – wäre, wenn das Modell korrekt wäre. Die Interpretation einer überzufälligen und damit *systematischen* Abweichung ist simpel: Das Modell ist so nicht korrekt. Bevor wir zu einer weiteren Diskussion über diesen Test kommen, wollen wir zuerst klären, worauf dieser Test beruht.

Die Grundlage eines jeden statistischen Tests ist die **Stichprobenverteilung** des betreffenden Parameters – z.B. der Kovarianz zwischen zwei Variablen X und Y. Sie gibt die empirisch ermittelten Kovarianzen an, die wir erhalten würden, wenn wir unendlich oft Stichproben aus einer Population ziehen würden. Da eine Stichprobenziehung ein stochastischer Prozess ist, wird man nicht immer die Populationskovarianz zwischen X und Y in den Daten finden – stattdessen bekommt man eine Verteilung der Stichprobenkovarianzen von X und Y. Wenn man nun in jeder der Stichproben nicht nur eine Kovarianz berechnet, sondern zusätzlich eine weitere (z.B. diejenige zwischen X und Z), bekommt man eine *bivariate* Verteilung, die alle möglichen Kombinationen beider Kovarianzen enthält. Dieser Gedanke lässt sich jetzt weiterführen, bis man zu einer Verteilung für eine gesamte Kovarianzmatrix kommt (der sog. Wishart-Verteilung, vgl. Hayduk, 1987, für eine detailliertere Erklärung). Nun ist der Gedanke der folgende: Wenn in der Population das Modell korrekt ist, dann ist die Populationskorrelationsmatrix Σ auch gleichzeitig die modell-implizite Kovarianzmatrix (meist $\Sigma(\theta)$ genannt). Wenn wir nun aus dieser Population unendlich viele Stichproben ziehen, bekommen wir die Stichprobenverteilung der Kovarianzmatrix aller Modellvariablen, die den Bereich an Matrizen-Diskrepanzen absteckt, den man per Zufall erwarten würde, obwohl das Modell korrekt ist. Damit beschreibt die **Wishart-Verteilung** die Wahrscheinlichkeit dafür, dass eine vorliegende Stichproben-Kovarianzmatrix S aus einer Population mit der Kovarianzmatrix $\Sigma(\theta)$ stammt. Wenn nun aber die Auftretenswahrscheinlichkeit der tatsächlich vorliegenden Kovarianzmatrix S in dieser Verteilung unter dieser angenommenen Populationskovarianzmatrix sehr gering ist ($p < .05$), lässt sich vermuten, dass das in den Daten spezifizierte Modell nicht dem Populationsmodell entspricht: Wie oben angemerkt, definiert die Stichprobenverteilung denjenigen Bereich an Abweichungen zwischen $\Sigma(\theta)$ und S, den man aufgrund des Zufalls erwarten würde. Jenseits dieses Bereichs müssen also irgendwelche systematische Probleme (entweder mit den Daten oder im Modell) bestehen, die nicht durch den Zufall erklärt werden können.

Die Parameter im Modell werden jetzt so geschätzt, dass die modell-implizite Matrix $\Sigma(\theta)$ so ähnlich wie möglich der Stichprobenmatrix S wird. Hierbei wird $\Sigma(0)$ – die ja auf den Parameterschätzungen basierend auf der aktuellen Stichprobe beruht – als stichprobenfehlerbehafteter Repräsentant der Populationsmatrix Σ benutzt. Die Parameterschätzung geschieht mit der **Maximum-likelihood-Fit-**

Funktion. Mit ihrer Hilfe werden die Parameter so geschätzt, dass die Diskrepanz zwischen S und $\Sigma(\theta)$ minimiert wird.

$$F_{ML}[S; \Sigma(\theta)] = tr[S\Sigma(\theta)^{-1}] + [\log |\Sigma(\theta)| - \log |S|] - q \quad (8)$$

Die Funktion sieht auf den ersten Blick kompliziert aus. Daher soll im Folgenden auf einfachem Wege erläutert werden, was hier passiert: Wir wollen dies an einem Beispiel illustrieren, das ein fehlspezifiziertes Modell zeigt. Abb. 11 zeigt ein links partielles Mediatormodell. Wie zu sehen, hat X1 sowohl einen Effekt auf Y, der z.T. über M vermittelt wird, als auch einen direkten Effekt. In dem hypothetisierten Modell fehlt dieser direkte Effekt jedoch. Der/die Forscher/in nimmt stattdessen fälschlicherweise an, der Effekt von X1 auf Y würde vollständig mediiert.

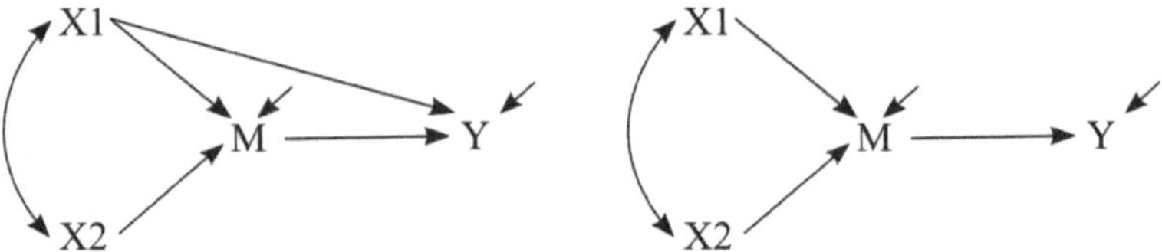

Abb. 11: Darstellung eines Populationsmodells (links) und des fehlspezifizierten Stichprobenmodells

Die Höhe der Effekte soll im Moment irrelevant sein – was interessiert, ist der Vergleich der beiden Matrizen:

Stichproben-Kovarianzmatrix S

	Y	M	X1	X2
Y	3.367			
M	2.042	1.898		
X1	1.405	0.878	1.040	
X2	0.861	0.905	0.454	1.198

Modell-implizite Matrix $\Sigma(\theta)$

	Y	M	X1	X2
Y	3.353			
M	2.033	1.890		
X1	0.940	0.874	1.036	
X2	0.969	0.901	0.452	1.193

Links ist die auf einer Stichprobe von N = 250 beruhende Kovarianzmatrix, die aus der Population gezogen wurde, in der das partielle Mediatormodell gilt – rechts die implizite Matrix, die aus der finalen Schätzung der Parameter folgt. Wie man sieht, gibt es über die verschiedenen Zellen hinweg unterschiedlich deutliche Abweichungen. Der Chi-Quadrat-Test gibt uns nun eine Zusammenfassung der Abweichungen sowie eine Information, ob diese Abweichungen gravierend genug sind, um an der Korrektheit des Modells zu zweifeln. Wenden wir nun die Fit-Funktion sukzessive an:

- **tr[SΣ(θ)$^{-1}$]:** In der eckigen Klammer werden die S und die invertierte implizite Kovarianzmatrix multipliziert (eine Matrix mit einem „-1"-Exponenten zeigt an, dass hier diese Matrix invertiert wird). Je ähnlicher S und Σ(θ) ist, umso mehr nähert sich das o.g. Produkt einer Identitätsmatrix an. Dies ist eine Matrix, die nur Nullen enthält – außer auf der Hauptdiagonalen, auf der Einsen stehen. Für das o.g. Beispiel sieht das Produkt folgendermaßen aus:

```
          Y       M    X1      X2
[1,]  1.004 -0.255 0.740 -0.181
[2,]  0.000  1.003 0.001  0.000
[3,]  0.395 -0.425 1.004  0.000
[4,] -0.096  0.104 0.000  1.004
```

  Die Buchstaben „tr" stehen für „trace" (Spur), was bedeutet, dass hier die Zahlen auf der Hauptdiagonalen aufsummiert werden – in unserem Beispiel also 1.004+1.003+1.004+1.004 = 4.015.
- **[log |Σ(θ)| - log |S|] :** In diesem Term werden ganz einfach die Determinanten beider Matrizen logarithmiert und voneinander subtrahiert. Es ist offensichtlich, dass je ähnlicher sich beide Matrizen sind, sich das Ergebnis Null annähert. Angewandt auf das o.g. Beispiel ist die Determinante von Σ(θ) = 1.059975; diejenige von S = .745574. Logarithmiert und subtrahiert man beide, erhält man .352.
- **q:** Dies ist die Anzahl der Variablen, die das Modell – und damit die Matrizen enthält – in unserem Fall ist q = 4

Setzt man die Bausteine nun zusammen, erhält man

$$F_{ML} = 4.015 + .352 - 4 = .367$$

Den Chi-Quadrat-Wert für dieses Modell erhält man nun, wenn man diesen Wert mit N-1 multipliziert:

$$\chi^2 = F_{ML} \times (N - 1)$$

Für das hier getestete Modell wäre das ein Chi-Quadrat-Wert von 91.4. Um die Signifikanz zu beurteilen, müssen wir die zu diesem Modell passende Chi-Quadrat-Verteilung auswählen. Dafür benötigt man die sie charakterisierende Anzahl von **Freiheitsgraden** (df, degrees of freedom). Natürlich macht all dies die Software, dennoch ist es sehr hilfreich, die Freiheitsgrade auch selbst berechnen zu können. Die Freiheitsgrade ergeben sich durch die Subtraktion der *geschätzten Modellparameter* von der *Anzahl der Elemente in S.* In unserem Fall enthält S vier Varianzen (für X1, X2, M und Y) und sechs Kovarianzen. Diese mag man sich in unserem Fall noch leicht vorstellen können – bei größeren Matrizen ist das nur noch schwer möglich. Die Anzahl lässt sich aber leicht durch die Formel $q \times (q + 1)/ 2$ berechnen, wobei q die Anzahl der Variablen ist – also $[4 \times (4+1)] / 2 = 10$.

Die Parameter des Modells muss man allerdings zählen. In unserem Beispiel ist das die Kovarianz zwischen X1 und X2, die beiden Effekte auf M, der Effekt von M auf Y, die beiden Varianzen von X1 und X2 sowie die beiden Fehlerterm-Varianzen von M und Y – insgesamt 8 zu schätzende Parameter. Somit ergeben sich für dieses Modell 2 df. Die anfangs gestellte Frage, mit welcher Wahrscheinlichkeit die Daten (d.h. S) aus einer Population stammen, in der das in Abb. 11 rechts abgebildete Modell gilt, entspricht der Frage nach der Auftretenswahrscheinlichkeit eines Chi-Quadrat-Wertes größer/gleich 91.4 in einer Chi-Quadrat-Verteilung mit 2 df. Das Ergebnis ist signifikant mit einem p-Wert kleiner .001 – die Nullhypothese, dass S aus einer Population mit dem vollständigen Mediatormodell kommt, muss verworfen werden. Dies ist eine andere Formulierung der Nullhypothese, dass die Populationskovarianzmatrix (Σ), die durch S geschätzt wird, identisch mit $\Sigma(\theta)$ – der modellimpliziten Kovarianzmatrix ist.

Wie bereits erwähnt, schätzt das Programm die Parameter so, dass sich $\Sigma(\theta)$ sukzessive S annähert. Begonnen wird mit Startwerten, die dann in sog. iterativen („sich annähernden“) Schritten soweit verändert werden, bis sie ein Optimum erreicht haben, das dem Minimum der Fit-Funktion entspricht. Ist das Modell korrekt, wird dieses Optimum implizieren, dass sich S und $\Sigma(\theta)$ nur in einem Ausmaß unterscheiden, wie dies der Stichprobenfehler nahelegt.

In unserem Beispiel hätte dies folgende Konsequenzen: Zunächst ist aufgrund der Pfadregeln (und dem d-separation-Kriterium) ersichtlich, dass die Kovarianz zwischen X1 und Y durch die Auslassung des direkten Effekts unterschätzt wird. Dies wird deutlich in der Diskrepanz zwischen S und $\Sigma(\theta)$ (1.405 vs. .940). Die Software versucht nun aber, dies zu kompensieren, indem sie die übrigen Effekte, die bei der Generierung dieser Kovarianz beteiligt sind, *über*schätzt. Und hier können die Pfadregeln helfen, zu verstehen, warum: Nach den Pfadregeln ergibt sich Kor(X1,Y) aus dem indirekten und direkten Effekt von X1 plus dem Produkt aus Korrelation mit X2 und dessen indirektem Effekt. Fehlt nun der direkte Effekt, versucht das Programm, die Kovarianz zu reproduzieren, indem es einige dieser anderen Parameter nach oben verzerrt. Zur Verfügung stehen a) der Effekt von M auf Y, b) der Effekt von X1 auf M und c) der Effekt von X2 auf M. Letztere wirken sich allerdings direkt auf die Kovarianzen von X1 bzw. X2 und M aus – daher wird nur der Effekt von M auf Y nach oben verzerrt. Je nachdem, wie stark der fehlende direkte Effekt von X1 auf Y ist, wird diese Verzerrung mehr oder weniger stark sein. Dies zeigt, dass wir bei Vorliegen von nicht-fittenden Modellen nie wissen, ob, wo und wie stark es zu Verzerrungen von Modellparametern kommt. Somit kommt dem Test des Modells und einer möglichen Respezifikation eine wichtige Rolle zu.

2.4.2 Akzeptanz des Chi-Quadrat-Tests und Fit-Indizes

In den letzten Jahrzehnten hat es sich als Praxis eingebürgert, das Ergebnis des Chi-Quadrat-Tests in wissenschaftlichen Veröffentlichungen zwar zu berichten, ihn aber bei der Evaluation des Modells eher zu vernachlässigen. Stattdessen ist es zur Praxis geworden, sogenannte **Fit-Indizes** zur Evaluation heranzuziehen (Schermelleh-Engel, Moosbrugger, & Müller, 2003). Gründe für diese Ablehnung des Tests sind zum einen, dass sich bestimmte Auffassungen über den Chi-Quadrat-Test verbreitet haben, die zwar einen wahren Kern haben, aber in ihrer Generalisierung und konkreten Ausformung schlicht falsch sind. Der zweite Grund ist eher psychologisch: SEM-Anwender haben oft – wie alle Menschen – zum einen Schwierigkeiten damit, Falsifikationen ihrer Vorstellungen zu akzeptieren und zum anderen oft nach einem fehlgeschlagenen Modell zu wenig Strategien, um eine Alternative zu entwickeln. Im Folgenden wird zunächst auf diese Gründe der mangelnden Akzeptanz des Chi-Quadrat-Tests eingegangen, im nachfolgenden Abschnitt auf den produktiven Umgang mit einem nicht-fittenden Modell und den Implikationen einer Respezifikation.

Kommen wir zunächst zu den üblichen Auffassungen über den Test. Ein praktisches Problem mit dem Chi-Quadrat-Test ist tatsächlich, dass er zwar darauf testet, dass das Modell in *irgendeiner Weise* fehlspezifiziert ist, man aber nicht weiß, wie groß das Problem ist und worin es genau besteht. D.h. es kann eine Form der Fehlspezifikation sein, die entweder völlig *trivial* oder aber *fundamental* ist. Insbesondere bei großen Stichproben wächst die **statistische Power** des Tests, auch bei kleinen und trivialen Fehlspezifikationen signifikant zu werden. Dies hat zum Mantra beigetragen, dies sei quasi ein Problem des Tests und er sei „übermäßig sensibel bzgl. der Stichprobengröße". Als Extremform herrscht sogar die Auffassung vor, die Stichprobengröße verursache den misfit (d.h. es wäre gar nicht möglich, einen akzeptablen Fit zu erhalten). Diese Auffassungen haben einen wahren Kern, ihre Ausformungen sind allerdings falsch.

Es ist tatsächlich ein Problem, dass der Test bei größeren Stichproben auch auf triviale Fehlspezifikationen reagiert, die das eigentliche theoretische Modell unberührt lassen, aber dies ist kein „Problem“ des Tests. Der Chi-Quadrat-Test ist ein Test wie jeder andere auch, daher steigt mit der Stichprobe die statistische Power. Der Hintergrund dieses Verhaltens ist, dass mit steigender Stichprobengröße der Stichprobenfehler sinkt und die Kovarianzmatrix in immer stärkerem Ausmaß als präzise Repräsentantin der Populationsmatrix behandelt werden kann. Ergo können Probleme, die Kovarianzmatrix durch das Modell zu reproduzieren, immer weniger durch Stichprobenfehler erklärt werden sondern müssen als konsistente und systematische Probleme des Modells angesehen werden. Genau dies – und nur dies – signalisiert der Chi-Quadrat-Test. Ob dies triviale oder fundamentale Probleme sind, muss eruiert werden. Es ist aber nicht die Stichprobengröße, die den misfit verursacht, sondern sie reduziert quasi das „Rauschen“, womit ein klarerer Blick auf das Modell möglich wird. Die Rolle der Stichprobengröße für den Test wird durch die oben besprochene Formel der Chi-Quadrat-Statistik ($\chi^2 = F_{ML} \times [N - 1]$) deutlich: Sie zeigt, dass die Stichprobengröße lediglich eine Gewichtung des Maximum likelihood Funktionswertes verursacht. Da korrekte Modelle einen Funktionswert nahe 0 haben, wird somit auch ein großes N zu keinem hohen Chi-Quadrat-Wert führen.

Der zweite Fehler besteht darin, die Tatsache, dass bei großen Stichproben auch triviale Fehlspezifikationen zu einem signifikanten Chi-Quadrat-Test führen, umzudrehen und daraus die Überzeugung abzuleiten, dass ein signifikanter Chi-Quadrat-Test in einer großen Stichprobe durch triviale Fehlspezifikationen verursacht wurde ("converse error", McIntosh, 2012). Stattdessen kann aus einem signifikanten Chi-Quadrat, egal bei welcher Stichprobengröße, nicht abgeleitet werden, dass die Fehlspezifikation irrelevant ist.

Weitere Probleme des Chi-Quadrat-Tests sind, dass Daten selten die **Annahmen** erfüllen, die er voraussetzt, z.B. multinormal-verteilte, kontinuierliche Variablen und eine große Stichprobe (Schermelleh-Engel et al., 2003). Diese Probleme können allerdings mittlerweile **korrigiert** werden. Möglichkeiten speziell der Korrektur des Tests bei Verletzungen der Verteilungs- und Skalierungsvoraussetzungen werden in diesem Buch behandelt. Zudem bieten Herzog und Boomsma (2009) eine Korrekturmöglichkeit des Tests bei kleinen Stichproben, bei denen die Wahrscheinlichkeit, ein korrektes Modell zurückzuweisen, erhöht ist. Die Autoren offerieren auch einen R-code, mit der diese sogenannte **swain-Korrektur** durchgeführt werden kann (siehe im Detail Abschnitt 8.5).

Aufgrund der genannten Auffassungen (und schlicht, weil Modelle oft nicht fitten), wurden in den letzten Jahrzehnten sogenannte **Fit-Indizes** entwickelt (Bentler, 1990; Bentler & Bonet, 1980; Browne & Cudeck, 1993). Während der

Chi-Quadrat-Test die Frage beantwortet, *ob* sich die empirische und modellimplizite Matrix systematisch, d.h. jenseits des Zufalls unterscheiden, versuchen Fit-Indizes, das *Ausmaß der Abweichung* zu quantifizieren. Der Hintergrund ist die Annahme, dass das Ausmaß der Abweichung gleichzeitig das *Ausmaß der Fehlspezifikation* kennzeichnet (im Sinne des „je besser die Fit-Indizes, um so korrekter das Modell"). Etabliert haben sich dabei der **Root mean square error of approximation** (RMSEA; Steiger, 1990), **Comparative fit index** (Bentler, 1990) und **Standardized root mean square residual** (SRMR; Jöreskog & Sörbom, 1981). Hu und Bentler (1999) empfehlen basierend auf einer berühmten Simulationsstudie, Modelle mit einem RMSEA < .06, CFI nahe oder über .98 und einem SRMR < .08 als „akzeptabel" zu erachten.

Das Problem dieser Empfehlungen ist, dass diese Werte **Daumenregeln** sind, die in der speziellen Simulation der Autoren als ein akzeptables Ausmaß zurückgewiesener Modelle sprachen. Daumenregeln suggerieren aber feste Grenzen, die es faktisch nicht gibt. So argumentieren aufgrund eigener Simulationsstudien Marsh, Hau und Wen (2004) gegen eine Generalisierung dieser „golden rules".

Das fundamentalere Problem von Fit-Indizes liegt aber in ihrer zugrundeliegenden Annahme, dass eine Quantifizierung des Abstandes der Kovarianzmatrizen (S und $\Sigma(\theta)$) als Ausmaß der Fehlspezifikation interpretiert werden kann. Genau dies aber ist eine falsche Annahme, da es fundamental falsche Modelle geben kann, die insbesondere durch die Schätzung unsinniger Parameter dennoch zu guten Fit-Indizes führen können (Hayduk, Cummings, Boadu, Pazderka-Robinson, & Boulianne, 2007; McIntosh, 2007). Sicherlich gibt es ein „Ausmaß von Fehlspezifikation" (d.h. Modelle können mehr oder weniger Fehler haben), aber jenseits eines Alarmsignals, *dass* es Fehler gibt, haben wir nichts zur Verfügung, um das *Ausmaß* dieses Fehlers zu evaluieren. Daher ist es nicht verwunderlich, dass eine immer größer werdende Anzahl von Autoren für eine stärkere Beachtung des Chi-Quadrat-Wertes plädieren (Barrett, 2007; Hayduk, Cummings, et al., 2007; Kline, 2011; McIntosh, 2007; Mulaik, 2009; Shipley, 2000).

Abgesehen von den besprochenen Problemen der Fit-Indizes berichtet McIntosh (2012) eine interessante Anekdote über die eigentlichen ursprünglichen Hintergründe für die Entwicklung des ersten Fit-Indizes: *"A telling anecdote in this regard comes from Dag Sorböm, a long-time collaborator of Karl Joreskög, one of the key pioneers of SEM and creator of the LISREL software package. In recounting a LISREL workshop that he jointly gave with Joreskög in 1985, Sorböm notes that: ''In his lecture Karl would say that the Chi-square is all you really need. One participant then asked 'Why have you then added GFI [goodness-of-fit index, i.e., an approximate fit index]?' Whereupon Karl answered 'Well, users threaten us saying they would stop using LISREL if it always produces such large Chi-squares. So we had to invent something to make people happy. GFI serves that purpose' (p. 10)''.*

Schließlich soll noch erwähnt werden, dass gegen den Chi-Quadrat-Test häufig ins Feld geführt wird, dass fehlspezifizierte Modelle oft einen nichtsignifikanten Chi-Quadrat-Test haben oder in anderen Worten: Selbst wenn man ein Modell fittet, weiß man nicht, ob es korrekt ist. Auch wenn dies korrekt ist, ist dies ein Dilemma eines jeden Hypothesentestenden wissenschaftlichen Ansatzes: nämlich, dass – wenn die Evidenz für eine Hypothese spricht – man dennoch niemals weiß, ob die Hypothese wirklich korrekt ist, weil andere Hypothesen ebenfalls mit den Daten übereinstimmen könnten (**fallacy of affirming the consequent**). Daraus folgt, dass wir lediglich testen können, ob es Kontra-Evidenz gegen das Modell gibt. Stellen wir keine Evidenz gegen das Modell fest, wird dieses dadurch niemals abschließend verifiziert, sondern bleibt einfach weiter „im Rennen". Stellen wir Evidenz gegen das Modell fest, ist dies kein Fehlschlag, sondern sollte als Grundlage für die Chance gesehen werden, das Modell zu verbessern und so einen wissenschaftlichen Lernprozess zu stimulieren.

2.4.3 Was tun wenn das Modell nicht fittet?

Wie weiter oben angemerkt, resultiert eine mangelnde Berücksichtigung des Chi-Quadrat-Tests häufig auch auf einem verständlichen Unbehagen gegenüber der Tatsache, dass ein plausibles Modell nicht so zu funktionieren scheint wie angenommen. Die Tatsache, dass der Veröffentlichungsdruck in den letzten Jahren zugenommen hat und Verlage von Fachzeitschriften vornehmlich Erfolgsgeschichten veröffentlichen (und ein nicht-fittendes Modell als Fehlschlag und als nicht veröffentlichungswürdig angesehen wird), hat einen verstärkenden Einfluss. Ein weiterer, vielleicht ausschlaggebender Grund liegt allerdings darin, dass Forscher/innen oft schlicht ratlos sind und sich hilflos fühlen, wenn ihr Modell nicht fittet. Folglich empfinden sie den signifikanten Chi-Quadrat-Test als „Todesurteil" für ihr Modell. Und wenn keine Strategien vorhanden sind, um mit dem misfit umzugehen, dann ist es menschlich verständlich, dass man eine Tendenz verspürt, den misfit zu ignorieren.

Die wissenschaftlich sinnvollere Alternative ist, zu explorieren, wo der Fehler liegen könnte. Denn anstatt den misfit als Zeichen eines Fehlschlags zu sehen, sollte er als Anlass und Möglichkeit zur Korrektur der eigenen Theorien interpretiert werden. Dies wird häufig nicht erfolgreich sein und der Chi-Quadrat-Test bleibt und bleibt signifikant. In diesem Fall wird man ein Modell präsentieren, das das Beste darstellt, was man erreichen konnte, aber welches eben mit Zweifeln behaftet bleibt. In vielen Fällen jedoch wird man Erfolg haben und eine Respezifikation wird u.U. eine radikale Veränderung der eigenen Sichtweise bewirken.

Dies wird oft als „data crunching" oder „Schrauben am Modell" kritisiert. Allerdings ist hier nicht die Rede von einem blindem Freisetzen von Parametern, bis ein fit erreicht ist, sondern von einer *theoriebasierten Respezifikation des Modells*. Dies entspricht dem wissenschaftstheoretischen Prinzip der **Abduktion**, was bedeutet, dass plausible theoretische Erklärungen gesucht werden, die ein empirisches Phänomen erklären können (Haig, 2005). Angewandt auf SEM bedeutet dies, dass eine Alternativerklärung gesucht wird, die die Kovarianzmatrix und damit zusammenhängend den misfit (Diskrepanz zwischen der vormals spezifizierten Struktur und der Kovarianzmatrix) erklären kann. Dies passiert durch Heranziehen von modellimpliziten, **diagnostischen Informationen** (z.B. standardisierte Residuen), Pfadregeln und deskriptiven Statistiken (siehe Abschnitte 4.6 und 5.2). Bei der Modifikation von Faktormodellen kommt die wichtige Inspektion der Item-Formulierungen hinzu. Wir werden später ein beispielhaftes Modell kennenlernen, das einen signifikanten Chi-Quadrat-Test, aber akzeptable Fit-Indizes hatte und wo die Inspektion der Korrelationsmatrix und der standardisierten Residuen zu einer alternativen kausalen Struktur führte.

Das Ergebnis einer **Respezifikation** ist eine neue Hypothese. Da nämlich die Respezifikation teilweise durch die Daten gespeist wird, besteht die Gefahr, ein artifizielles Modell zu entwickeln, das speziell für die vorliegende Stichprobe gilt, aber nicht darüber hinaus. Allerdings beweist auch eine erfolgreiche Replikation nicht, dass dies auch das kausal korrekt spezifizierte Modell ist. Entwickelt man nämlich ein fittendes aber falsches Modell, stellt dies eine systematische Fehlspezifikation da, die über weitere Stichproben stabil wäre (d.h. in neuen Stichproben wieder fitten würde). Die einzige Lösung dieses Problems stellt daher eine Erweiterung / Veränderung des Modells dar – insbesondere das Spezifizieren von Restriktionen, die gelten müssen, wenn das Modell korrekt ist. Instrumentalvariablen, wie sie in Kapitel 7 behandelt werden, sind dabei eine gute Möglichkeit.

Im Zusammenhang mit der Respezifikation ist auch zu beachten, dass der Chi-Quadrat-Test des veränderten Modells seine Funktion als statistischer Test verliert, wenn die Veränderungen aufgrund von Informationen der Daten oder der

Ergebnisse des ersten Tests vorgenommen wurden. Daher ist es eine gute Strategie, bereits vor dem ersten Test quasi worst-case-Szenarien zu entwickeln, was an dem favorisierten Modell falsch sein könnte. Dies betrifft z.B. auch das genaue Überprüfen der Indikatoren von latenten Variablen (vgl. Kap. 3). Kommen in den Indikatoren der verschiedenen latenten Variablen gemeinsame Begriffe vor? Dies kann bereits dazu führen, dass beide eine gemeinsame Kovarianz ausbilden. Betonen Indikatoren einen Aspekt, der eine vom Rest des Modells abweichende Kausalstruktur ermöglicht? Wenn ja, könnte das auf die Liste der Respezifikationen kommen. Ein Beispiel dazu: Ich hatte einmal ein Modell, in dem ein Effekt der Leistungsmotivation auf Job Involvement vorgesehen war. Ein Indikator der Leistungsmotivation lautete „ich arbeite sehr hart". Mir kam plötzlich der Gedanke, dass dies nicht einfach nur ein Indikator für Leistungsmotivation sein könnte, sondern ein Indikator einer spezifischeren latenten Variable „Fleiß auf der Arbeit" und diese eine *kausale Folge* von Job Involvement sein könnte. Allein wegen dieser Möglichkeit erschien mir dieser Indikator als ungeeignet für jene latente Variable (Leistungsmotivation als eine allgemeine Persönlichkeitseigenschaft), deren Wirkung ich im Modell testen wollte.

Eine weitere Form des worst-case Szenarios ist, unnötige und unplausible Restriktionen (z.B. auf null fixierte Pfade) im Modell zu spezifizieren, ohne sich dies explizit klar zu machen. So fixieren z.B. viele Anwender/innen Fehlerterm-Kovarianzen zwischen mehreren abhängigen Variablen automatisch auf null, ohne zu reflektieren, dass dies die Annahme ausdrückt, dass alle gemeinsamen Ursachen dieser Variablen im Modell sein müssen. Schauen Sie in ihrem Modell auch auf die Effekte oder Beziehungen, die sie *nicht* schätzen wollen, da diese letztendlich den misfit verursachen.

3

Modelle mit latenten Variablen

3.1 Latente Variablen

Das Ziel bei der Anwendung von SEM ist das Testen von theoretisch abgeleiteten Kausalhypothesen zwischen Variablen. Diese Variablen repräsentieren in der Regel **hypothetische Konstrukte** bzw. deren zugeordnete empirische Phänomene und werden wiederum repräsentiert durch Beobachtungen, Befragungen oder Experimente. Eine Variable kann dabei direkt repräsentiert werden durch eine Frage (sog. „Item“) z.B. in einem Fragebogen oder durch die Summe oder den Mittelwert mehrerer Items. Der Sinn einer solchen Summe beruht auf der Idee, dass der Summen- oder Mittelwert verschiedener Items, die dasselbe messen, einen geringeren Messfehler hat, als die einzelnen Items. Dies ist vorteilhaft, weil Messfehler (in unabhängigen Variablen) die Effektschätzungen der Variablen verzerren.

In den vergangenen beiden Kapiteln wurden Modellvariablen implizit mit ihren Messungen gleichgesetzt. Eine Stärke von SEM ist nun allerdings, dass die Beziehung zwischen der/den Messung/en und sogenannten **latenten Variablen** modelliert werden können. Diese repräsentieren das empirische Phänomen von Interesse, für das man kausale Effekte hypothetisiert. Allerdings hat man keinen direkten und absoluten Zugang zu diesem Phänomen, sondern kann seine Existenz und Effektivität nur über **beobachtbare Indikatoren** erfassen. Durch die explizite Spezifikation von latenter Variable und zugehörigen Indikatoren erhält man selbst wieder ein Kausalmodell – ein **Messmodell** – das den kausalen Einfluss der latenten Variablen auf die Indikatoren beschreibt. Dies hat zwei Vorteile: a) die Parameter, die die latenten Variablen betreffen (Varianzen, Kovarianzen und Struktureffekte) sind messfehlerbefreit und lösen das o.g. Verzerrtheitsproblem; b) mit dem Testen der Beziehungen zwischen latenten und gemessenen Variablen erhält man die wichtige Möglichkeiten, das Messmodell selbst und damit sogenannte **Brückenhypothesen** (auxilliary hypotheses) zu spezifizieren und zu testen. Brückenhypothesen sind Annahmen darüber, wie hypothetische Konstrukte, welche die latenten Variablen repräsentieren sollen, mit beobachtbaren Messungen zusammenhängen. Die Spezifizierung der Beziehungen zwischen beiden innerhalb eines Messmodells bietet daher einen wichtigen Ansatz zur Lösung des **Korrespondenzproblems,** welches die grundsätzliche Kluft zwischen Theorie und Empirie betrifft.

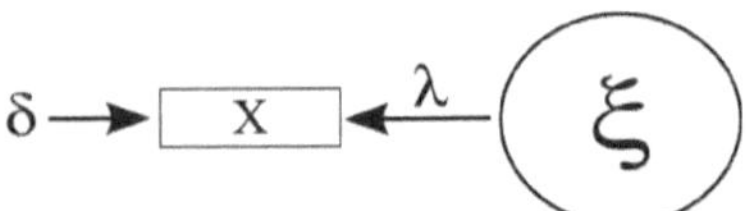

Abb. 12: Ein einfaches Messmodell

Abb. 12 zeigt ein einfaches Messmodell. Wie in der Beschreibung von Modellen üblich, werden griechische Symbole zur Beschreibung von Variablen und Parametern benutzt. ξ ("Ksi") ist die latente Variable; sie repräsentiert das empirische Phänomen von Interesse, für das das entsprechende Konstrukt den theoretischen Kontext, Namensgebung und Bedeutung liefert. Als Beispiel orientieren wir uns an Edwards und Bagozzi (2000): so könnte ξ „Arbeitszufriedenheit" sein. Das empirische Phänomen ist das Ausmaß, in dem eine Person positiv oder negativ über ihren Arbeitsplatz urteilt. Voraussetzung für die Sinnhaftigkeit ist hier, dass es plausibel ist, dass sich Personen in der Population entlang dieser empirischen wenn auch unbeobachtbaren Dimension anordnen lassen. Dennoch ist „Arbeitszufriedenheit" ein Konstrukt – d.h. ein fiktionaler Begriff, der von Menschen erfunden wurde, um einem Phänomen einen Namen und eine Bedeutung zu verleihen.

X wäre in dem Modell ein beobachtbarer und damit messbarer Indikator der Arbeitszufriedenheit und könnte „im allgemeinen bin ich zufrieden mit meiner Arbeit" lauten – mit den Antwortmöglichkeiten „trifft überhaupt nicht zu", „trifft eher nicht zu", „trifft teils teils zu", trifft eher zu" und „trifft völlig zu". Das Messmodell unterstellt, dass die latente Arbeitszufriedenheit kausal mit der Stärke λ auf den Indikator wirkt, d.h. die Antwort auf das Item „X" *reflektiert* das zugrundeliegende Ausmaß an Arbeitszufriedenheit. Daher nennt man X auch einen **reflektiven Indikator.**

Dennoch besteht zwischen ξ und X keine 1:1-Beziehung – d.h. X ist in der Regel ein messfehlerbehafteter Indikator. Das Symbol δ bzw. ein Pfeil ohne einen Ursprung ist ein Symbol dafür. Der Messfehler in SEM besteht dabei aus einer Zusammenfassung aller zufälliger als auch systematischer weiterer Ursachen für einen Wert in dem Indikator. Zufällige Einflüsse können z.B. mangelnde Motivation und wahlloses Ankreuzen, mangelnde Konzentriertheit durch Hitze, Müdigkeit etc. oder unklare Formulierungen des Indikators oder der Antwortkategorie sein; systematische Einflüsse dagegen soziale Erwünschtheit, Antworttendenzen (z.B. Aquieszenzbias) etc. Diese Mitberücksichtigung systematischer Faktoren unterscheidet den Messfehler in einem Messmodell vom Konzept des Zufallsmessfehlers in der klassischen Testtheorie, wie er in dessen zentralem Gütemaß **„Reliabilität“** zum Ausdruck kommt (Borsboom, 2005).

3.2 Konstrukte vs. latente Variablen

Eine häufige Fehlerursache in SEM rührt daher, dass latente Variablen mit Konstrukten gleichgesetzt werden. In der Folge werden aus der Literatur sog. „Skalen“ mit mehreren Items entnommen, die „das“ Konstrukt messen sollen, und als reflektive Indikatoren einer latenten Variable spezifiziert. Solche Modelle fitten meistens nicht, weil der Begriff „Konstrukt“ nicht gleichzusetzen ist mit dem Begriff der latenten Variablen und die Spezifikation eines Messmodells etwas zusätzliche Theoriearbeit voraussetzt. Dies soll nun etwas ausgeführt werden.

Wie oben beschrieben, repräsentieren Variablen in Modellen Konstrukte. Ein Problem dabei ist, dass man in der wissenschaftlichen Praxis so sehr mit dem Konstrukt (als der menschlichen Konstruktion) beschäftigt ist, dass man etwas aus den Augen verliert, was das empirische Phänomen dahinter ist. Dazu kommt, das in vielen Fällen nicht klar genug spezifiziert ist, ob sich das Konstrukt nur auf ein Phänomen bezieht, oder ein Sammelbegriff für eine Reihe ähnlicher Phänomene ist. So wurde das Konstrukt „Arbeitszufriedenheit“ am Anfang des Kapitels als eine empirische Dimension verstanden, auf der sich Personen entlang ihrer allgemeinen Bewertung ihrer Arbeit unterscheiden. In diesem Fall würde der theoretische Begriff durch eine einzige latente Variable **konzeptualisiert**. Ein/e ander/e Forscher/in könnte dagegen Arbeitszufriedenheit als Sammelbegriff für die Summe aller spezifischen Zufriedenheiten verstehen (d.h. Zufriedenheit mit der Aufgabe, dem/der Vorgesetzten, den Weiterbildungsmöglichkeiten etc.). Beide Konstrukte „Arbeitszufriedenheit“ unterscheiden sich fundamental bzgl. der Phänomene, die sie beschreiben und implizieren völlig andere Vorgehensweisen, wenn man sie messen und in ein SEM einfügen will. Der Grund ist, dass – will man in einem SEM wirklich einen Ausschnitt aus der Realität modellieren – man sich auch bzgl. der Variablen in dem SEM an in der Realität natürlich vorkommenden empirischen Phänomenen orientieren sollte. Dies bedeutet, dass das Konstrukt bzgl. seiner „Anteile" bzw. Sub-Dimensionen präzisiert werden sollte, welche die einzelnen empirischen Phänomene repräsentieren. Diese Dimensionen sollten dann als latente Variablen in dem SEM modelliert werden.

Oft ist es aber so, dass das Konstrukt auf der theoretischen Ebene eher unscharf gehalten wird, dann aber über die Einbeziehung einer Vielzahl heterogener Messungen die Komplexität des Konstrukts wieder zu adressieren versucht wird. Zum Beispiel könnte der/die Forschende „Arbeitszufriedenheit" hinsichtlich der zu repräsentierenden Phänomene unzureichend konzeptionalisiert haben und intuitiv Standardskalen mit Items benutzen, die sich auf verschiedene Aspekte der Arbeit beziehen („ich bin zufrieden mit … meinen Kollegen/innen, Vorgesetzten, Weiterbildungsmöglichkeiten" etc.). Wird nun ein Messmodell spezifiziert, das *eine* latente Variable vorsieht, auf dem diese Items laden, stellt dies eine Fehlspezifikation dar (MacKenzie, Podsakoff, & Jarvis, 2005), denn Arbeitszufriedenheit ist hier keine latente zugrundeliegende Ursache der einzelnen Facetten, sondern ein Sammelbegriff.

Hayduk und Kolleginnen (Hayduk & Beckie, 1997; Hayduk & Pazderka-Robinson, 2007) weisen in diesem Kontext darauf hin, dass oft Facetten oder Dimensionen eines Konstrukts eher Prädiktoren oder Konsequenzen als Bestandteile sind. So macht es z.B. einen bedeutsamen Unterschied ob „Arbeitszufriedenheit" konzeptionalisiert wird als Summe spezifischer Zufriedenheiten oder ob diese Determinanten bzw. Einflussfaktoren einer latenten Variable „Arbeitszufriedenheit" sind. Ersteres impliziert eine konstruktivistische Perspektive der Arbeitszufriedenheit (d.h. das Konstrukt als ein Sammelbegriff); innerhalb dieser Perspektive gibt es Arbeitszufriedenheit nicht unabhängig von seiner Definition als Gesamtheit der Einzelzufriedenheiten. Die zweite Perspektive erlaubt dagegen eine realistische Perspektive, innerhalb derer Arbeitszufriedenheit als eine einzelne latente Variable konzeptualisiert wird, die kausal von den Einzelzufriedenheiten beeinflusst wird. Es ist dabei wichtig, dass diese latente Variablen genauso eindimensional und spezifisch ist wie jede andere (latente) Variable und durch reflektive Indikatoren gemessen werden kann.

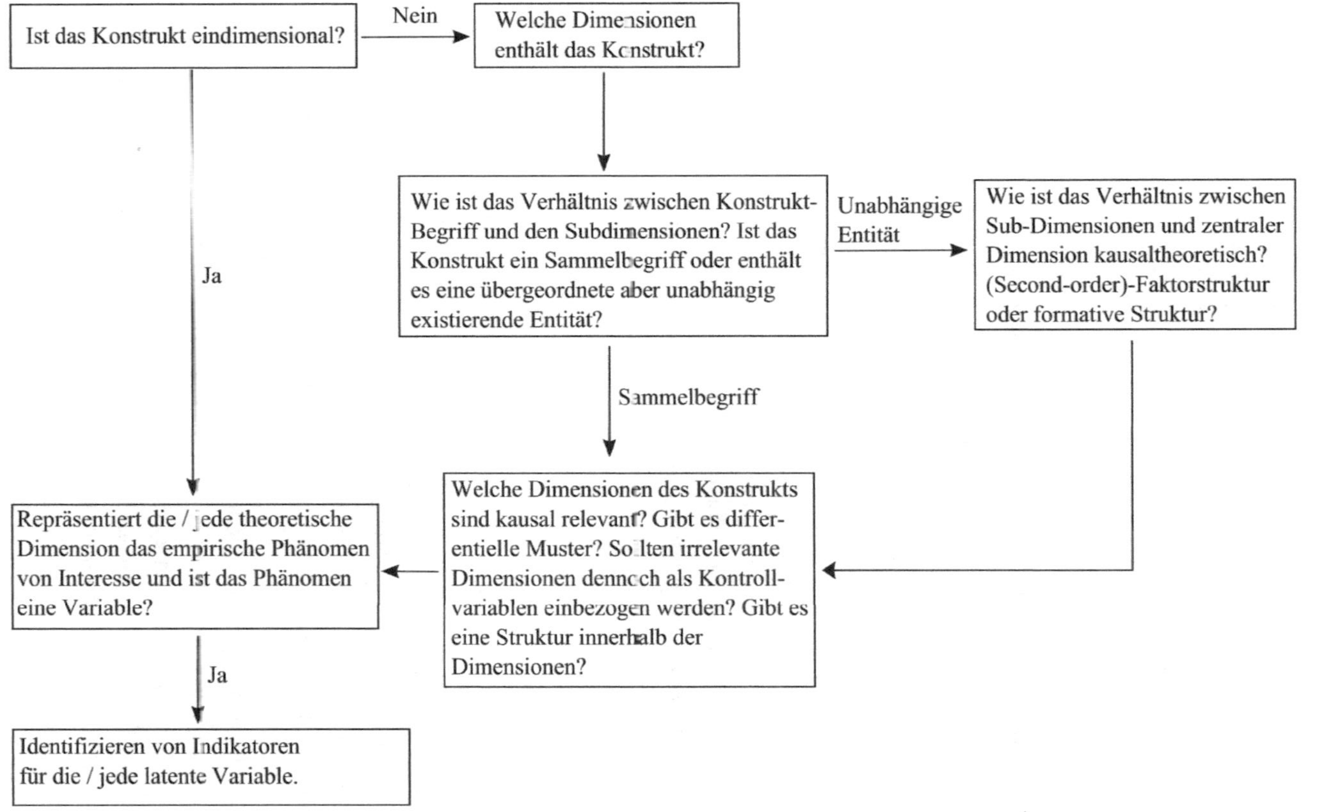

Abb. 13: Zu klärende Fragen bei der Übersetzung von theoretischen Konstrukten in Modelle mit latenten Variablen.

3.3 Schritte bei der Übersetzung eines Konstrukts in ein Modell

Abb. 13 enthält die Abfolge der Fragen, die SEM-Anwender/innen auf dem Weg der Übersetzung eines Konstruktes in ein SEM beantworten müssen. Hierbei sei vorangestellt, dass die Beantwortung dieser Fragen v.a. das theoretische Verständnis des Konstruktes und seine empirische Repräsentanz betrifft, welches der/die Anwender/in hat. Im Folgenden werden die zentralen Fragen diskutiert und erklärt.

3.3.1 Klärung der Dimensionalität des Konstrukts

Die primäre Frage ist zunächst, ob das Konstrukt ein- oder mehrdimensional ist. Diese Frage lässt sich meist durch folgendes Gedankenexperiment klären: Angenommen, Sie hätten nur ein einziges Item, mit dem sie das Konstrukt messen könnten und dieses Item hätte keinen Messfehler. Welches würden Sie wählen? Sind Sie nun der Meinung, dass dieses eine Item zu wenig ist, weil bestimmte Facetten, Aspekte oder Bereiche fehlen, ist dies ein Hinweis auf Multidimensionalität, denn jede dieser Facetten etc. ist eine Dimension. Daher sollte nun die Aufgabe darin bestehen, auf der theoretischen Ebene (und nicht durch die Entwicklung eines breiten Item-pools) zu re-konzeptualisieren, was diese Facetten sind.

3.3.2 Klärung des Verhältnisses zwischen den Subdimensionen und dem Konstrukt

Hat man geklärt, welche Subdimensionen bzw. Facetten das Konstrukt umfasst, sollte man klären, was genau das Konstrukt nun ausmacht. So kann das Konstrukt ein simpler Sammelbegriff der Dimensionen sein (siehe Edwards, 2001, für eine Vertiefung). Beispielsweise könnte die Definition des Konstrukts „Intelligenz“ lauten, dass sie ein Sammelbegriff für die Summe von spezifischeren Intelligenzen, z.B. verbale Intelligenz, ist (vgl. Abb. 14, links). In diesem Fall existiert Intelligenz nicht als eigenständige Entität, da sie an die der Definition von Intelligenz zugrundeliegenden Einschlusskriterien der Subdimensionen gebunden ist. Ändert man diese Kriterien, ändert sich folglich das Konstrukt.

In anderen Fällen kann es durchaus sein, dass das Konstrukt zwar ein Sammelbegriff für die Dimensionen ist, es aber dennoch eine zentrale Dimension gibt (quasi der Kern des Komplexes).

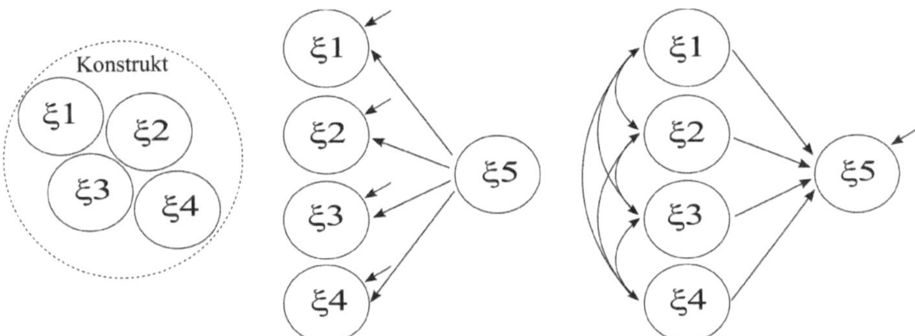

Abb. 14: Strukturen multidimensionaler Konstrukte: a) als Sammelbegriff für mehrere Dimensionen, b) als Second-Order-Faktor, c) als formative Struktur

Hiervon gibt es zwei Versionen: Das Second-Order-Faktor-Modell und das formative Modell. Das Second-Order-Faktor-Modell postuliert, dass den verschiedenen Subdimensionen eine zentrale latente Variable zugrunde liegt und somit die Erklärung dafür ist, dass die Subdimensionen miteinander korrelieren. In Abb. 14 beschreibt das mittlere Pfaddiagramm diese Struktur. Eine Implikation dieses Modells ist, dass die Korrelationen zwischen den Subdimensionen verschwinden, wenn der Second-Order-Faktor statistisch konstant gehalten wird (vgl. Abschnitt 2.3). So wurde z.B. die Inter-Korrelationen der verschiedenen spezifischen Intelligenzen (z.B. verbale und mathematische Intelligenz, räumliches Denken, logisches Schlussfolgern) als Hinweis dafür genommen, dass es eine zentrale Intelligenz gibt (den „G-Faktor"), die für diese Korrelationen verantwortlich ist (Neisser et al., 1996). Lee und Cadogan (2013) argumentieren allerdings, dass Second-Order-Faktorstrukturen nur dann Sinn machen, wenn die Primärfaktoren konzeptionell sehr ähnlich sind (siehe auch Abschnitt 8.9 über die Gefahr des Reifikationsfehlers bei Second-Order-Faktoren). Je konzeptionell unähnlicher sich die Primärfaktoren sind, um so unplausibler ist es, dass sie alle eine gemeinsame Ursache haben und umso unklarer die Bedeutung eines Faktors, der diese Ursache repräsentieren soll.

Das formative Modell postuliert, dass die zentrale latente Variable nicht Ursache, sondern Folge des kausalen Einflusses der Subdimensionen ist. Das am meisten zitierte Beispiel für solche eine Struktur ist der soziale Status, der sich z.B. aus dem Einkommen der Person, ihrer Bildung und dem beruflichen Status ergibt (Bollen, Glanville, & Stecklov, 2001). Es sei darauf hingewiesen, dass in einem Großteil der Literatur über formative Konstrukte nicht die Subdimensionen selbst als Ursache der zentralen latenten Variable konzeptualisiert werden, sondern gemessene Indikatoren. Zum Beispiel würde der Effekt von „Bildung“ nicht von der latenten/tatsächlichen Bildung ausgehen, sondern vom Fragebogen-Item „welchen Schulabschluss haben Sie“. In diesem Zusammenhang spricht man oft von *kausalen Indikatoren* und grenzt diese von reflektiven Indikatoren ab, wie sie weiter oben behandelt wurden (vgl. Abb. 12). Obwohl es allerdings sicher nicht auszuschließen ist, dass Indikatoren einen kausalen Einfluss haben, ist es in den meisten Fällen aber plausibler, dass die latenten Variablen, die hinter diesen Indikatoren sind, formative Determinanten des Konstrukts sind. Die Gemeinsamkeit von Second-Order-Faktor und formativem Modell ist, dass das Konstrukt durchaus ein Sammelbegriff sein kann – die zentrale Variable aber selbst als eine eigenständige, latente Dimension anzusehen ist (siehe dazu auch Abschnitt 8.8 über die vermeintliche Breite von latenten Variablen).

3.3.3 Klärung der kausalen Rolle der Dimensionen

Die bisherigen Ausführungen haben gezeigt, dass hinter dem „Konstrukt“ eine Menge unterschiedlich spezifischer vs. multidimensionaler Strukturen stecken können. Es muss also eine Menge Konzeptualisierungsarbeit geleistet werden, um das eigene Verständnis des Konstrukts zu klären. Es sei darauf hingewiesen, dass dies keine Frage von richtig oder falsch ist – sondern eine derjenigen plausiblen theoretischen Konzeptualisierung, die vor dem Hintergrund der Forschungsfrage Sinn macht.

Hat man die Struktur konzeptualisiert, sollte man klären, wie das Konstrukt in das Gesamtnetz von kausalen Beziehungen eingeordnet werden kann, die im letztendlich interessierenden SEM modelliert werden sollen. Dies ist v.a. bei multidimensionalen Konstrukten der Fall. Hier muss geklärt werden, welche latenten Variablen überhaupt einen kausalen Effekt haben sollen. Fasst man das Konstrukt beispielsweise als Sammelbegriff mehrerer latenter Variablen auf, sollte man sich fragen, ob denn alle Subdimensionen im SEM einen Effekt haben oder von anderen Variablen beeinflusst sein sollten. Wenn das Konstrukt Zusammenhänge mit mehr als nur einer Variablen haben soll, könnte die Überlegung entstehen, dass es ein differenziertes Muster von Effekten und Zusammenhängen gibt. Unter Umständen kommt man zu dem Schluss, dass nur eine der Subdimensionen überhaupt relevant ist und damit gemessen werden muss. So könnte der/die Forschende in einer Untersuchung zur Wirkung des organisationalen Commitments lediglich Interesse an „affektivem commitment“ – d.h. der emotionalen Bindung der Person an die Firma haben – aber nicht an den anderen beiden Dimensionen

„normatives commitment“ (fühlt sie sich der Firma verpflichtet) und „kalkulatorisches commitment“ (hat sie so viel Investitionen in die Firma getätigt, dass eine Kündigung eine negative Nutzenbilanz ergäbe).

Im Falle eines **Second-Order-Faktor-Modells** oder des **formativen Modells** (vgl. Abb. 14) stellt sich ebenfalls die Frage, welche der implizierten latenten Variablen kausal wirksam ist. Beim Second-Order-Faktor-Modell besteht die Auffassung, dass die Primärfaktoren durch den Second-Order-Faktor beeinflusst werden. Daher sollte in vielen Fällen auch dieser Faktor als zentrale kausale Größe für die Effekte auf andere Variablen im Modell angenommen werden (andernfalls macht seine Modellierung wenig Sinn).

Formative Modelle werden üblicherweise als unabhängige Variable modelliert. Hierbei ist die wesentliche Annahme häufig, dass der Einfluss der Primär-Faktoren auf abhängige Variablen über die formative Variable mediiert wird. Aus Gründen, die hier nicht vertieft werden können, empfehlen Howell, Breivik, und Wilcox (2007), die formative Variable direkt über reflektive Indikatoren zu messen, da dies die statistische Identifikation und Beurteilung der Bedeutung der formativen Variable sicherstellt (s.u.). Als Beispiel nennen sie den sozioökonomischen Status einer Person, der meist als eine Konsequenz von Ursachen wie Bildung, Einkommen und beruflichem Status konzeptualisiert wird. Wenn dieses Modell aus Sicht einer kausaltheoretischen als auch wissenschaftstheoreti schen realistischen Perspektive korrekt ist (d.h. der Effekt dieser Variablen wird über eine latente Variable „sozioökonomischen Status“ auf abhängige Variablen mediiert und diese latente Variable hat einen eigenen von den Einflüssen unabhängigen Existenzstatus), dann sollte diese latente Variable auch durch reflektive Indikatoren messbar sein. Howell et al. schlagen hier z.B. solche Items wie „wie hoch in der sozialen Leiter befinden Sie sich?“ vor, die eine (messfehlerbehaftete) Messung des sozioökonomischen Status darstellen dürfte.

Schließlich kann das Ergebnis dieses Schrittes sein, dass die Subdimensionen selbst miteinander kausal verbunden sind. Dies mag vor dem Hintergrund der eigentlichen Forschungsfrage zwar u.U. irrelevant sein, führt aber ebenfalls zur Frage, wie dann dieses kausale Subsystem in das Gesamtmodell integriert werden soll. Als Beispiel lässt sich wieder das organisationale commitment heranziehen. Nach einer populäre Definition des affektiven commitments (Mowday, Porter, & Steers, 1982) besteht affektives commitment aus a) einem starken Glauben und der Akzeptanz der Werte der Organisation (kurz: „Akzeptanz“), b) der Bereitschaft, sich für die Belange der Organisation anzustrengen („Anstrengung“) und c) einem starken Wunsch, in der Organisation zu bleiben („Bindung“). Bei einem solchen Konstrukt zumindest die Multidimensionalität zu erkennen, ist bereits ein Fortschritt (gegenüber undifferenzierten eindimensionalen Konzeptualisierungen). Allerdings ist es hier wenig plausibel, commitment als Second-Order-Faktor zu konzeptualisieren; eher als Sammelbegriff. Was die Beziehungen dieser drei Facetten anbelangt, so ist die Struktur „Akzeptanz“ → „Anstrengung“ → „Bindung“ sinnvoll. Sollte nun im originären Modell beispielsweise der Effekt von Arbeitszufriedenheit und Leistungsmotivation auf „commitment“ untersucht werden, könnte eine sinnvolle Respezifikation darin bestehen, dass Arbeitszufriedenheit auf „Akzeptanz“ wirkt, während Leistungsmotivation sowohl auf „Akzeptanz“ als auch direkt auf „Anstrengung“ wirkt. Auch wenn andere Strukturen denkbar sind, soll dieses Beispiel verdeutlichen, wie problematisch eine undifferenzierte eindimensionale Konzeptualisierung eines Commitment-Faktors ist.

3.4 Das Common-Factor-Modell

Gleichgültig, wie das Ergebnis der Konzeptualisierungsarbeit ausfällt, die im vorherigen Abschnitt beschrieben wurde: sie endet mit einer Reihe von spezifischen latenten Variablen. Der nächste und letzte Schritt (vgl. Abb. 13) ist daher die Identifikation oder Auswahl von Indikatoren (meist Fragebogen-Items), die die entsprechende latente Variable reflektieren und damit messen (vgl. das grundlegende reflektive Messmodell in Abb. 12). In der Regel hat es Vorteile, mehrere Indikatoren zu wählen. Dies erlaubt die Schätzung und damit Modellierung des Messfehlers, mit dem jedes Item behaftet ist. Das daraus folgende Messmodell ist das sogenannte Common-Factor-Modell, welches das am häufigsten angewendete Messmodell zumindest in der Psychologie ist. Konfirmatorische Faktorenanalysen (confirmatory factor analyses, CFA) dienen dazu, die Kausalstruktur, die für eine bestimmte Menge von Indikatoren postuliert wird, zu testen.

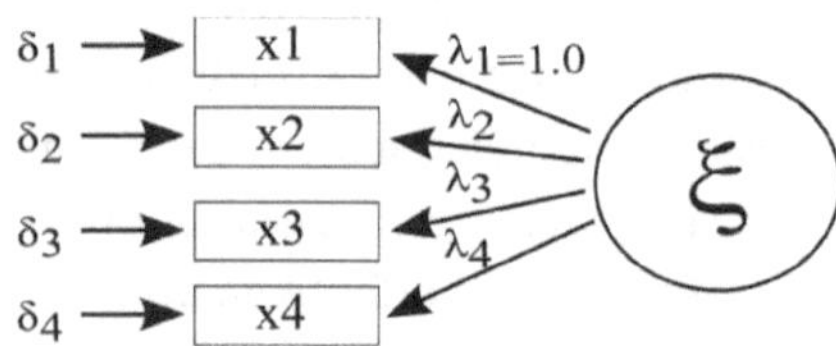

Abb. 15: Das Common-Factor-Modell

Die zentrale Idee dieses Modell ist es, dass die Indikatoren alle eine gemeinsame Ursache haben – die latente Variable (ξ). Einflüsse auf die Indikatoren werden üblicherweise mit λ_i notiert und in der Regel mit dem Begriff „**Ladung**“ bezeichnet. Wie zu sehen, entspricht das Common-Factor-Modell der common-cause-Struktur wie sie in Abschnitt 2.1, Abb. 5, besprochen wurde. Dies heißt nicht, dass ξ die einzige Ursache seiner Indikatoren (x1-x4) ist. Jeder Indikator wird von weiteren systematischen als auch zufälligen Einflussgrößen beeinflusst, die zusammen durch den **Messfehler** δ_i repräsentiert werden – nur haben die Indikatoren keine weiteren *gemeinsamen* Ursachen. Ist dies doch der Fall, ist das Modell fehlspezifiziert und müsste angepasst werden. Wie in Abschnitt 2.1 ausführlich erläutert, hat dieses Modell Implikationen für die Daten – allem voran für die Kovarianzen (bzw. Korrelationen) zwischen den Indikatoren. Diese lassen sich wie bei allen Strukturen durch die in Abschnitt 2.1 beschriebenen Pfadregeln bestimmen. So ergibt sich beispielsweise die Kovarianz zwischen x1 und x2 aus $\lambda_1 Var(\xi)\lambda_2$. Mit l sind hier die sogenannten Ladungen notiert, d.h. die angenommenen Effekte der latenten Variablen auf ihre Indikatoren. Alle weiteren Kovarianzen ergeben dasselbe Muster des Produkts zwischen Ladungen und Varianz der latenten Variable. Wie leicht zu erkennen ist, kommen die einzelnen Parameter in mehreren Gleichungen vor, was uns zu einem wichtigen Kriterium der Schätz- und Testbarkeit des Modells bringt: der **Identifizierbarkeit**.

Ein Gleichungssystem (und damit auch dasjenige, das ein Modell repräsentiert) ist dann identifiziert, wenn die Anzahl der „Bekannten“ gleich der der „Unbekannten“ ist. Zum Beispiel ist die Gleichung „13 = 2a + 3b“ nicht identifiziert, da es *eine* Bekannte (13) und *zwei* Unbekannte gibt (a und b). Es gibt somit keine Möglichkeit, die Gleichung nach a oder b aufzulösen. Dies ändert sich aber dann, wenn eine weitere Gleichung dazu kommt: „24 = a + 1.5b“. Nun ist sie **„gerade-identifiziert“**. Dadurch lässt sich genau eine Lösung für a und b finden. Angewendet auf SEM sind die „Bekannten“ die empirischen Varianzen und Kovarianzen der Indikatoren und die „Unbekannten“ die zu schätzenden Parameter (λ, δ, und Var(ξ)). Was im Rahmen der Algebra vorteilhaft ist (nämlich genau eine Lösung zu finden) ist bei SEM ein Nachteil, denn erst die **Über-Identifzierbarkeit** ermöglicht den Test der kausalen Struktur. Grund ist, dass das Vorkommen eines Parameters in mehreren Gleichungen den Maximum-likelihood-Algorithmus zwingt, einen Kompromiss zu finden, der nur haltbar ist (im Rahmen des Modellfits), wenn die Struktur korrekt ist. Ist das Modell gerade-identifiziert, ist dies nicht der Fall. Selbst völlig falsche Modelle liefern Parameterschätzungen, aber keinen Test der Struktur. Der Grad an Über-Identifzierbarkeit reflektiert somit den Grad an Restriktionen, angesichts derer das Modell fitten muss.

Wie weiß man nun, ob das Modell identifiziert ist? Die Antwort ergibt sich aus

$$df = s - q$$

wobei df die Anzahl der **Freiheitsgrade** (degrees of freedom) des Modells sind, s die Anzahl der Elemente der empirischen Kovarianzmatrix S und q die Anzahl der zu schätzenden Parameter. Ist das Modell gerade-identifiziert, ist df = 0; ist es überidentifiziert, sind die df > 0. Die Anzahl für s bekommt man bei kleinen Matrizen noch relativ einfach (durch Zählen), bei größeren ist die simple Formel

$$s = \frac{(k \times k + 1)}{2}$$

sehr hilfreich (k ist die Anzahl der Variablen in der Kovarianzmatrix). Angewendet auf Common-Factor-Modelle ergibt sich daraus, dass ein Modell mit 2 Indikatoren nicht identifiziert ist: Es hat 3 Elemente in S (zwei Varianzen und eine Kovarianz), aber es sollen 4 Parameter geschätzt werden (2 Messfehlervarianzen, $Var(\delta_i)$, eine latente Varianz und eine Ladung). Ein Modell mit drei Indikatoren hätte null Freiheitsgrade und hätte keinerlei testbare Restriktionen. Dies bedeutet, dass man nahezu jedes beliebige Set von inter-korrelierenden Variablen als reflektive Indikatoren einer gemeinsamen Ursache modellieren könnte und bekäme eine perfekte und vollständige Replikation der Kovarianzmatrix (d.h. S und $\Sigma(\theta)$ wären nahezu identisch – und das, obwohl das Modell falsch ist). Ein Modell mit 4 oder mehr Indikatoren ist dagegen überidentifiziert und hat eine positive Anzahl von Freiheitsgraden.

Eine Ausnahme von dieser „Vier-Item-Regel“ als Voraussetzung für ein überidentifiziertes Modell besteht dann, wenn das Messmodell in einen Kontext anderer Variablen integriert wird (sog. „**empirische Über-Identifikation**“). Hier kommen weitere Kovarianzen in S hinzu (nämlich zwischen den Indikatoren der latenten Variablen und diesen anderen Variablen). So hat ein Modell mit zwei latenten Variablen, die jeweils durch zwei Indikatoren gemessen werden, 1 df und ermöglicht einen Test dieser Struktur. Da Messmodelle *immer* in einem Kontext integriert und getestet werden sollten, sollte sich in der Praxis kein Problem mit der Identifikation dieser Modellstrukturen ergeben. Isolierte Messmodelle, die lediglich das Set an Indikatoren und „ihre“ latente Variablen enthalten, sind schwache Tests dieser Struktur. Daher stellt ein fit solcher Modelle keine besonders starke Evidenz für diese Struktur dar.

Neben der nötigen Anzahl muss eine Ladung innerhalb eines Messmodells **fixiert** sein (in der Regel auf 1), um die latente Variablen zu **skalieren**. Dies bedeutet, dass diese Ladung nicht geschätzt wird, sondern ihr Parameter festgelegt wird. Der Grund dafür ist, dass latente Variablen kontinuierliche Dimensionen sind und für quantitative Auswertungen eine Metrik bekommen müssen, damit ihre Varianzen, Kovarianzen oder Effekte geschätzt werden können. Eine Fixierung einer Ladung auf 1 impliziert, dass die Skala des betreffenden Indikators auf die latente Variable übertragen wird. Diese Ladung entspricht einer Regressionsgeraden, die eine Steigung von 45° hat; somit entspricht der Unterschied zwischen zwei Einheiten in der latenten Variable dem Unterschied zwischen zwei Einheiten in der Indikatorvariablen (vgl. die herkömmliche Interpretation eines Regressionskoeffizienten).

Eine wichtige zusätzliche Sichtweise auf das Common-Factor-Modell liefert die Graphentheorie (Pearl, 2009), speziell mit Aussagen zu **bedingten Unabhängigkeiten**, die sich als Implikationen aus einem Modell ergeben (siehe hierzu Abschnitt 2.3 über **d-separation**). Danach impliziert das Common-Factor-Modell nicht nur, dass die Varianzen und Kovarianzen der Indikatoren sich durch die Faktorvarianz, Ladungen und Messfehler ergeben, sondern auch, dass die Kovarianzen der Indikatoren *dann* 0 sind, wenn der Faktor statistisch konstant gehalten oder auspartialisiert würde. Es sind diese Restriktionen, auf die das Modell im Falle einer Abweichung mit einem misfit reagiert und die einen Hinweis liefern, dass das Modell in dieser Form fehlspezifiziert ist. Wir werden auf diese Implikationen des Common-Factor-Modells noch einmal in Kapitel 5 stoßen, wo es um die Diagnostik von fehlspezifizierten Modellen geht, und dies etwas vertiefen.

Als Ergebnis des Schätzprozesses bekommt man als Parameterschätzungen die Faktorladungen (λ_i), die Varianz der latenten Variable(n) und Messfehlervarianzen ($Var(\delta_i)$). Die Faktorladungen sind von zentraler Bedeutung, da sie als **Validitätskoeffizienten** dienen (Bollen, 1989). Sie geben den empirischen Zusammenhang zwischen dem Indikator und der latenten Variablen an und sollten in einer Höhe sein, wie sie dem theoretischen vermuteten Zusammenhang zwischen dem Indikator und *derjenigen latenten Variable entspricht, die konzeptualisiert wurde.* Es kann nämlich durchaus sein, dass das Faktormodell zwar fittet, der Faktor aber nicht derjenigen latenten Variable entspricht, die man konzeptualisiert hat.

Die **Messfehlervarianz** (Var(δ)) kennzeichnet den Teil der Indikatorvarianz, der durch Messfehler verursacht ist. Auch wenn es üblich ist, Messfehler in einem SEM mit den klassischen Zufallsfehlern zu assoziieren und folglich Konzepte (wie Reliabilität) aus der klassischen Testtheorie anzuwenden, so ist das strenggenommen inkorrekt, da der Messfehler in SEM für alle nicht modellierten Einflüsse des Indikators steht und somit **zufällige** als auch **systematische Messfehler** beinhaltet (Scherpenzeel & Saris, 1997).

Eine zentrale Implikation eines Common-Factor-Modells ist die **Austauschbarkeit** der Indikatoren. Wenn beispielsweise eine latente Variable durch k Indikatoren gemessen wird (vorausgesetzt, das Modell ist korrekt spezifiziert), dann wird dieselbe latente Variable auch durch jede mögliche Teilmenge k – j gemessen (wobei j < k ist). Die Auswahl und Anzahl der Indikatoren verändert nicht die Bedeutung der latenten Variablen, genauso wenig, wie die Wahl einer Körpergewichtswaage das tatsächliche Körpergewicht eines Menschen verändert. Modelle, bei denen dies der Fall ist, sind vor einigen Jahren unter dem Begriff des **formativen Messmodells** populär geworden, bei denen die latente Variable als eine Folge oder Funktion der sog. *kausalen Indikatoren* konzeptualisiert wird. Wie schon unter Punkt 3.3 diskutiert, macht es hier allerdings wenig Sinn, anzunehmen, dass die latente Variable durch die manifesten Indikatoren selbst verursacht wird, sondern eher von deren latenten Entsprechungen (also die tatsächliche Bildung als Ursache der latenten Variable und nicht der in der Befragung gemessen Indikator der Bildung, s. Abb. 14c). Weiterhin sehen es einige Forscher kritisch, hier von Messmodellen zu sprechen, weil eine Ursache (d.h. der kausale Indikator) nur schlecht seine Wirkung (d.h. die latente Variable) messen kann (Edwards, 2011). Es ist also in diesem Fall sinnvoller, das „formative Messmodell" als ein latentes Regressionsmodell zu spezifizieren, in dem eine latente abhängige Variable durch mehrere latente Ursachen beeinflusst wird, man diese latenten Ursachen aber durch ihre gemessenen Indikatoren ersetzt.

Das Common-Factor-Modell ist quasi das Standardmodell, das nahezu allen Messtheorien zugrunde liegt. Aber gerade weil es so häufig angewendet wird, ist es fast schon zu einem Reflex geworden, Indikatoren eines „Konstrukts" (vgl. die Diskussion im vorherigen Kapitel), unreflektiert als Indikatoren eines common factors zu modellieren. Die meisten Probleme, die SEM-Anwender/innen haben und die am häufigsten Grund für misfit und Fehlspezifikationen sind, haben mit diesem Reflex zu tun. Daher sollte man sich bei dem Test einer solchen Struktur nicht vor Alternativen verschließen, auf die u.U. ein misfit des Modells hinweist. Wir werden in Kapitel 5 einen solchen Fall kennenlernen. Zudem sei auf den Artikel von Hayduk (2014) verwiesen, in dem dieser beispielhafte Alternativen illustriert, die alle zu ähnlichen oder identischen Datenanpassungen wie das Common-Factor-Modell führen würden.

3.5 Anzahl der Indikatoren und Single-Indicator-Variablen

Wie schon angesprochen, ist es durchaus vorteilhaft, mehrere Indikatoren zur Messung einer latenten Variable zu nutzen. Hat man mindestens 4 Indikatoren, ist das entsprechende Messmodell isoliert (d.h. ohne Verbindung zu anderen im Modell befindlichen Messmodellen) überidentifiziert; bei 3 Indikatoren ist es gerade identifiziert, bei 2 Indikatoren ist es nur identifiziert, wenn die latente Variable mit anderen Variablen (manifest oder latent) signifikant in Verbindung steht (über Effekte oder Kovarianzen). Hat man nur einen Indikator, kann die Messfehlervarianz nicht geschätzt werden, sondern muss fixiert werden. Weiterhin können mehrere Indikatoren mit starken Ladungen eine geringe Stichprobengröße kompensieren (Boomsma & Hoogland, 2001).

Diese Vorteile sowie die v.a. in der Psychometrik entwickelte Tradition, mehrere Items zu nutzen, um psychologische Konstrukte zu messen, haben dazu geführt, dass grundsätzlich Messmodelle mit mehreren Indikatoren von vielen Forschern/innen bevorzugt werden, selbst wenn das Modell nicht fittet. Häufig sollen durch eine Menge von heterogenen Indikatoren die „Breite" des Konstrukts adressiert werden, die eigentlich besser auf der latenten Ebene abgebildet werden sollte (im Rahmen der Multidimensionalität des Konstrukts). Daher soll hier nochmals betont werden, wie wichtig es ist, dass das Messmodell valide ist – d.h. die Kausalstruktur adäquat spezifiziert wurde und die Restriktionen (d.h. Erklärung der Kovarianz durch die latente Variable) haltbar sind. Dies wird bei der durch die Nutzung jedes weiteren Indikators immer unwahrscheinlicher, da mit der zwangsläufig induzierten sprachlichen Heterogenität schnell eine latente Heterogenität einhergeht, d.h. unterschiedliche Entitäten gemessen werden. Ist das Messmodell fehlspezifiziert, muss das nicht unbedingt problematisch sein, kann u.U. aber so weit gehen, dass die modellierte latente Variable ein Artefakt ist und kein reales Phänomen mehr repräsentiert. Daher ist es in vielen Fällen nicht nur realistischer, sondern auch inhaltlich präziser, wenige aber valide (im Sinne der Konzeptualisierung) Indikatoren einzubeziehen.

Hayduk und Litvay (2012) sowie Hayduk et al. (2007) empfehlen, nur einen einzigen Indikator zu benutzen – und zwar einen **„Golden-Standard"-Indikator**, der aufgrund theoretischer Erwägungen als auch der Kenntnis und

Erfahrung mit seinen Messeigenschaften am geeignetsten erscheint. Da in diesem Fall die Messfehlervarianz nicht empirisch geschätzt werden kann, empfehlen die Autoren, die Messfehlervarianz zu fixieren – und zwar auf einen Wert, der aufgrund theoretischer Erwägungen einem prozentualen Anteil der Varianz des Indikators entspricht. Dieser dient auch dazu, den theoretisch erwünschten Bedeutungsgehalt der latenten Variablen zu *bestimmen.* Will man z.B. eine latente Variable in das Modell einfügen, die konzeptionell sehr nahe am semantischen Inhalt des Indikators angesiedelt ist, würde man die Messfehlervarianz auf einen geringeren Wert fixieren, als wenn die latente Variable eine größere konzeptionelle Distanz zum Indikator aufweist. Allerdings bleibt hier unklar, auf welchen Wert man die Messfehlervarianz bestimmen soll.

Eine Möglichkeit, die in einer Reihe von Arbeiten verwendet wird, um die Messfehlervarianz eines Indikators zu fixieren, ist **Reliabilitätsinformationen** zu nutzen (z.B. Lin, Sher, & Shih, 2005). In diesem Sinne werden dann entweder Maße der internen Konsistenz wie Cronbach's alpha oder retest-Reliabilitäten benutzt und die Messfehlervarianz auf $\text{Var}(x_i) \times (1\text{-Rel})$ fixiert, wobei „$\text{Var}(x_i)$" die beobachtete Varianz des Indikators x_i und „Rel" das jeweilige Reliabilitätsmaß. Das Problem hierbei ist, dass bei Gebrauch von Cronbach's alpha mehrere Indikatoren vorliegen müssen und somit diese auch zum Test der Faktorenstruktur herangezogen werden müssen. Die Nutzung der Re-test-Reliabilität setzt voraus, dass die Veränderung der beobachteten Werte über die Zeit messfehlerbedingt ist und keine Veränderungen der latenten Variablen vorliegen. Und schließlich sei daran erinnert, dass Reliabilität die Abwesenheit von Zufallsmessfehlern angibt, während der Messfehler in Messmodellen mehr als nur Zufallsmessfehler enthält. Daher unterschätzt die Reliabilität u.U. den tatsächlichen Messfehler.

Auch wenn das Fixierungsproblem nicht befriedigend gelöst werden kann und somit Annahmen in der Modell gesteckt werden, die man letztlich nicht testen kann, haben Single-Indicator-Variablen den Vorteil, dass man sich bei der Auswahl des „Golden-Standard"-Indikators genaue Gedanken über die gewünschte Bedeutung der latenten Variablen von Interesse machen muss. Dies wird durch den Gebrauch multipler Indikatoren häufig vermieden. Schließlich betonen Hayduk et al. (2007) korrekterweise, dass das zentrale Merkmal einer validen Messung die Korrektheit der kausalen Spezifikation ist. Modelle mit vielen Indikatoren induzieren häufig eine sprachliche Heterogenität und damit mögliche Ursachen einer Fehlspezifikation. Daher ist der erste Adressat bei einem misfit immer der Fragebogen und die darin enthaltenen Item-Formulierungen.

4

Schritt für Schritt: SEM in R

4.1 R und lavaan

Bis vor kurzem war die Analyse von SEM nur mittels kommerzieller Software-Pakete wie LISREL, Mplus, EQS oder AMOS möglich. Dies war vielleicht ein Grund, dass im Zweifel insbesondere Studierende auf exploratorische Faktorenanalysen, wie sie mit jeder gängigen Software durchführbar sind, zurückgriffen. Mit der Software R, die von der Website des (www.cran.r-project.org) frei heruntergeladen werden kann, und insbesondere mit dem Paket lavaan (www.lavaan.org) ist nun eine Möglichkeit vorhanden, kostenlos und mit einfachen Mitteln konfirmatorische Faktorenanalysen und SEM zu rechnen. Alternative Pakete, mit denen SEM spezifiziert werden können, sind openMX und das sem-Paket.

R ist eine Statistiksoftware, die eine große Auswahl an analytischen Prozeduren bereits beim Installieren bereitstellt und durch über 1000 sog. Pakete erweitert werden kann. Lavaan ist eines dieser Pakete. Nach der Installation von R stellt sich dabei meist etwas Ernüchterung ein, weil der oder die Nutzer/in lediglich eine leere Konsole vorfindet, in der Befehle eingetippt werden müssen – im Gegensatz zu Programmen wie EXCEL oder SPSS, die über komplexe Menüleisten und Icons verfügen, über die analytische Prozeduren recht einsteigerfreundlich durchgeführt werden können. Dies sollte jedoch nicht abschrecken, weil die R-Befehle nach einem klaren Prinzip aufgebaut sind, das sich leicht erlernen lässt. Und selbst, wenn Sie R nicht über die Analyse von SEM hinaus zur allgemeinen Datenanalyse verwenden möchten, werden Sie in diesem Buch alle dazu nötigen Prozeduren – angefangen vom Einlesen der Daten bis zur Interpretation der Parameter und des Modellfits – kennenlernen.

Anstatt dabei die einzelnen Befehle direkt in die Konsole zu tippen, bietet es sich an, dies in einem **Editor** zu tun. Dies hat den Vorteil, dass alle Analysen gespeichert werden können.

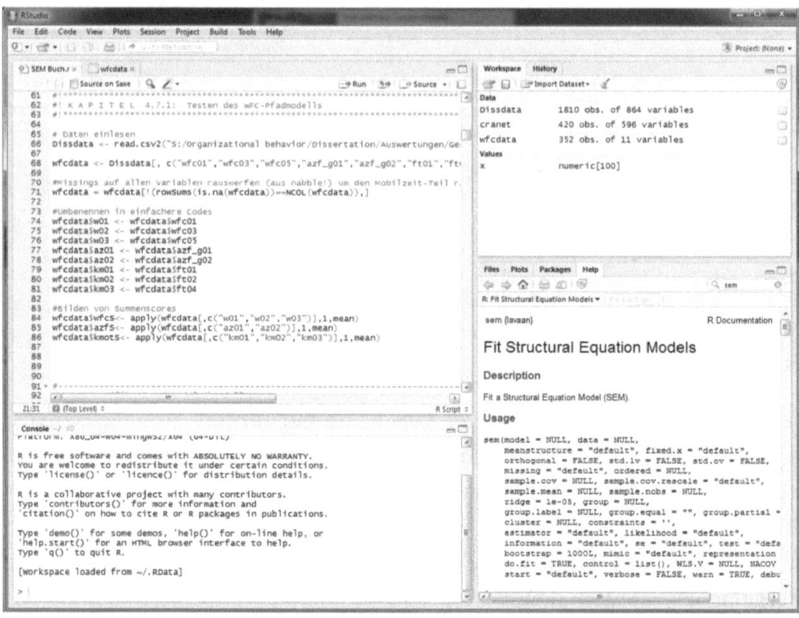

Abb. 16: RStudio

Aus dem Editor können dann die Befehle ausgeführt werden – der Befehl wird dabei an die Konsole geschickt und die Ergebnisse in der Konsole selbst ausgeworfen. Als Editor bietet sich dabei **RStudio** an (s. Abb. 16). Dieses kostenlose Programm (www.rstudio.com) integriert die Konsole und einen Editor zum Schreiben und Sichern aller Analyseschritte (siehe Abschnitt 4.5). Wie in der Abbildung zu sehen, bietet RStudio vier Fenster an, die in ihrer Größe und Anordnung vollständig angepasst werden können. Zentral ist das Fenster „Source" (oben links), das als Editor dient. Hier werden alle Kommandos aufgelistet, die dann über „STRG-R" oder einen Klick auf „run" in der obersten Zeile des Editors zur Konsole geschickt werden (unten links). Im Editor können eine oder mehrere „Skripte" angelegt und gespeichert werden, die wie Registerkarten in EXCEL funktionieren. Zusätzlich zu den eigentlichen Befehlen können und sollten die einzelnen Inhalte dokumentiert und kommentiert werden – dies passiert, indem man vor den Kommentar eine Raute (#) setzt. Ergebnisse werden in der Konsole abgebildet.

RStudio bietet die Möglichkeit, innerhalb von **Projekten** den Inhalt der Sitzung inklusive der Objekte (allem voran der Datensatz) und mehrerer Skripte zu speichern und zu laden. Um ein neues Skript zu erstellen, klickt man auf „File/New File“ und dann auf „R Script“. Über das Anlegen verschiedener Skripte lässt sich das gesamte Projekt strukturieren. Beispielsweise könnte man ein Skript anlegen, in dem alle explorativen Analysen und Veränderungen der Daten aufgelistet sind und ein separates, in dem alle SEM-Analysen enthalten sind. Die verschieden Skripte kann man über die Registerkarten sehen und ansteuern. Speichert man das Projekt, fragt RStudio, welche Skripte und ob der Inhalt des Workspace gespeichert werden soll; lädt man es, werden die Skripte und der Workspace geladen.

Um Pakete herunterzuladen, klicken Sie in der Menüleiste von R auf Pakete/Installiere Pakete, entscheiden Sie sich für einen mirror und wählen dann das entsprechende Paket aus – dies kann aber auch in RStudio geschehen, indem sie im rechten, unteren Fenster „Packages“ „install Packages“ anklicken. Ist das Paket installiert, müssen Sie allerdings in jeder Sitzung dieses Paket aktivieren (mit `library(Paketname)` oder indem Sie im RStudio das Paket anklicken).

Die Arbeit mit R, die für die Durchführung von SEM für uns von Belang sind und in diesem Buch behandelt werden, entfallen auf die Bereiche

- Einlesen der Rohdaten
- Prozeduren zur Veranschaulichen der Daten („data screening“) und Überprüfen bestimmter Annahmen
- SEM im engeren Sinne
- Heranziehung diagnostisch hilfreicher Informationen zur Respezifikation des Modells.

Allerdings werden die Wurzeln für das zu testende Modell weit früher im Forschungsprozess gelegt. Daher ist es nötig, diese grundlegenden Arbeitsschritte mit einzubeziehen. Oft kommt es zu einer Diskrepanz zwischen der theoretischen Vorstellung des/der Forscher/in und der tatsächlichen Datenerhebung, die dann erst später, im Rahmen der Konfrontation mit einem nicht-fittenden Modell, auffällt. Die Übersetzung eines meist eher vagen Konstrukts in empirisch messbare Variablen wurde in Kapitel 3 ausführlich erörtert. Hier sollen eher die praktischen Aspekte und typischen Fehler und Probleme diskutiert werden. Wir beginnen daher mit dem ersten Schritt – der möglichst präzisen Konzeptualisierung des Modells und der darin enthaltenen Variablen.

4.2 Spezifizierung des Forschungsmodells

Ausgangslage für ein Modell sind theoretische Überlegungen, welche kausalen Beziehungen zwischen denjenigen empirischen Phänomenen bestehen, für die die theoretische Konstrukte bedeutungsstiftend sind. Kapitel 3 beschrieb bereits, dass es einen zusätzlichen theoretischen Aufwand erfordert, ein Konstrukt in ein empirisch testbares Modell zu übersetzen. Konstrukte können eindimensional und damit direkt als einzelne latente Variable repräsentierbar sein oder mehrdimensional – was die Frage aufwirft, welche Dimensionen für das Modell relevant sind und welche nicht.

Ein/e Forscher/in ist daher schlecht beraten, sich auf sog. **validierte Skalen** zu verlassen. Skalen sind Summen von Items, für die in vielen Fällen ein unzureichender Beleg für ein Messmodell mit der Struktur eines Common-Factor-Modells (vgl. Abschnitt 3.4) vorliegt. Das heißt, dass die Items dieser Skalen nicht unbedingt eine latente Variable messen müssen und somit häufig heterogener sind, als vermutet. Oft sind solche Skalen mittels **exploratorische Faktorenanalysen** entstanden. Für solche Faktorenanalysen existieren nur unklare Entscheidungskriterien (z.B. Kaiser-Guttman-Kriterium) für die Auswahl der Faktoren, was die Auswahl von Faktoren eher zu einer Frage der „Adäquatheit" oder „Handhabbarkeit" macht, als zu einer Frage nach dem korrektem Messmodell (vgl. 3.1). D.h., die Faktorenanalyse extrahiert mehrere Faktoren, aber es wird nur ein einziger berücksichtigt, auf dem dann alle Items mehr oder weniger stark laden.

Zudem ist eine Faktorenanalyse nicht in der Lage, Items zu identifizieren, die als einzige eine latente Variable messen (vgl. Single-Indicator-Variablen, Abschnitt 3.5), da ein Faktor nur extrahiert wird, wenn mehrere Items korrelieren. Sollten daher Items im Datensatz sein, die eine eigene latente Variable messen, werden sie einem der anderen Faktoren zugewiesen. Dies hat den Nachteil, dass unter Umständen eine theoretisch wie praktisch bedeutsame latente Variable unentdeckt bleibt und das Item wegen der niedrigen Ladung sogar eliminiert wird.

In vielen Fällen der Skalenentwicklung wurde sogar nur eine **Hauptkomponentenanalyse** (principal component analysis, PCA) durchgeführt. Obwohl jahrzehntelang fälschlicherweise als Unterart der exploratorischen Faktorenanalyse angesehen, identifiziert diese gar keine latenten Faktoren, sondern konstruiert gewichtete Summenindizes der Items (vgl. Borsboom, 2006, für eine diesbezügliche Kritik der Persönlichkeitspsychologie).

Wegen all dieser Aspekte überrascht es auch nicht, dass CFA-Modelle häufig nicht fitten, wenn Sie auf Skalen angewendet werden. Während Vertreter der Skalen dies meist als ein Zeichen sehen, dass CFA-Modelle eine inadäquate Methode zur Überprüfung der Skalen sind (Borsboom, 2006), stellt dies meist für CFA-erfahrene Forscher/innen keine Überraschung dar, da traditionell bei der Konstruktion von Skalen darauf geachtet wird, eine möglichst konzeptionelle Breite der Item-Formulierungen zu erreichen um die „Breite des Konstrukts" „abzudecken" (siehe dazu Abschnitt 8.8). Wie bereits in Abschnitt 3.1 diskutiert, impliziert diese Sichtweise allerdings ein multidimensionales Konstrukt bestehend aus mehr als nur einer latenten Dimension. Es kann also davon ausgegangen werden, dass viele Skalen, die ein Konstrukt messen, dennoch eine **heterogene Faktorstruktur** haben, was bedeutet, dass die Items unterschiedliche latente Variablen messen.

Aus diesen Gründen ist es bei in der Konzeptualisierung des Modells unabdingbar, die Konstrukte des Modells in Termini von spezifischen latenten Dimensionen zu spezifizieren, die ein empirisch existierendes Phänomen betreffen. Folgendes Gedankenexperiment kann dabei helfen: Bei der Auswahl von Indikatoren für ein beliebiges Konstrukt darf man nur einen einzigen Indikator auswählen. Welchen würde man wählen? Der „Zwang" zur Wahl eines solchen *Golden-Standard -Indikators* hat zwei Konsequenzen:

Erstens hinterfragt man die Dimensionalität des Konstrukts: Hat man nämlich den Eindruck, dass das Konstrukt durch diesen einen Indikator nicht ausreichend repräsentiert werden kann, ist das ein Hinweis auf Multidimensionalität des Konstrukts. Diese sollte explizit gemacht werden durch die Beantwortung der Frage, welche *latenten Dimensionen* das Konstrukt repräsentieren. D.h. die Breite des Konstrukts sollte auf latenter und damit Konzept-Ebene geklärt werden und nicht nachträglich durch eine Konstruktion heterogener Items.

Zweitens ist man gezwungen, die kausale Rolle „des" Konstrukts zu präzisieren. Es geht also nicht darum, das Konstrukt an sich umfassend abzubilden, sondern *genau diejenigen latenten Dimensionen zu wählen, die im Rahmen des Forschungsmodells und damit entsprechend der Theorie des/der Forschers/in relevant sind.* Kommt im Rahmen dieses Gedankenexperiments heraus, dass das Konstrukt mehrere Dimensionen hat, stellt sich die Frage, ob alle Dimensionen relevant sind (d.h. jede einzelne Dimension hat oder empfängt Effekte), oder nur einige. Die Beantwortung all dieser Fragen und entsprechende Item-Auswahl bildet eine realistische Grundlage für das geplante Forschungsprojekt. Hat man den Golden-Standard-Indikator identifiziert, der theoretisch nahezu perfekt die latente

Variable von Interesse repräsentiert, sollten 2-3 weitere Indikatoren ausgewählt werden, die exakt diese latente Variablen in gleicher Weise reflektieren (im Sinne des Common-Cause-Modells).

4.3 Pretesting von Items

Bevor die Datenerhebung startet, ist es ratsam, dass Verständnis der Items zu überprüfen (Krosnick, 1999; Sudman, Bradburn, & Schwarz, 1996). Dieses pretesting ist in der Demoskopie und den Sozialwissenschaften üblich – in anderen Verhaltenswissenschaften (z.B. Psychologie) nahezu unbekannt (Willis, 2005). Insbesondere **kognitive Interviews** bieten die Möglichkeit zu überprüfen, ob Befragte Schwierigkeiten beim Verständnis haben, die durch die Item-Formulierung erzeugt wurden (z.B. unklare Formulierungen, oder Fehler wie doppelte Verneinungen, unverständliche Begriffe). Darüber hinaus bekommt man einen Eindruck, ob die Items das messen, was sie messen sollen. Bei einem kognitiven Interview wird eine Person aus der Population von Interesse mit ausgewählten Items aus dem Fragebogen konfrontiert. Dabei gibt es eine Reihe von Techniken, um das Verständnis der Items zu prüfen. Beispiele sind:

- **Thinking aloud:** Der/die Befragte wird aufgefordert, beim Lesen und Beantworten der Frage „laut zu denken". Ziel ist es, herauszufinden, welche Gedanken und Schlussfolgerungen auftauchen.
- **Paraphrasing:** Hierbei wird die Person aufgefordert, die Frage in eigenen Worten wiederzugeben.
- **Probing:** Hierbei werden Fragen zu einigen Aspekten des Items gestellt, z.B.: zum Verständnis bestimmter Begriffe (*comprehension probing*), zu den Gründen ihrer Wahl einer Antwortkategorie („Warum haben Sie die „4" angekreuzt?", *category selection probing*), oder auf welchem Wege sie zu ihrer Antwort gelangt ist (*information retrieval probing*)
- **Confidence rating:** Oft mag es bedeutend sein, z.B. Ereignisse oder die Häufigkeit von Verhalten genau abzuschätzen. In diesem Fall kann man durch Fragen, wie sicher sich die Person bei Ihrer Antwort ist, abschätzen, ob eine genaue Messung möglich ist.

Je nach Messinstrument kann eine Interview-Studie mit einem Dutzend Personen bereits ausreichend sein, um gravierende Probleme mit der Item-Formulierung auszuräumen. Ist alles im Vorfeld geleistet, kann die Studie ins Feld gehen.

4.4 Einlesen der Daten

Meist werden die Daten eines Forschungsprojekts mithilfe anderer Programme eingegeben und gespeichert (z.B. SPSS, Excel) oder liegen als ascii-Dateien vor. Vor der Verwendung in R müssen sie also eingelesen werden. Eine einfache Methode ist, die Datei im csv-Format abzuspeichern und die entsprechende Datei einzulesen:

```
wfcdata <- read.csv2("c:\\... \\dataset.csv", header=T)
```

Hierdurch wird mittels der Funktion `read.csv2` die csv-Datei „dataset.csv" eingelesen und dem R-Objekt „wfcdata" zugewiesen. Dieses Objekt gehört zur Klasse „dataframe". In späteren Operationen (z.B. SEM) wird immer auf dieses Objekt verwiesen werden. Die Funktion `read.csv2` liest Datensätze ein, die Semikola zur Trennung der Variablen enthalten. Sollten die Variablen stattdessen mit Kommata getrennt sein, benutzt man `read.csv`. Das Argument `header=T` teilt R mit, dass die erste Zeile (der „header") die Namen der Variablen enthält.

4.5 Projektmanagement und Housekeeping

Der R-Workspace enthält nun ein Objekt – den Datensatz. Diesen sieht man im RStudio-Fenster „workspace". Im Laufe der weiteren Arbeit werden viele weitere Objekte hinzukommen. Man sollte sich gelegentlich durch die Funktion `ls()` einen Überblick verschaffen, was sich alles angesammelt hat, bevor der Arbeitsspeicher zur Müllhalde verkommt. Beim Benutzen von RStudio sieht man die Objekte im „Workspace"-Fenster. Objekte löschen kann man mit `rm(Objekt1,Objekt2)`. Ich neige dazu, nur den Datensatz zu behalten und alle anfallenden Objekte (wie Ergebnisse oder Grafiken) nicht als Objekte zu behalten, sondern stattdessen die R-Befehle zu ihrer Erzeugung in einem separaten Editor zu speichern. Dieser Editor enthält alle Befehle, die im Projekt anfallen – angefangen vom Befehl, der das Einlesen der Daten beinhaltet, über die Darstellung deskriptiver Statistiken und Grafiken bis zur eigentlichen Analyse. Zusätzlich kann man jede Analyse kommentieren (was man ausgiebig tun sollte, wenn man sich teilweise Monate später noch erinnern will, was man wozu getan hat).

4.6 Exploratorische Datenanalyse

Sind die Daten eingelesen und die Maßnahmen zum Projektmanagement ergriffen, besteht das erste Ziel einer jeden Arbeit mit Daten darin, die Stichprobe und den Datensatz kennenzulernen. Diese ersten exploratorischen Analysen haben 3 Ziele:

a) **Auffinden von Fehlern** – sowohl von Fehlern in der primären Datenquelle (z.B. Fragebogen) als auch von Fehlern bei der Eingabe.
b) **Überprüfen von Annahmen**, die das Analyseverfahren von Interesse (z.B. SEM) macht oder Implikationen des unterstellten Modells. So ist es z.B. sinnvoll, sich die Kovarianzmatrix und Korrelationsmatrix anzuschauen um sich ein Bild über Zusammenhänge zwischen den Variablen zu machen. Oft fallen einem dabei bereits Dinge auf, die später (d.h. bei der Modifikation eines nicht-fittenden Modells) helfen können. Beispielsweise sollten Items, die im Rahmen eines Faktormodells (vgl. Abschnitt 3.4) modelliert werden sollen, möglichst homogen und hoch miteinander und in ähnlicher Ausprägung mit anderen korrelieren. Abweichungen können darauf hinweisen, dass die Items doch verschiedene latente Variablen messen.
c) **Kenntnis über die Stichprobe** und Identifikation interessanter Fragestellungen, Zusammenhangsmuster und evtl. Subgruppen, die voneinander abweichen (in Mittelwerten oder Zusammenhängen zwischen Variablen). Wenngleich das Hypothesentesten das dominante Paradigma in der gegenwärtigen Wissenschaft ist, hat die exploratorische Arbeit mit Daten einen wichtigen (wenn nicht den wichtigsten) Einfluss auf die Forschung (sog. abductive reasoning, vgl. Haig, 2005, 2008; Locke, 2007; Rozeboom, 1997).

Im Folgenden werden einige wichtige R-Befehle exemplarisch aufgelistet, die bzgl. der o.g. Ziele behilflich sein können.

4.6.1 Auffinden von Fehlern

Zum Auffinden von Fehlern sollten Sie zunächst den structure-Befehl (`str(wfcdata)`) eintippen. Angezeigt werden die Stichprobengröße, Anzahl der Variablen, die Variablennamen, ihr **Modus** (ob numerisch, character – d.h. Textvariable – oder kategorial) und die ersten 10 Fälle. Hier ist v.a. von Interesse, ob der Modus der Variablen so zugewiesen wurde, wie beabsichtigt. Vor allem sollten Items, die später als Indikatoren im SEM verwendet werden sollen, den Modus „int“ (für integer“), oder „num“ (für numerisch) – nicht aber „Factor“ enthalten. Diese Modi sind zwar im Grunde nicht korrekt, weil Rating-Skalen strenggenommen kategoriale Variablen sind; praktischerweise behandelt man diese aber als approximativ metrisch (zur Diskussion von kategorialen Indikatoren und den Möglichkeiten der Berücksichtigung dieser Klasse siehe Abschnitt 4.11).

Als nächstes sollte man sich anschauen, ob Werte falsch eingegeben wurden. Eine gute Funktion dafür ist `describe(wfcdata)` aus dem psych-Paket. Sie liefert alle wichtigen deskriptiven Statistiken. Hier sind zunächst die „min"- und die „max"-Spalte interessant. Manchmal werden Zahlen versehentlich doppelt eingetippt (also „55" anstatt „5"). Solche Fehler sind sehr gravierend und können leicht erkannt werden, wenn der Zahlen außerhalb des plausiblen Bereichs vorkommen. Will man die Häufigkeiten aller Werte jeder Variablen des gesamten Datensatzes auf einen Schlag, tippt man z.B. `lapply(wfcdata,table)`. Häufigkeiten spezifischer Variablen bekommt man mit `CrossTable(wfcdata$wl01)` aus dem gmodels-Paket – hier für die Variable wl01 – oder deren grafische Veranschaulichung `plot(table(wfcdata$wl01))`.

R Console

Datei Bearbeiten Verschiedenes Pakete Windows Hilfe

```
> describe(mydata)
       var   n   mean     sd median trimmed    mad    min    max  range  skew kurtosis   se
code     1 134 275.31 115.11 267.00  273.30 152.71 101.00 476.00 375.00  0.13    -1.34 9.94
wl01     2 128   3.12   1.02   3.00    3.26   1.48   0.00   4.00   4.00 -0.86    -0.23 0.09
wl02     3 127   2.36   1.26   2.00    2.45   1.48   0.00   4.00   4.00 -0.32    -0.82 0.11
wl03     4 125   3.22   0.93   4.00    3.32   0.00   1.00   5.00   4.00 -0.69    -0.65 0.08
sst01    5 127   1.59   1.43   1.00    1.50   1.48   0.00   4.00   4.00  0.41     1.15 0.13
sst02    6 128   0.70   1.17   0.00    0.47   0.00   0.00   4.00   4.00  1.47     1.06 0.10
sst03    7 127   1.20   1.17   1.00    1.07   1.48   0.00   4.00   4.00  0.71    -0.35 0.10
sst04    8 126   0.48   0.99   0.00    0.23   0.00   0.00   4.00   4.00  2.20     4.36 0.09
cn01     9 127   0.76   1.10   0.00    0.53   0.00   0.00   4.00   4.00  1.44     1.36 0.10
cn02    10 127   0.65   1.03   0.00    0.44   0.00   0.00   4.00   4.00  1.75     2.74 0.09
cn03    11 127   1.18   1.17   1.00    1.04   1.48   0.00   4.00   4.00  0.80    -0.15 0.10
```

Abb. 17: Tabelle mit deskriptiven Statistiken

Fälle, bei denen stark abweichende Werte auftreten, kann man gut mittels der Kombination von `plot(wfcdata$wl01)` und nachfolgendem `identify(wfcdata$wl01)` identifizieren. Durch das erste Kommando wird eine Grafik erstellt, in der die einzelnen Werte auf der Ordinate entlang ihrer Position im Datensatz auf der Abzisse angezeigt werden. Durch das zweite Kommando verwandelt sich der mouse-Zeiger in ein Kreuz, mit dem man auf einzelne Punkte in der Grafik klicken kann. Beendet man diesen Vorgang mit der Taste „escape", bekommt man deren Fallnummer in der Grafik angezeigt.

Neben dieser univariaten Strategie der Fehlersuche sollte man sich Gedanken machen, welche *Kombinationen von Werten* noch plausibel sind und welche auf Fehler hinweisen. So ist ein 15jähriger Studienteilnehmer nichts Auffälliges und eine Person mit 50.000€ Jahreseinkommen auch nicht – aber ein 15jähriger mit 50.000€ schon. Auch beim Vorliegen von Filterfragen sollten bestimmte Kombinationen von Werten schlicht nicht vorkommen – z.B. die Kombination von „nein“ auf die Frage „Haben Sie Kinder“ und eine positive Zahl bei der nachfolgenden Frage „wie viele Kinder haben Sie“. Kombinationen lassen sich gut mit Kreuz- oder Kontingenztabellen anzeigen, z.B. `CrossTable(x,y, prop.chisq=F)`. Die Variablen x und y sollen hier stellvertretend für zwei substantielle Variablen stehen. Das Argument „prop.chisq“ fügt der Tabelle den in der Regel uninteressanten Beitrag der Zelle zum Chi-Quadrat-Test auf Gleichverteilung hinzu und kann durch „=F“ für „false“ unterdrückt werden.

Schließlich sollte das Augenmerk auf die Definition von **missing data** (d.h. fehlenden Werten) gerichtet werden. Fehlende Werte können in dem Programm, mit dem die Daten eingegeben wurden (z.B. EXCEL), entweder als leere Zellen oder sogenannte „missing codes“ enthalten sein. Letztere sind in der Regel meist Zahlen, die in der betreffenden Variablen nicht vorkommen können (z.B. „999“) und kennzeichnen, dass es sich bei dem Zelleintrag um einen fehlenden Wert handelt. Liest man die Daten dann in R ein (mit dem o.g. `read.csv2`-Befehl), muss R unbedingt mit `"na.strings=999"` klargemacht werden, dass dies ein fehlender Wert ist. Unterbleibt dies, behandelt R diesen Wert wie jede andere Zahl, was zu extremen Ausreißern und ggf. starken Verfälschungen der Daten führen kann. Durch die Definition der missing codes fügt R ein „NA“ für „not available“ ein. Ein Teil der exploratorischen Analyse besteht somit darin, zu elaborieren, ob die missing codes richtig vergeben wurden. Falls dies nicht der Fall war, fällt dies bei den oben besprochenen univariaten Häufigkeitsanalysen auf.

Abgesehen davon besteht ein wichtiges Ziel der exploratorischen Datenanalyse darin, sich ein Bild zu machen, wie viel fehlende Werte für jede Variable vorliegen. Tiefergehende Analysen betreffen die Frage der Zufälligkeit vs. Systematik fehlender Werte und möglicher Muster (siehe die exzellente Einführung von McKnight, McKnight, Sidani, & Figueredo, 2007). So kann man in einer neuen Variable codieren, ob eine Person in einer bestimmten Variablen überhaupt einen Wert hat (mit „1“) oder dort ein Eintrag fehlt (mit „0“). Indem man nun diese neue, binäre, Variable auf andere Variablen im Datensatz mittels einer logistischen Regression regressiert, bekommt man Informationen darüber, welche Variablen im Datensatz erklären können, *ob* sie einen Wert in der Zielvariablen eingetragen hat.

4.6.2 Kennenlernen des Datensatzes

Wie anfangs erwähnt, gibt es eine Vielzahl von Funktionen, mit denen sich die Daten veranschaulichen lassen. Tabelle 1 enthält eine Auflistung einiger dafür sinnvoller Funktionen.

Tabelle 1: Hilfreiche Funktionen zur Beschreibung des Datensatzes

Deskriptive Statistiken*	
`mean(x), median(x)`	Mittelwerte, Median der Variablen „x"
`var(x), sd(x), range(x)`	Varianz, Standardabweichung und Bereich von x
`cov(x,y), cor(x,y)`	Kovarianz und Korrelation zwischen x und y
Univariate Verteilungen	
`CrossTable(x)`	Häufigkeiten der Variablenwerte von x (Paket gmodels)
`hist(x, nclass=#)`	Histogramm von x. `nclass=#` bestimmt die Anzahl der gewünschten Säulen.
`plot(table(x))`	Balkendiagramm bei Faktoren
`barplot(table(x))`	Balkendiagramm bei Faktoren
`boxplot(x)`	Boxplot von x.
`qqnorm(x)`	QQ-Plot zur Prüfung der Normalverteilung. Punkte sollten einer Diagonalen folgen.
Bivariate Verteilungen	
`plot(y~x)`	Streuungsdiagramm von x und y. Tippt man anschließend `abline(lm(y~x))` ein, wird eine Regressionsgerade abgebildet. Durch das Argument „jittering" wird die Punktewolke etwas plastischer dargestellt. Dabei wird jedem Fall ein kleiner Zufallsfehler hinzugefügt: `plot(jitter(y,amount=.05)~jitter(x,amount=.05))`
`scatterplot (y~x,data=wfcdata)`	Ebenfalls ein Streuungsdiagramm aus dem car-Paket, das aber a) sofort eine Regressionsgerade und b) zusätzlich eine nicht-parametrische Kurve in die Punktewolke legt. Diese Funktion ist sehr gut geeignet, um nicht-lineare Effekte zu identifizieren. Gibt man hinter „y~x" noch einen vertikalen Strich („|") und einen Faktor ein (z.B. Geschlecht) bekommt man getrennte Punktewolken für die Kategorien des Faktors.

*Enthält der Datensatz fehlende Werte, muss hier na.rm=T als zusätzliches Argument in den Befehl eingegeben werden (z.B. `mean(x, na.rm=TRUE)`)

Leitfragen beim Check der Daten sind:

(1) **Haben die Indikatoren eine ausreichende Varianz** (d.h. Informationsgehalt oder Differenzierungsfähigkeit)? In manchen Fällen hat man Variablen, die eine extrem hohe Varianz haben (z.B. eine freie Abfrage der Wochenarbeitszeit). Es mag nötig sein, solche Variablen durch eine Konstante zu teilen, weil SEM-Programme häufig Probleme bekommen. Dies wird durch eine Fehlermeldung aber angezeigt.

(2) **Ist die Verteilung adäquat?** Wie wir noch sehen werden, macht SEM die Annahme, dass nicht die Indikatoren einzeln normalverteilt sind, sondern alle zusammen einer multinormalen Verteilung folgen sollten. Bevor Sie aber *statistische Tests* zur Überprüfung von Verteilungsannahmen anwenden, bedenken Sie: Solche Tests werden nahezu immer eine signifikante Abweichung von der Multi-Normalverteilungsannahme liefern. Wir werden später (Abschnitt 4.8.2) Möglichkeiten kennenlernen, dass lavaan verschiedene Schätzer für solche Situationen anbietet. Vielmehr sollte das Augenmerk darauf gerichtet sein, was z.B. eine extremschiefe Verteilung für die inhaltliche Bedeutung eines Indikators bedeutet. Deutlich zu wenige Werte im oberen Skalenbereich implizieren beispielsweise, dass Befragte hier nicht zugestimmt haben. Bedeutet das, dass die Frage zu „streng" formuliert war? Wie lässt sich das vor dem Hintergrund der gewählten Konzeption der latenten Variable von Interesse interpretieren? Passt dieses Item (in dieser Formulierung) dann zu der latenten Variablen?

(3) **Wie ist das Ausmaß fehlender Werte in den Indikatoren?** Dazu gehört zunächst die Überprüfung, ob die missing codes (s.o.) korrekt sind. Einfache missing-data-Analysen lassen sich z.B. durchführen, in dem der Status („vorhanden" vs. „nicht vorhanden") in eine neue Variable codiert wird und andere Variablen im Datensatz im Rahmen einer logistischen Regression (mit der Status-Variablen) als Prädiktoren getestet werden. Dies kann Aufschluss geben, ob es systematische Gründe für das Nicht-Ausfüllen gibt und wieder um Rückschlüsse auf die Bedeutung liefern.

(4) **Sind Varianzen und Kovarianzen der Indikatoren wie theoretisch erwartet?** Man sollte sich die Kovarianzmatrix und Korrelationsmatrix der in Frage kommenden Variablen genau anschauen. Folgende Aspekte *können* ein Indiz sein, dass die theoretische Konzeption eines oder mehrere Messmodelle problematisch ist: Wenn Indikatoren eine gemeinsame Ursache haben (d.h. die latente Variable), sollten in der Regel ihre Varianzen in etwa gleich hoch sein. Ausnahmen kann es z.B. geben, wenn Indikatoren ein unterschiedliches

Ausmaß an Messfehlervarianz aufweisen. Diese Erklärungsmöglichkeit sollte in Anbetracht der Indikator-Formulierungen Sinn machen. Wenn die Indikatoren ähnlich formuliert sind, macht die Annahme, dass ein bestimmter Indikator einen höheren Messfehler hat, wenig Sinn. Analog sollten die Korrelationen der Indikatoren untereinander ähnlich hoch sein und Korrelationen mit anderen Modellvariablen (z.B. Indikatoren anderer latenter Variablen) sollten das gleiche Muster haben. Abweichungen müssen nichts bedeuten, können aber Ausgangspunkt für eine konzeptionelle Überprüfung der Indikatoren sein. In dem praktischen Beispiel, das später berichtet werden wird, war die Korrelationsmatrix ein Hinweisgeber (vgl. Abschnitt 5.3).

4.7 Spezifizieren und Testen von Modellen in lavaan

Sind die Daten überprüft, geht es an das eigentliche Modell. Der Vorgang des Modelltestens läuft in lavaan über drei sequenzielle Schritte. Das mag umständlich klingen, entspricht aber dem objekt-basierten Arbeiten in R: jeder Schritt liefert ein Objekt, indem je nach Interesse mehr oder weniger weitere Informationen entnommen werden. Dies hat den Vorteil, dass man nicht als Grundeinstellung eine Masse von Informationen geliefert bekommt, wenn man nur an einem Teil davon interessiert ist.
Die drei Schritte sind:

(1) **Das Spezifizieren des Modells:** Hier legen Sie aufgrund von Erwartungen fest, welche Parameter das Modell enthalten soll und welche nicht.
(2) **Das Rechnen des Modells**: Hier werden die Parameter geschätzt.
(3) **Anzeige des Modellfits und der Parameterschätzungen:** Das Ergebnis des zweiten Schritts hat ein Objekt ergeben, das viele Informationen enthält. Primär sind hier der Modellfit und die Parameterschätzungen und ihre Teststatistiken von Interesse. Darüber hinaus lassen sich weitere wichtige Informationen extrahieren, die v.a. im Rahmen der Modell-Respezifikation wichtig sind, wie Modifikationsindizes oder standardisierte Residuen (siehe Kap. 5).

Als Illustration dieser Schritte soll das Modell geprüft werden, das in
Abb. **18** abgebildet ist. Das Modell postuliert eine partielle Mediation des Effektes von work-family conflict über Arbeitszufriedenheit auf die Kündigungsmotivation. Work-family conflict bezeichnet wahrgenommene Probleme einer Person, Arbeits- und Familienleben zu vereinbaren, was durch Arbeitsanforderungen, Arbeitszeit und Stress verursacht sein kann. Das Modell sieht nun vor, dass dies die Arbeitszufriedenheit reduziert und dies wiederum die Kündigungsmotivation

erhöht. Zudem wird angenommen, dass work-family conflict direkt zu einer höheren Kündigungsmotivation beiträgt. Dies bedeutet, dass ein Zusammenhang zwischen beiden Variablen auch dann noch besteht, wenn Arbeitszufriedenheit auspartialisiert bzw. statistisch kontrolliert wird (vgl. Abschnitt 2.3). Eine Begründung dieses Effekts besteht darin, dass work-family conflict die Kündigungsmotivation auch aus faktischen Gründen erhöht (z.B. weil er sich auf durch die Arbeit bedingten Probleme mit der Kinderbetreuung bezieht), ohne dass damit immer eine Reduktion der Arbeitszufriedenheit einhergehen muss. Es sollte angemerkt werden, dass unter realistischen Umständen dieses Modell zu einfach wäre, weil wichtige Kovariate der Modellvariablen fehlen. Dies wird Gegenstand von Kapitel 7 sein.

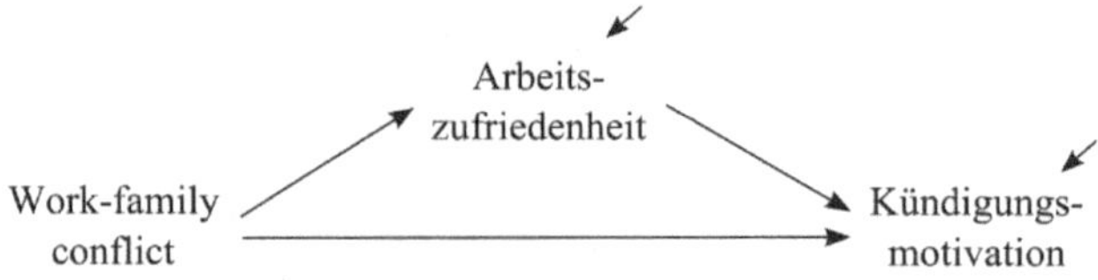

Abb. 18: Theoretisches Modell

Für jedes der drei Konstrukte wurden mehrere Items benutzt. Stichprobe waren 387 Erwerbstätige aus einer Vielzahl von Branchen oder Berufen (Steinmetz, 2007). Die Items zur Messung der Konstrukte sind in Tabelle 2 dargestellt.

Tabelle 2: Items zur Messung der Konstrukte

w01	Meine beruflichen Anforderungen behindern mein Privat- und Familienleben
w02	Wegen meiner beruflichen Anforderungen kann ich Dinge, die ich zu Hause erledigen möchte, nicht tun.
w03	Aufgrund meines Berufs muss ich oft familiäre Pläne ändern.
az01	Wie sehr entspricht Ihre Arbeit insgesamt Ihrer Vorstellung, wie sie sein sollte?
az02	Alles in Allem: Wie zufrieden sind Sie mit Ihrer Arbeit?
km01	Wie häufig kommt Ihnen der Gedanke, zu kündigen?
km02	Wie häufig haben Sie sich in letzter Zeit nach einem anderen Arbeitsplatz erkundigt (z.B. Stellenanzeigen gelesen, Bekannte gefragt etc.)
km03	Wie wahrscheinlich ist es, dass Sie tatsächlich innerhalb des nächsten Jahres kündigen werden?

4.8 Spezifizieren von Pfadmodellen

Zunächst soll das oben abgebildete Modell in einem **Pfadmodell** illustriert werden.

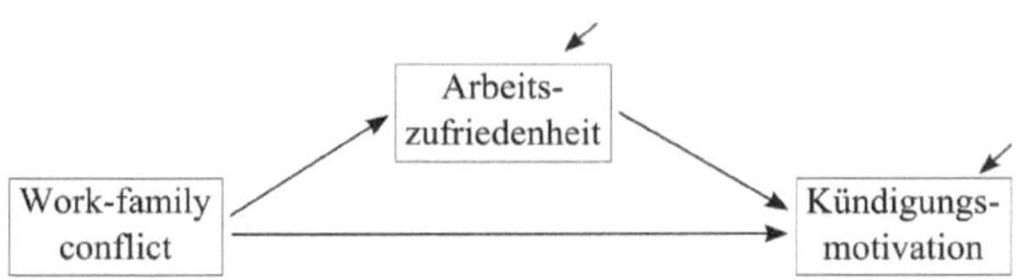

Abb. 19: Ein Pfadmodell zur Wirkung von work-family conflict

In einem Pfadmodell wird das Messmodell jeder Variablen – also die explizite Modellierung der Kausalstruktur, die jede Modellvariable mit ihren gemessenen Indikatoren verbindet – nicht spezifiziert. Stattdessen wird jede Variable im Modell direkt durch die gemessene(n) Variable(n) repräsentiert. Dies wird durch die Kästchen um die Variablennamen kommuniziert. Somit kann das Pfadmodell als eine Erweiterung der Regressionsanalyse aufgefasst werden. Vorteile der Pfadanalyse gegenüber der Regression sind, a) dass mehr als nur eine endogene Variable modelliert werden kann und b) dass ein komplexes kausales Netz inkl. Restriktionen getestet werden kann (z.B. fixierte Pfade). Nachteil gegenüber einem SEM mit latenten Variablen ist, dass dieses Vorgehen impliziert oder voraussetzt, dass die Variablen messfehlerfrei sind. Ist diese Annahme falsch (was sie in der Regel ist), sind die geschätzten Effekte abhängig von der Höhe der Messfehler mehr oder weniger stark verzerrt. Eine Ausnahme ist die Einbeziehung von Instrumentalvariablen, mit deren Hilfe auch in Regressions- und Pfadmodellen bei Vorliegen von Messfehlern, unverzerrte Effekte geschätzt werden können. Darauf werden wir später noch ausführlicher eingehen. Ein weiterer Nachteil ist, dass häufig die Konstruktspezifikation leidet – v.a. wenn mehrere gemessene Items zur Verfügung stehen – d.h. es unklar bleibt, ob diese Items reflektive Messungen einer einzigen latenten Variable sind, oder ob sie verschiedene Facetten eines multidimensionalen Konstrukts sind.

Im Fall des hier getesteten Modells gehen wir davon aus, dass wir für jede der drei Modellvariablen mehrere Items haben. Daher bilden wir zunächst **Variablenwerte** für jede Variable, in dem die betreffenden Items einfach gemittelt werden:

```
wfcdata$wfcS<- apply(wfcdata[,c("w01","w02","w03")],1,mean)
wfcdata$azfS<- apply(wfcdata[,c("az01","az02")],1,mean)
wfcdata$kmotS<- apply(wfcdata[,c("km01","km02","km03")],1,mean)
```

Diese drei Zeilen wenden die "`apply()`"-Funktion auf den Datensatz "wfcdata" an – und zwar auf den Teil, der durch die Nennung der Variablennamen determiniert wird. Der Begriff „mean" und die Zahl „1" sagen R, dass es den Mittelwert über die Zeilen der genannten Items berechnen soll. Eine „2" würde bedeuten, dass es den Mittelwert spaltenweise (also für jedes Item) berechnen würde. Das Ergebnis dieser Mittelwertsbildung wird als neue Variable (z.B. wfcS) in den Datensatz geschrieben. Wichtig hierbei ist, dass der Datensatz inkl. Dollar-Symbol (wfcdata$) vor den Variablennamen geschrieben wird, weil ansonsten die Variable als vom Datensatz separiertes eigenes Objekt angelegt wird.

Um Modelle mit lavaan zu spezifizieren, muss zunächst das lavaan-Paket aktiviert werden:

```
library(lavaan)
```

4.8.1 Spezifikation des Modells

Die Spezifikation des in Abb. 19 abgebildeten Modells geschieht mit dem Befehls-Block

```
wfcmodel <- '
  kmotS ~ wfcS + azfS
  azfS ~ wfcS
  '
```

Durch diesen Block wird das Objekt „wfcmodel" generiert, das die postulierte Struktur enthält. Die Variablen in diesem Modell sind diejenigen, die oben durch die apply-Funktion berechnet wurden. Beachtet werden sollte hier v.a. dass die Befehle in zwei Oberstriche eingefasst sind, die den Anfang und das Ende des Blocks kennzeichnen. Wie zu sehen, ist die Syntax sehr simpel und benötigt lediglich eine Nennung der beiden Strukturgleichungen

Kündigungsmotivation = γ_1*work-family conflict* + γ_2*Arbeitszufriedenheit* + ε_1
und
Arbeitszufriedenheit = γ_3*work-family conflict* + ε_2

Wie zu sehen, wird durch die Tilde (~) spezifiziert, dass die jeweilige Variable auf ihre(n) unabhängige(n) Variable(n) regressiert wird.

Charakteristisch für lavaan ist, dass es Parameter gibt, die automatisch geschätzt werden, und die somit in der Syntax nicht explizit aufgeführt werden müssen. Dies sind die Fehlervarianzen Var(ε_1) und Var(ε_2). Hat ein Modell mehrere endogene Variablen, die nicht durch einen Effekt miteinander verbunden sind, werden zusätzlich automatisch deren Fehlerkovarianzen geschätzt. Dies ist auch sinnvoll – es sei denn, man geht davon aus, dass alle gemeinsamen Ursachen von Y1 und Y2 bereits im Modell enthalten sind. Dennoch kann man diese Parameter in der Syntax ansprechen – z.B. wenn man sie fixieren möchte. Varianzen und Kovarianzen spricht man in der Spezifikation mit einer doppelten Tilde (~~) an: Stehen vor dieser Doppeltilde zwei verschiedene Variablen, repräsentiert sie die Kovarianz dieser Variablen; steht die selbe Variable sowohl vor als auch hinter der Doppeltilde, ist es die Varianz dieser Variablen (die Varianz einer Variablen ist die Kovarianz dieser Variablen mit sich selbst).

Will man einen Parameter **fixieren**, muss man den Wert, auf den er fixiert werden soll, gefolgt von einem Sternchen hinter diesen Parameter (und vor die betreffende Variable) schreiben. Tabelle 3 enthält alle Syntax-Symbole zur Spezifikation von Parametern in Pfadmodellen und zwei Beispiele für Fixierungen. Es sei angemerkt, dass lavaan die Schätzung (und damit Ansprache) von Varianzen und Kovarianzen *von exogenen beobachteten Kovariaten* per default nicht vorsieht („fixed-x"-Ansatz). Das führt allerdings dazu, dass ihre Kovarianzen mit latenten Variablen im Modell nicht geschätzt werden, was in den meisten Fällen eine Fehlspezifikation darstellt. Daher sollte man sich zur Gewohnheit machen, die Kovarianzen *aller* exogenen Variablen (egal ob latent oder beobachtet) explizit zu spezifizieren und das Argument `"fixed.x=FALSE"` hinzuzufügen. Dies geschieht, in dem man diejenigen Variablenpaare, die man korrelieren lassen möchte, durch ein „~~" verbindet

Tabelle 3: Syntax zur Spezifikation von Parametern in Pfadmodellen

`~`	Pfad- bzw. Struktureffekt
`~~`	Varianz oder Kovarianz
`Y ~ 1.5*X`	Fixieren des Effektes von X auf Y auf den Wert 1.5
`Y1~~.5*Y2`	Fixieren der Fehlerkovarianz von Y1 und Y2
`Y1~~.7*Y1`	Fixieren der Fehlervarianz von Y1

4.8.2 Rechnen des Modells

Als nächster Schritt werden die Parameter geschätzt. Dies geschieht durch:

```
fit <- sem(wfcmodel,data=wfcdata, fixed.x=FALSE, missing="fiml",
       estimator="mlr")
```

Ähnlich zum ersten Schritt wird wieder ein Objekt mit beliebigem Namen angelegt (hier „fit"), in dem alle Ergebnisse des Schätzprozesses (Parameterschätzungen, Chi-Quadrat-Statistik, Fit-Indizes, standardisierte Residuen etc.) enthalten sind. Wie zu sehen, wird hierbei die Funktion „sem" auf das vormals generierte Objekt „wfcmodel" angewendet. Die Daten zur Schätzung werden dabei aus dem dataframe „wfcdata" gewonnen. Hat man keine Rohdaten, sondern eine Kovarianzmatrix, dann kann diese stattdessen genutzt werden. Hierbei muss das Argument `sample.cov=CM, sample.nobs=#N#` in die obigen Klammer eingefügt werden, wobei CM ein Objekt ist, dass die Kovarianzmatrix enthält und #N# die Größe der Stichprobe.

Schätzungen mittels Maximum Likelihood setzen voraus, dass die Rohdaten **multivariat normalverteilt** sind. Ist dies nicht der Fall, wird der Chi-Quadrat-Wert überhöht und die Standardfehler der Parameter unterschätzt (Finney & DiStefano, 2006). Die Abweichung von der Multi-Normalverteilung ist aber die Regel. Lavaan bietet in diesen Fällen Möglichkeiten zur robusten Schätzung mittels **„Robust maximum likelihood"**. Die beiden Schätzer MLM und MLR sind Erweiterungen der ML-Methode, die sowohl eine Korrektur der Chi-Quadrat-Statistik als auch die Schätzung robuster Standardfehler bewirkt. Dabei führt `e-stimator="mlm"` neben der **Satorra-Bentler-Korrektur** (Satorra & Bentler, 1994) des Chi-Quadrat-Werts zur Schätzung klassischer robuster Standardfehler, während `estimator="mlr"` die **Yuan-Bentler-Korrektur** und Schätzung von nach Huber-White korrigierten Standardfehlern bewirkt. Allerdings können statt des estimator-Arguments die Korrektur und robuste Schätzung auch getrennt voneinander vorgenommen werden (Argumente `test="Satorra-Bentler"` und `"Yuan-Bentler"`, bzw. `se="robust.mlm"` und `"robust.mlr"`). Daneben kann sowohl für die Chi-Quadrat-Statistik als auch für die Standardfehler **bootstrapping** vorgenommen werden. Auf letzteres werden wir bei der Schätzung indirekter Effekte noch einmal zurückkommen (vgl. Abschnitt 4.8.4).

Weiterhin bietet lavaan die Möglichkeit, bei Vorliegen von **fehlenden Werten** (missing data) das Modell mittels **Full information maximum likelihood (FIML)** zu schätzen (Abraham & Russell, 2004; Enders, 2001; Enders & Bandalos, 2001; Graham, 2009). FIML schätzt basierend auf der Annahme der multivariaten Normalverteilung der Variablen die Parameter des Modells, indem es sowohl die Fälle benutzt, die in den Modellvariablen vollständige Werte haben als auch diejenigen *implizierten Werte* der fehlenden Fälle, die sich durch die Werte in den vollständigen Fällen voraussagen lassen (Schlomer, Bauman, & Card, 2010, p. 5). Um das Modell mit FIML zu schätzen, fügt man der o.g. Funktion das Argument `missing="fiml"` hinzu. Liegen fehlende Daten in Kombination mit nicht-normverteilten Daten vor, kann `missing="fiml"` mit `"estimator=mlr"` (Yuan-Bentler-Korrektur mit nach Huber-White geschätzten Standardfehlern) kombiniert werden. Andere Schätzer sowie die Satorra-Bentler-Korrektur sind mit FIML inkompatibel. In der o.g. Syntax wurde daher FIML mit der Yuan-Bentler-Korrektur kombiniert. Tabelle 4 enthält die wichtigsten optionalen Argumente, die in die `sem()` – oder `cfa()`-Funktion zusätzlich eingefügt werden können. Aus Platzgründen werden diese nur aufgelistet. Weitere Schätzer finden sich im lavaan-Manual (siehe. www.lavaan.org).

Tabelle 4: Optionale Argumente der Schätz-Funktion `sem()` oder `cfa()`

`sample.cov=<#>,` `sample.nobs=<#>`	Verwendung einer Kovarianzmatrix als Datengrundlage. Die #'s enthalten den Namen der Matrix und die Nennung der Stichprobengröße

`estimator =`
`"gls"`
`"uls"`
`"wls"`
`"ml"`
`"mlm"`
`"mlr"`
`...`
	Auswahl der Schätzer
	• Generalized least squares (GLS)
	• Unweighted least squares (ULS)
	• Weighted least squares (WLS) / Asymptotically distribution free (ADF)
	• Maximum likelihood (ML; Grundeinstellung)
	• Robust maximum likelihood (Satorra-Bentler Korrektur und robuste Standardfehler, MLM)
	• Robust maximum likelihood mit einer Yuan-Bentler Korrektur der Chi-Quadrate-Statistik und nach Huber-White geschätzten Standardfehlern (MLR)

`missing="fiml"`	Schätzung mittels full information maximum likelihood bei fehlenden Werten

`test =`
`"Satorra-Bentler"`
`"Yuan-Bentler"`
`"boot"`
	Korrektur der Chi-Quadrat-Statistik
	• Satorra-Bentler-Korrektur
	• Yuan-Bentler-Korrcktur
	• Bollen-Stine bootstrap

`se =`
`"robust.mlm"`
`"robust.mlr"`
`"boot"`
	Anforderung robuster Standardfehler (SE)
	• Konventionelle robuste SE (robust.mlm)
	• Huber-White SE (robust.mlr)
	• Bootstrapping der SE (boot)

4.8.3 Anzeige des Modellfits und der Parameterschätzungen

Nachdem das Modell geschätzt wurde, schauen wir uns die Parameterschätzungen und die Fitstatistiken an. Dies geschieht dadurch, dass die Funktion `summary()` auf das oben angelegte Objekt „fit" angewendet wird:

```
summary(fit,standardized=T)
```

Wie zusehen, fordern wir als zusätzliches Argument zunächst lediglich die standardisierten Koeffizienten an. Als Ergebnis erhalten wir folgenden Output:

```
lavaan (0.5-11) converged normally after  18 iterations

  Number of observations                         352

  Number of missing patterns                       4

  Estimator                                       ML      Robust
  Minimum Function Test Statistic              0.000       0.000
  Degrees of freedom                               0           0
  P-value (Chi-square)                         0.000       0.000
  Scaling correction factor                                   NA
    for the Yuan-Bentler correction

Parameter estimates:

  Information                                 Observed
  Standard Errors                   Robust.huber.white

                 Estimate  Std.err  Z-value  P(>|z|)   Std.lv  Std.all
Regressions:
  kmotS ~
    wfcS            0.106    0.048    2.210    0.027    0.106    0.119
    azfS           -0.339    0.055   -6.194    0.000   -0.339   -0.367
  azfS ~
    wfcS           -0.279    0.051   -5.463    0.000   -0.279   -0.289


Intercepts:
    kmotS           0.763    0.101    7.519    0.000    0.763    0.918
    azfS            1.072    0.082   13.142    0.000    1.072    1.191
    wfcS            1.462    0.050   29.331    0.000    1.462    1.570

Variances:
    kmotS           0.570    0.056                      0.570    0.826
```

```
    azfS        0.743    0.061                               0.743    0.917
    wfcS        0.867    0.060                               0.867    1.000
```

Wie zu sehen, wird für dieses Modell kein Chi-Quadrat-Test durchgeführt, da die Freiheitsgrade gleich 0 sind. Es enthält somit keine testbaren Restriktionen. Die darunter enthaltene Tabelle zeigt die **Pfadkoeffizienten** (Estimate), ihre **Standardfehler** (Std.err), und den **z-Wert** – d.h. den Wert der Prüfverteilung unter der Nullhypothese, dass der Pfadkoeffizient gleich 0 ist. Dieser ergibt sich, indem man den Pfadkoeffizient durch seinen Standardfehler teilt. P(>|z|) ist die Auftretenswahrscheinlichkeit des (Stichproben)-Pfadkoeffizienten gegeben der Populationskoeffzient ist gleich 0 – also $p(b \mid \beta = 0)$. Ist diese Wahrscheinlichkeit kleiner .05, ist der Koeffizient signifikant – d.h. überzufällig – von 0 verschieden. Die nächsten beiden Spalten geben zwei Arten **standardisierter Koeffizienten** an und gehen auf den Aufruf des Arguments `standardized=T` zurück. „Std.all" ist hier lediglich relevant und entspricht der vollständigen Standardisierung („completely standardized solution"). Es soll hierbei kurz erwähnt werden, dass standardisierte Koeffizienten keineswegs dazu führen, eine Effektstärke unabhängig von der Metrik und Varianz der beteiligten Variablen zu erhalten und es so erlauben, die Effektstärke von unabhängigen Variablen zu vergleichen (Grace & Bollen, 2005; Greenland, Schlesselman, & Criqui, 1986; Kim & Mueller, 1976; King, 1986) – im Gegenteil. Während der unstandardisierte Regressions- oder Pfadkoeffizient den tatsächlichen kausalen Effekt angibt (vorausgesetzt, das Modell ist korrekt), konfundieren standardisierte Effekte die Effektstärke mit der Varianzen der Variablen. Folglich können zwei unabhängige Variablen mit faktisch demselben kausalen Effekt einen unterschiedlichen standardisierten Effekt bekommen, wenn sie unterschiedlich starke Streuungen haben.

Der Abschnitt „variances" enthält die Varianzen der unabhängigen Variablen und die Fehlerterm-Varianzen der abhängigen Variablen (nicht die Varianzen der abhängigen Variablen, da diese keine geschätzten Parameter des Modells sind). Wie zu sehen werden für die Varianzen kein z- und p-Wert ausgegeben. Man kann diese aber leicht manuell berechnen, indem man (s.o.) den Schätzwert der Varianz durch seinen Standardfehler teilt. Varianzen müssen signifikant sein, weil man nur dann überhaupt von einer Variable sprechen kann. Nichtsignifikante Varianzen (unabhängig davon, ob sie Variablen oder Fehlerterme betreffen) deuten meist auf ein fehlspezifiziertes Modell hin.

Weitere optionale Argumente in der `summary()` – Funktion sind `fit.measures=T`, was dazu führt, dass die **Fit-Indizes** CFI (comparative fit index), TLI (Tucker-Lewis-Index), RMSEA (root mean square error of approximation), AIC (Akaike information criterion) und BIC (Bayesian information criterion) berichtet werden (siehe die kritische Diskussion in Abschnitt 2.4.2). Zusätzlich zum RMSEA erhält man einen p-Wert für die Wahrscheinlichkeit, dass in der Population eine hohe Approximation erreicht würde (RMSEA < .05) und sein Konfidenzintervall. Weitere Argumente sind in Tabelle 5 enthalten. Eine sparsame Ausgabe aller Parameterschätzungen mitsamt der Konfidenzintervalle enthält man durch die Funktion `parameterEstimates()`; Argument dafür ist das im vorherigen Schritt geschätzte Objekt (in unserem Falle „fit").

Tabelle 5: Optionale Argumente der summary-Funktion

`rsquare=T`	Anzeigen des R-Quadrats für alle abhängigen Variablen
`fit.measures=T`	Anzeigen der Fitindizes
`standardized=T`	Anzeigen der standardisierten Koeffizienten
`modindices=T`	Anzeigen der Modifikationsindizes

4.8.4 Test von indirekten und totalen Effekten

Die bisherigen Syntaxbefehle haben ein Modell spezifiziert, das Modell geschätzt und sowohl den Modellfit als auch Parameterschätzungen für die direkten Effekte in dem Modell geliefert. Häufig beziehen sich Hypothesen allerdings auch auf **indirekte Effekte** oder **Mediatoreffekte**. Wie in Abschnitt 1.2 beschrieben, kann z.B. der indirekte Effekt, den work-family conflict auf die Kündigungsmotivation *über Arbeitszufriedenheit* hat, manuell leicht berechnet werden über das Produkt der beiden Teileffekte „work-family conflict →Arbeitszufriedenheit" × „Arbeitszufriedenheit → Kündigungsmotivation". Schwieriger ist es allerdings, die Signifikanz dieses indirekten Effekts zu beurteilen. Dazu benötigt man den Standardfehler dieses indirekten Effekts. Lavaan bietet hierzu alle modernen Möglichkeiten, diesen Standardfehler zu schätzen. Zusätzlich zu dem indirekten Effekt lassen sich sogar der **totale Effekt** (d.h. die Summe aller Einflüsse von work-family conflict auf die Kündigungsmotivation) sowie dessen Standardfehler berechnen. Zwei Möglichkeiten stehen zur Schätzung dieser komplexeren Effektarten und ihrer Standardfehler zur Verfügung: a) die sog. „Delta-Methode" (auch "Sobel-Test" genannt; Sobel, 1982) und b) das bootstrapping. Bei der Delta-Methode wird der indirekte Effekt geteilt durch seinen Standardfehler, der sich aus den direkten Effekten und deren Standardfehlern berechnen lässt:

$$SE_{ab} = \sqrt{a^2 SE_b{}^2 + b^2 SE_a{}^2}$$

Ein Problem der Delta-Methode ist, dass sie annimmt, dass die Stichprobenverteilung von „ab" normalverteilt ist. Dies ist aber nicht der Fall – da die Verteilung eine positive Schiefe und Kurtosis hat. MacKinnon et al. (2007) argumentieren aufgrund eigener Simulationsstudien allerdings, dass in Modellen mit einem indirekten Effekt der Sobel-Test ab N=50 einen nur geringen bias zeigt und in Modellen mit mehreren indirekten Effekten ab einem N=100-200. Da nun aber bootstrapping als weitere Option zur Verfügung steht, sollte diese genutzt werden. Dabei werden aus dem Datensatz Stichproben gezogen, in denen das Modell und der indirekte Effekt geschätzt werden. Anzahl der Ziehungen ist per default 1000. Als Ergebnis erhält man eine normalverteilte Stichprobenverteilung mit einem Standardfehler, der im lavaan-output ausgegeben wird.

Um beide Methoden anzuwenden, muss man lavaan mitteilen, welchen indirekten Effekt (inkl. seines Standardfehlers) man schätzen möchte. Dies funktioniert, indem den beteiligten direkten Effekten Label gegeben werden, auf die anschließend verwiesen werden kann, um den indirekten Effekt zu definieren:

```
wfcmodel <- '
  kmotS ~ c*wfcS + b*azfS
  azfS ~ a*wfcS
  ab := a*b
 '
```

Wie zu sehen, wurden mittels `a*` und `b*` die einzelnen direkten Effekte benannt. Dies sollte nicht verwechselt werden mit der Fixierung von Effekten, die ebenfalls mit solchen Sternchen erfolgt, weil hier Buchstaben verwendet werden. Die Zeile „ab := a*b" sorgt dafür, dass ein neuer Parameter mitsamt Standardfehler geschätzt wird, der als Produkt der beiden Teileffekte a und b definiert wird („:="). Will man zusätzlich den totalen Effekt, definiert man diesen mittels „c + a*b". Wird das Modell jetzt geschätzt, wird per default die Delta-Methode angewendet. Alternativ kann man als weiteres Argument in der Schätz-Funktion `"se=bootstrap"` nennen. Daraufhin werden 1000 bootstrap-Stichproben gezogen. Möchte man die Anzahl der bootstraps verändern kann man das Argument `"boostrap=#"` wählen, bei dem einfach die gewünschte Anzahl eingegeben wird. Beachtet werden muss, dass der bootstrap inkompatibel mit MLM und MLR ist (vgl. Tabelle 4), da diese ja robuste Schätzungen der Standardfehler initiiert (was die Alternative zum bootstrapping ist). Auch setzt MLM vollständige Daten voraus, wodurch FIML damit inkompatibel ist. Eine Lösung besteht darin, FIML zu benutzen, den Chi-Quadrat-Test durch die Yuan-Bentler-Korrektur zu korrigieren (`test="Yuan-Bentler"`) und robuste Standardfehler durch bootstrapping zu erhalten:

```
fit <- sem(wfcmodel,data=wfcdata, fixed.x=FALSE, missing="fiml",
       se="boot", test="Yuan-Bentler")
```

und erhalten durch `summary(fit)`

```
                     Estimate  Std.err  Z-value  P(>|z|)
Regressions:
  kmotS ~
    wfcS                0.108    0.048    2.228    0.026
    azfS       (b)     -0.340    0.055   -6.147    0.000
  azfS ~
    wfcS       (a)     -0.289    0.051   -5.696    0.000

Variances:
    kmotS               0.576    0.058
    azfS                0.750    0.063
    wfcS                0.882    0.063

Defined parameters:
    ab                  0.098    0.023    4.182    0.000
```

Die letzte Zeile gibt uns die Schätzung des indirekten Effekts, seines Standardfehlers und die Teststatistiken. Wie zu sehen, ist der indirekte Effekt signifikant.

4.9 Spezifizieren von konfirmatorischen Faktormodellen

4.9.1 Einführung

Im vorherigen Abschnitt wurden Pfadmodelle behandelt, in denen hypothetisierte Effekte zwischen gemessenen Variablen geschätzt werden. Anstatt die Variablen in einem solchen Modell mit ihren Messungen gleichzusetzen, kann man stattdessen Messmodelle spezifizieren, in denen die Effekte von latenten Variablen ausgehen. Dabei wird die Beziehung zwischen einer latenten Variablen und ihrer Messung ein spezifizierbarer und damit testbarer Teil des Gesamtmodells (anstatt eine Messung mit ihrer latenten Ursache gleichzusetzen).

Konfirmatorische Faktoranalysen (confirmatory factor analyses, CFA) sind eine Art Sonderfall eines vollständigen SEM, bei denen es hauptsächlich um das Spezifizieren und Testen dieser Messmodelle geht, die Effekte der latenten Variablen aber in den Hintergrund treten. In der Regel spezifiziert man daher eine saturierte Kovarianzstruktur unter den latenten Variablen – d.h. es werden alle möglichen Kovarianzen der latenten Variablen geschätzt und es gibt keine Effekte. CFA spielen in der Forschung zwei Rollen: Zum einen werden sie in der Entwicklung von Messinstrumenten benutzt, in denen latente Variablen durch mehrere Indikatoren gemessen werden sollen (Floyd & Widaman, 1995; Gerbing & Anderson, 1988). Zum anderen werden sie in der Regel als eine Art Vortest in einem Forschungsvorhaben durchgeführt, in dem es final um das Testen eines gesamten Strukturgleichungsmodells (s.o.) geht.

Populär ist in diesem Zusammenhang der **„two step approach“** von Anderson und Gerbing (1988) geworden. Dabei wird im ersten Schritt das Messmodell getestet, bei dem alle latenten Variablen miteinander kovariieren dürfen, ohne dass hierbei irgendwelche Restriktionen spezifiziert werden. Dieses Modell testet, ob die postulierte Kausalstruktur, die Effekte der latenten Variablen auf „ihre“ Indikatoren vorsieht, haltbar ist. Im zweiten Schritt werden die Kovarianzen zwischen den latenten Variablen ersetzt durch das meist sparsamere Set von Struktureffekten der latenten Variablen aufeinander. Sowohl der absolute Fit dieses Modells als auch eine signifikante Verschlechterung gegenüber dem Messmodell deutet auf eine Fehlspezifikation des Strukturmodells hin. Dabei kommt dem ersten Schritt eine fundamentale Bedeutung zu, da hier Hinweise erhältlich sind, ob ggf. die Anzahl und Bedeutung der latenten Variablen falsch ist. Es versteht sich von selbst, dass eine falsche Anzahl latenter Variablen oder die Modellierung der falschen (d.h. nicht theoretisch intendierten) latenten Variablen den Sinn des zweiten Schrittes obsolet macht.

Ein Nachteil des CFA-Ansatzes ist, dass er die Vorstellung, dass latente Variablen per se nur mittels mehrerer Indikatoren modelliert werden können, zu einem allgemeinen Paradigma gemacht hat. Innerhalb dieses Paradigmas liegen alternative Modelle kaum noch im Blickwinkel möglicher Respezifikationen (z.B. mit Single-Indicator-Variablen, vgl. Abschnitt 3.5). Anwender/innen, die mit nicht-fittenden Modellen konfrontiert werden, reagieren in der Folge meist hilflos, wenn ihr postuliertes Multi-Indikator-Modell nicht fittet. Wir werden die Implikationen und diagnostischen Möglichkeiten in solchen Situationen in Abschnitt 5.2 etwas vertiefen.

Ein Nachtteil der CFA innerhalb des two step approach ist, dass das Strukturmodell, in dem zu einem gewichtigen Maß die (kausale) Theorie der latenten Variablen und damit ihre Bedeutung modelliert wird, eliminiert wird (Fornell & Yi, 1992a, 1992b; Hayduk & Glaser, 2000). Dadurch werden theoretisch bedeutsame Restriktionen über das Muster von Kausaleffekten durch eine theorielose Kovarianzstruktur, die keinerlei Restriktionen erhält, ersetzt. Dies kann sich auf die Schätzung der Messparameter (d.h. Faktorladungen und Messfehlervarianzen) auswirken, da diese im Verbund mit allen zu schätzenden Parameter (inkl. Kovarianzen versus Struktureffekten der latenten Variablen) geschätzt werden. Enthält das Modell viele theoretisch nicht-fundierte Parameter, können die Schätzungen der Messparameter sich von denen des theoretisch fundierteren Modells unterscheiden, weil alle Parameter bei der Reproduktion der empirischen Kovarianzen eine Rolle spielen (vgl. die Pfadregeln, Abschnitt 2.2). Aus praktischer Sicht ist eine isolierte CFA allerdings in der Regel hilfreich, weil in einem nicht-fittenden vollständigen SEM meist unklar ist, wo das Problem des Modells zu suchen ist.

4.9.2 Spezifikation des Modells

Bei der Spezifikation einer CFA in lavaan kommt ein zentrales neues Symbol gegenüber den reinen Pfadmodellen hinzu („=~“), das die Ladung eines Indikators auf einer latenten Variablen kennzeichnet. So bedeutet beispielsweise die Zeile „`WFC =~ w01+w02+w03`“, dass die latente Variable „WFC“ durch die Indikatoren w01, w02 und w03 gemessen wurde – bzw. diese auf ihr laden sollen. Die Fixierung des jeweils ersten Indikators zur Spezifizierung der Metrik der latenten Variablen wie auch das Schätzen der Messfehlervarianzen muss nicht explizit vorgenommen werden. Dies macht die Spezifikation des CFA-Modells sehr einfach.

In unserem Beispiel sollen nun die Variablenwerte für work-family conflict, Arbeitszufriedenheit und Kündigungsmotivation, die in Abschnitt 4.8 aus den Indikatoren dieser Variablen manuell berechnet wurde, ersetzt werden durch latente Variablen. Dies erfordert die Spezifikation von Messmodellen, in denen diese drei Konstrukte als latente Variablen postuliert werden, die den jeweiligen Indikatoren kausal zugrunde liegen. Abb. 20 zeigt das Pfaddiagramm dieses Modells. Wir werden im Folgenden zunächst ignorieren, dass die Indikatoren mit fünf Antwortkategorien gemessen wurden, was die Annahme verletzt, dass Variablen in einem SEM kontinuierlich sein sollten. Auch wenn eine Verwendung in den meisten Fällen das Ziel der Modelltestung nicht wirklich in Frage stellen dürfte, sei hier bereits angemerkt, dass lavaan mit DWLS (Diagonally weighted least squares) einen Schätzer zur Verfügung stellt, der die ordinale Natur von Indikatoren mit Antwortkategorien explizit berücksichtigt (siehe ausführlich Abschnitt 4.11).

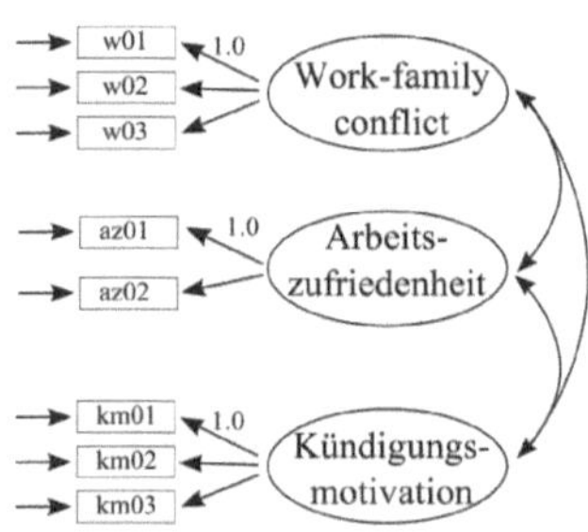

Abb. 20: Pfaddiagramm der konfirmatorischen Faktoranalyse

Die Syntax für dieses Modell ist

```
wfcCFA <- '
  WFC =~ w01+w02+w03
  AZF =~ az01 + az02
  KMOT =~ km01+km02+km03
  '
```

Wie zu sehen, muss lediglich bestimmt werden, welche Indikatoren auf welchen latenten Variablen laden sollen. Es müssen weder die Fixierungen der „marker-Indikatoren" (d.h. derjenigen Indikatoren, deren Ladung auf 1 fixiert werden soll), noch Varianzen oder Kovarianzen der latenten Variablen oder Messfehler der Indikatoren adressiert werden. Veränderungen dieser Basis-Syntax sind allerdings dann nötig, wenn man zusätzliche Restriktionen testen möche. Diese können sich z.B. auf die angenommene Gleichheit von Parametern (**Gleichheitsrestriktionen**, equality constraints), oder spezifische Werte von Parametern (**Fixierungen**) beziehen. Beide Formen von Restriktionen werden im nächsten Abschnitt behandelt. Tabelle 6 fasst alle möglichen Syntaxbefehle zur Spezifikation von konfirmatorischen Faktorenanalysen zusammen.

Tabelle 6: Syntax zur Spezifikation von Parametern in konfirmatorischen Faktoranalysen

`WFC =~ w01`	Ladung des Indikators w01 auf der latenten Variable WFC
`WFC ~~.8*WFC`	Die Varianz der latenten Variablen WFC wird auf den Wert .80 fixiert
`WFC ~~ 1.2*AZF`	Die Kovarianz zwischen den latenten Variablen WFC und AZF wird auf 1.2 fixiert
`w01~~.10*w01`	Die Messfehlervarianz des Indikators w01 wird auf .10 fixiert.
`w01~~az01`	Die Kovarianz der Messfehler von w01 und az01 wird frei geschätzt (diese ist als Grundeinstellung auf 0 fixiert). Ähnlich zu den Fixierungen in der vorherigen Zeile kann diese auch mit * auf einen bestimmten Wert fixiert werden.
`WFC =~ w01+a*w02+a*w03`	Die Ladungen von w02 und w03 werden mit einem Label „a" versehen und so mittels einer Gleichheitsrestriktion gleichgesetzt. Sollten diese Ladungen in der Population ungleich sein, führt dies zu einem misfit des Modells.

4.9.3 Rechnen des Modells

Ähnlich zur Schätzung des Pfadmodells im vorherigen Abschnitt, wird das Schätzen des CFA-Modells mittels der cfa()-Funktion vorgenommen.

```
fitCFA <- cfa(wfcCFA, data=wfcdata, missing="fiml",
         estimator="mlr")
```

So muss man sowohl das im ersten Schritt erzeugte Objekt nennen, in dem die Struktur enthalten ist (wfcCFA) als auch den dataframe, in dem die Rohdaten enthalten sind. Alle weiteren optionalen Argumente sind identisch zu denen, wie sie im Rahmen des Pfadmodells diskutiert wurden (vgl. Tabelle 4, S. 67), d.h. Benutzung der **Kovarianzmatrix** der Indikatoren als Grundlage des Modells (`sample.cov=` ; `sample.nobs=`), Wahl unterschiedlicher **Schätzer** durch `estimator=#` ("gls", "uls", "mlm", "wls", "mlr") oder von **full information maximum likelihood** (`missing=FIML`), und **robuster Schätzung der Standardfehler** `("robust.mlm","robust.mlr", "boot")`.

4.9.4 Anzeige der Parameterschätzungen und des Modellfits

Die Syntax zur Anzeige der Parameterschätzungen und des Modellfits ist identisch zur der im Beispiel des Pfadmodells. Diesmal fordern wir aber die Anzeige der Fitindizes an:

```
summary(fitCFA, standardized=T, fit.measures=T)
```

Da der output dieses Modells etwas umfangreicher ist, werden einzelne Abschnitte durch # und eine laufende Nummer indiziert und anschließend erläutert.

```
lavaan (0.5-12) converged normally after  37 iterations

Number of observations                          352

Number of missing patterns                        8

Estimator                                        ML      Robust #1
Minimum Function Test Statistic              39.980      36.349
Degrees of freedom                               17          17
P-value (Chi-square)                          0.001       0.004
Scaling correction factor                                 1.100
    for the Yuan-Bentler correction

Model test baseline model:                                      #2

Minimum Function Test Statistic            1235.588     962.707
Degrees of freedom                               28          28
P-value                                       0.000       0.000

Full model versus baseline model:
Comparative Fit Index (CFI)                   0.981       0.979
Tucker-Lewis Index (TLI)                      0.969       0.966

...

Number of free parameters                        27          27 #3
Akaike (AIC)                               6801.156    6801.156 #4
Bayesian (BIC)                             6905.474    6905.474
Sample-size adjusted Bayesian (BIC)        6819.819    6819.819

Root Mean Square Error of Approximation:

RMSEA                                        0.062    0.057     #5
```

```
90 Percent Confidence Interval          0.037  0.087   0.032 0.081
P-value RMSEA <= 0.05                          0.195   0.294

Standardized Root Mean Square Residual:

SRMR                                           0.040       0.040

...                                                              #6

           Estimate  Std.err  Z-value  P(>|z|)   Std.lv  Std.all
Latent variables:
  WFC =~
   w01      1.000                                 0.858    0.838
   w02      1.005    0.071   14.137    0.000     0.862    0.798
   w03      0.977    0.069   14.076    0.000     0.838    0.770
  AZF =~
   az01     1.000                                 0.814    0.827
   az02     1.083    0.108   10.036    0.000     0.881    0.964
  KMOT =~
   km01     1.000                                 0.784    0.783
   km02     1.069    0.129    8.313    0.000     0.838    0.802
   km03     0.746    0.107    6.963    0.000     0.585    0.631

Covariances:
  WFC ~~
   AZF     -0.228    0.055   -4.111    0.000    -0.326   -0.326
   KMOT     0.191    0.055    3.496    0.000     0.283    0.283
  AZF ~~
   KMOT    -0.326    0.071   -4.615    0.000    -0.510   -0.510

...

Variances:
   w01      0.311    0.047                        0.311    0.297
   w02      0.425    0.060                        0.425    0.364
   w03      0.482    0.056                        0.482    0.407
   az01     0.307    0.067                        0.307    0.317
   az02     0.060    0.068                        0.060    0.071
   km01     0.389    0.075                        0.389    0.387
   km02     0.391    0.092                        0.391    0.357
   km03     0.517    0.077                        0.517    0.602
   WFC      0.736    0.080                        1.000    1.000
   AZF      0.662    0.095                        1.000    1.000
   KMOT     0.615    0.108                        1.000    1.000
```

(1) Zunächst wird der Schätzer (ML), der Chi-Quadrat-Wert (39.980) und die Freiheitsgrade des Modells (17) ausgegeben. Daneben ist die robuste Form der Statistik zu finden, deren Berechnung durch `estimator="mlr"` veranlasst wurde. Der p-Wert des Chi-Quadrat-Wertes zeigt, dass es einen signifikanten misfit des Modells gibt. Wie in Abschnitt 2.4 erläutert, bedeutet dies, dass die implizite Kovarianzmatrix der Indikatoren dieser Modellstruktur von der tatsächlich vorhandenen in einem Ausmaß abweicht, das nicht durch Zufall erklärbar wäre. Im aktuellen Fall ist die Wahrscheinlichkeit für so einen Fall $p < .001$.

(2) In diesem Abschnitt des Outputs wird ein weiterer Chi-Quadratwert mitsamt Freiheitsgraden ausgegeben – und zwar der des sog. „**Nullmodells**“. Das Nullmodell ist dasjenige Modell mit der größtmöglichen Anzahl von Freiheitsgraden und dem schlechtesten fit. Dieses Modell hat als einzige geschätzte Parameter die Varianzen der Indikatoren. Ansonsten impliziert dieses Modell, dass alle Beziehungen zwischen den Indikatoren gleich Null sind. Benötigt wird das Nullmodell zur Berechnung der beiden darunter abgebildeten Fitindizes **CFI** und **TLI**. Diese kennzeichnen die prozentuale Verbesserung des sog. „target-Modells“ (also des theoretisch postulierten) mit eben dem schlechtest-fittenden Modell. Nach allgemeinen Standards (Hu & Bentler, 1999) wird ein Modell mit einem CFA und TLI nahe .95 als „akzeptabel“ bezeichnet. Es soll aber noch einmal darauf hingewiesen werden, dass im Fall eines signifikanten Chi-Quadrat-Tests von einer Fehlspezifikation ausgegangen werden muss, deren Bedeutsamkeit untersucht werden sollte.

(3) Hier ist die Anzahl der in dem Modell geschätzten Parameter genannt.

(4) In diesem Abschnitt werden die informationstheoretischen Fitindizes **Akaike Information Criterion (AIC)** und **Bayesian Information Criterion (BIC)** genannt. Sie sind bei der Beurteilung eines einzelnen Modells unerheblich. Beide werden gewöhnlich beim Vergleich alternativer Modellstrukturen herangezogen. Das Problem bei dem Vergleich zweier nach dem Chi-Quadrat-Test nicht-fittenden Modelle ist, dass aus dem Unterschied des Ausmaßes des misfits (d.h. dem Vergleich der AIC-Werte) nicht auf den Unterschied in der Fehlspezifikation der Modelle geschlossen werden kann. D.h. ein Modell mit einem niedrigeren AIC kann dennoch problematischer sein, als ein Alternativmodell mit einem höheren, weil es fundamentaler fehlspezifiziert ist, aber die geschätzten Parameter dennoch eine Annäherung von S und $\Sigma(\theta)$ ermöglichen. Auch hier sollte besser konstatiert werden, *dass* beide Mo-

delle in irgendeiner Form fehlspezifiziert sind und nach Möglichkeiten einer Verbesserung gesucht werden.

(5) Hier findet sich der **root mean square error of approximation (RMSEA)**, sein Konfidenzintervall als auch die Wahrscheinlichkeit, mit der das vorliegende geschätzte Modell aus einer Population stammt, in der ein RMSEA von kleiner/gleich .05 vorliegt. In unserem Fall ist der RMSEA gemessen an allgemeinen Standards (Hu & Bentler, 1999) akzeptabel, nach denen ein RMSEA $\leq$ .06 als Beleg für einen ausreichenden Modelfit gesehen werden kann. Unter dem RMSEA wird der **standardized root mean square residual (SRMR)** ausgegeben. Ähnlich zu den anderen o.g. Fitindizes existieren hier Empfehlungen, Modelle mit einem SRMR $\leq$ .08 zu akzeptieren.

(6) In diesem Abschnitt werden analog zum Output des Pfadmodells die Parameterschätzungen, ihre Standardfehler und Teststatistiken präsentiert. Den Anfang machen die Messmodelle mit den Faktorladungen; anschließend kommen die Kovarianzen der latenten Variablen und die Varianzen der Messfehler und latenten Variablen.

Insgesamt zeigt das Modell einen signifikanten misfit, während die Fitindizes für eine Akzeptanz des Modells sprechen würden. Eine genauere Diskussion der diagnostischen Hinweise für eine mögliche Verbesserung des Modells wird zusammenfassend im nächsten Abschnitt präsentiert.

4.10 Spezifizieren von Strukturgleichungsmodellen mit latenten Variablen

In diesem Abschnitt sollen nun die Pfadanalyse und das Faktormodell zusammengeführt werden. Dies führt zu einem vollständigen Strukturgleichungsmodell, in dem die Konstrukte durch latente Variablen und nicht durch Summenindizes der Indikatoren (wie im Pfadmodell) repräsentiert werden. Dabei sind die Messmodelle der latenten Variablen Teil des gesamten Modells. Der Unterschied zur konfirmatorischen Faktorenanalyse ist, dass hierbei die im Faktormodell vorliegende vollständige Kovarianzstruktur der latenten Variablen (d.h. jede latente Variable kovariiert mit jeder anderen) ersetzt über das theoretische Strukturmodell mit spezifischen geschätzten und fixierten Effekten (vgl. Abschnitt 2.3).

Dies sollte aber wieder nicht so verstanden werden, dass Faktormodelle notwendiger Bestandteil eines vollständigen SEM sein müssen. Stattdessen ist ein Faktormodell ein mögliches Modell, dass die Beziehung zwischen latenter Variable und mehreren Indikatoren beschreibt. Das in Abschnitt 3.5 diskutierte Single-Indicator-Modell ist ein weiteres, bei dem nur ein Indikator zur Messung der latenten Variable zur Verfügung steht. Das zentrale Merkmal des SEM ist daher, dass eine Gesamtstruktur von kausalen Beziehungen modelliert wird, in die sowohl hypothetisierte latente Variablen als auch beobachtete Variablen eingewoben sind. Abb. 21 zeigt das Pfaddiagramm unseres Modells. Wie zu sehen, kovariieren die drei latenten Variablen nicht frei – stattdessen wurden die theoretisch erwarteten Struktureffekte spezifiziert, wie sie bereits im Pfadmodell spezifiziert worden waren.

Mithilfe der Pfadregeln (vgl. Abschnitt 2.2) lassen sich nun die impliziten Korrelationen dieses Modells ableiten. So postuliert das Modell beispielsweise, dass die Korrelation zwischen w01 und km01 durch die kausalen Prozesse entstanden sind, in die die latenten Variablen work-family conflict, Arbeitszufriedenheit und Kündigungsmotivation und ihre Effekte aufeinander sowie auf diese beiden Indikatoren involviert sind (speziell das Produkt von Ladungen und Effekten der latenten Variablen). Dies zeigt, dass die Beziehung zwischen einem Indikator und „seiner" latenten Variable nie isoliert, sondern immer nur im Netzwerk *aller Indikatoren* betrachtet werden muss (eben, weil er mit ihnen über die latente Variable und deren Effekte verbunden ist), womit auch eine geringe Anzahl von Indikatoren eine enorme Bedeutung im Testen dieser kausalen Netzwerke zukommt.

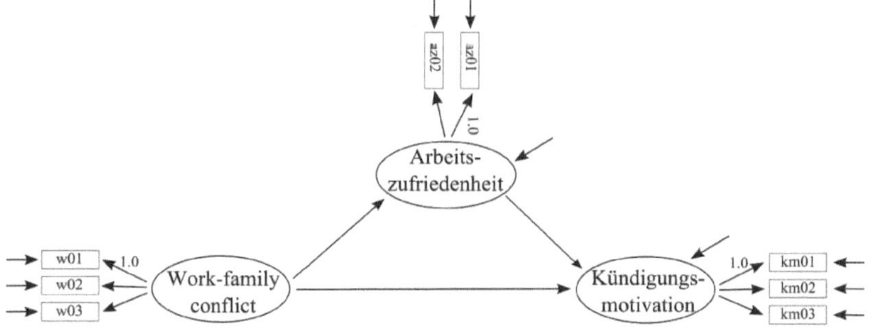

Abb. 21: Pfaddiagramm eines vollständigen SEM

4.10.1 Spezifikation des Modells

Bei der Spezifikation des vollständigen latenten Modells werden einfach alle Elemente wie sie in den Abschnitten 4.8.1 und 4.9.2 behandelt wurden, integriert. Die Syntax für das in Abb. 21 abgebildete Modell ist demnach:

```
wfcSEM <- '
  #Messmodell
  WFC =~ w01+w02+w03
  AZF =~ az01 + az02
  KMOT =~ km01+km02+km03
  #Strukturmodell
  KMOT ~ WFC + AZF
  AZF ~ WFC
  '
```

Zur besseren Strukturierung wurden in der Syntax Abschnitte benannt. Die Verwendung der Raute (#) führt dazu, dass R die betreffende Zeile ignoriert.

4.10.2 Rechnen des Modells und Anzeige der Ausgabe

Bei der der Schätzung des Modells nutzen wir wieder FIML und die robuste Schätzung der Chi-quadrat-Statistik:

```
fitSEM <- sem(wfcSEM, data=wfcdata,
                  missing="fiml",estimator="mlr")

summary(fitSEM, standardized=T, fit.measures=T)
```

Der Output des vollständigen SEM ist nahezu identisch mit der konfirmatorischen Faktoranalyse in Abschnitt 4.9.4 – mit der Ausnahme, dass die Ausgabe der freigeschätzten Kovarianzen zwischen allen drei latenten Variablen ersetzt wurde mit einer theoretisch angenommenen Struktur von kausalen Effekten. Dies entspricht dem Output des Pfadmodells in Abschnitt 4.8.3. Der betreffende Abschnitt im Output ist nun wie folgt.

```
           Estimate  Std.err  Z-value  P(>|z|)  Std.lv   Std.all
Regressions:
  KMOT ~
    WFC      0.119    0.066    1.818    0.069    0.131    0.131
    AZF     -0.451    0.089   -5.051    0.000   -0.468   -0.468
  AZF ~
    WFC     -0.310    0.066   -4.694    0.000   -0.326   -0.326
```

Vergleicht man diese Koeffizienten mit denen des Pfadmodells in Abschnitt 4.8.3, fällt auf, dass sie durchweg höher sind, was auf die Messerfehlerbefreiung durch die Modellierung latenter Variablen zurückgeführt werden kann. Der Modellfit ($\chi^2(17) = 36.35$, $p = .004$, RMSEA = .057, CFI = .98) ist identisch zu dem der CFA (Abschnitt 4.9.4), was dadurch bedingt ist, dass die latente Struktur in beiden Fällen **saturiert** ist (d.h. alle möglichen latenten Koeffizienten werden geschätzt) – nur dass bei der CFA alle Kovarianzen geschätzt werden, und im Strukturmodell alle Effekte. D.h. auf der latenten Ebene besitzt dieses Modell keinerlei kausale Restriktionen und die einzigen getesteten Restriktionen sind diejenigen, die die Messmodelle implizieren. Nach wie vor weist der der Chi-Quadrat-Test auf ein oder mehrere Probleme hin – während aufgrund der Fit-Indizes dieses Modell akzeptiert werden würde. Bevor wir uns im nächsten Kapitel detaillierter mit Formen von Fehlspezifikationen, deren Konsequenzen und möglichen diagnostischen Möglichkeiten, diese zu erkennen, beschäftigen, wird der folgende Abschnitt kurz auf die Verwendung kategorialer Items als Variablen in einem Modell eingehen.

4.11 Ordinale Indikatoren

Bislang wurden als Schätzer der SEM das maximum likelihood (ML) Verfahren oder deren robuste Versionen (MLM, MLR) benutzt. Diese liefern Statistiken, die asymptotisch Chi-Quadrat-verteilt sind und asymptotische unverzerrte Parameterschätzungen. Allerdings setzt ML ebenfalls voraus, dass die Daten kontinuierlich sind. Allerdings basieren Modelle mit latenten Variablen meist auf Indikatoren, die mit rating-Skalen gemessen wurden – und damit **Ordinal-Niveau** haben. Daher ist diese Annahme grundsätzlich verletzt. In der Folge werden die Chi-Quadrat-Werte etwas nach oben verzerrt, wodurch der Modelltest strenger ausfällt, als er nach traditionellen Richtlinien ausfallen würde (d.h. das Fehlerrisiko von 5% wird erhöht). Lavaan bietet nun mit Diagonally Weighted Least Squares (DWLS) einen Schätzer an, der die ordinale Natur der Indikatoren berücksichtigt (Flora & Curran, 2004). Sind die Indikatoren allerdings zusätzlich nicht-normal verteilt, kann eine Schätzung mit robuster maximum likelihood – Schätzung (`estimator="mlr"`) sinnvoller sein (Rhemtulla, Brosseau-Liard, & Savalei, 2012), woraus unten ausführlicher erläutert wird.

DWLS nutzt die **polychorische Korrelationsmatrix** der Indikatoren, bei der deren kontinuierlichen Korrelationen aufgrund der ordinalen Indikatoren geschätzt werden. Daneben nutzt DWLS im Schätzprozess als Gewichtungsmatrix lediglich die Diagonale der **asymptotischen Kovarianzmatrix** der Varianzen und Kovarianzen der Variablen (deshalb „diagonally weighted“). Asymptotische Varianzen sind Schätzungen der Streuung der Parameterschätzungen, wie sie bei wiederholter Stichprobenziehung auftreten würden. Beispielsweise hat die Variable X1 eine Varianz. Würde man jetzt eine sehr große Anzahl von weiteren

Stichproben (=asymptotisch) ziehen, hätte diese Varianz selbst eine Varianz. Genauso verhält es sich mit jeder anderer Parameterschätzung.

Bei der Modellierung eines ordinalen Indikators x_i unterstellt DWLS, dass diesem Indikator eine **kontinuierliche latente response-Variable** $*x_i$ zugrunde liegt. Zum Beispiel haben Befragte eine kontinuierliche Zustimmungstendenz ($*x_i$) zu der Aussage „ich bin zufrieden mit meiner Arbeit" (x_i), die von völliger Ablehnung bis zur völligen Zustimmung führt. Das Modell nimmt weiter an, dass $*x_i$ normalverteilt ist und dass es bestimmte Schwellen (thresholds) gibt, ab denen, wenn sie überschritten werden, die nächst höhere Antwortkategorie gewählt wird. Das geschätzte Modell enthält daher die Schätzung dieser Schwellen, während die geschätzte Ladung den Effekt der latenten Variable von Interesse auf die latente response-Variable darstellt (zur Schätzung siehe Yang-Wallentin, Jöreskog, & Luo, 2010).

Um den DWLS-Schätzer in lavaan zu benutzen, muss in der Schätz-Funktion lediglich mit dem Argument `ordered=c(...)`, eine Liste der ordinalen Indikatoren geliefert werden. Nachteil ist, dass in diesem Fall weder FIML benutzt werden kann, noch die zur Diagnostik wichtigen standardisierten Residuen ausgegeben werden. Die Syntax für das im vorherigen Abschnitt getestete Modell wäre damit

```
fitSEM <- sem(wfcSEM, data=wfcdata,
  ordered=c("w01","w02","w03","az01","az02,"km01","km02","km03"))
```

Wie oben angemerkt, wird die Situation durch eine nicht-normale Verteilung der Indikatoren verkompliziert. So nimmt DWLS zumindest an, dass die latenten response-Variablen normalverteilt sind. Ist dies nicht der Fall (was oft der Fall sein dürfte), wird unter Umständen die Nutzung der korrigierten Versionen des ML-Schätzers (`estimator= "mlr"` und `"mlm"`) günstiger sein, v.a., wenn die Indikatoren eine Kategorien-Anzahl über vier haben. So zeigte eine Simulationsstudie von Rhemtulla et al. (2012), dass sowohl Fit-Statistiken als auch Kovarianzen zwischen den latenten Variablen durch robust maximum likelihood besser geschätzt wurden als mit ULS (Unweighted least squares), die ähnlich zu DWLS zur Schätzung bei kategorialen Indikatoren verwendet werden kann (Forero, Maydeu-Olivares, & Gallardo-Pujol, 2009; Yang-Wallentin et al., 2010). Dies war v.a. der Fall, wenn zur Nicht-Normalverteilung auch noch eine kleine Stichprobengröße hinzukam. Ob das auch auf DWLS übertragen werden kann, konnte die Studie allerdings nicht beantworten.

DWLS ist allerdings auf jeden Fall angebracht, wenn die abhängige Variable kategorial, z.B. binär ist. In dem Fall sind die Ergebnisse analog zu denen einer probit Regression.

5

Diagnostik im Fall fehlspezifizierter Modelle

Nachdem wir uns jetzt mit den praktischen Aspekten bei der Spezifizierung von Pfadmodellen, konfirmatorischen Faktoranalysen und Strukturmodellen mit latenten Variablen beschäftigt haben, soll das Augenmerk auf die **Diagnostik** und **Respezifikation** und damit mögliche Verbesserung von Modellen gerichtet werden. Wie der Output sowohl der konfirmatorischen Faktorenanalyse als auch des latenten Strukturmodells zeigt, würde ein/e Forscher/in aufgrund der Fitindizes (RMSEA = .057, CFI=.98, und SRMR = .04) schließen, dass der Modellfit akzeptabel ist. Im Gegensatz dazu legt der Chi-Quadrat-Test ($\chi^2(17) = 36.35$, $p = .004$) nahe, dass das Modell in irgendeiner Weise fehlspezifiziert ist und die entsprechende Abweichung zwischen modell-impliziter und empirischer Kovarianz nicht durch den Stichprobenfehler erklärt werden kann. Abgesehen von diesen allgemeinen Evaluationskriterien, sollte in *jedem* Fall (d.h. auch wenn das Modell fittet) untersucht werden, ob es Signale für eine mögliche Fehlspezifikation gibt. Das beginnt mit dem Überprüfen von Parameterschätzungen, die auf ihre Konsistenz mit theoretischen Erwägungen beurteilt werden sollten und reicht bis zur Inspektion der Rohdaten, der Korrelationen und der sogenannten standardisierten Residuen. Um zu Hypothesen zu gelangen, braucht es allerdings ein Grundverständnis darüber, wie sich ein Misfit sowohl in den impliziten Kovarianzen als auch in der Verzerrung von Parametern niederschlägt.

5.1 Formen der Fehlspezifikation und ihre Konsequenzen

Die Diagnostik eines nicht-fittenden Modells kommt einer Detektivarbeit gleich. Nötige Kompetenzen beziehen sich v.a. auf die Kenntnis der Pfadregeln und d-separation-Implikationen, wie sie in Kapitel 2 ausführlich behandelt wurden. Immerhin versucht der ML-Algorithmus, mithilfe der freigeschätzten Parameter, die empirischen Kovarianzen zu reproduzieren. Scheitert der Fit, gibt es mehrere Erklärungsmöglichkeiten: So können beispielsweise die Annahmen, die das Modell an die Daten stellt, verletzt sein oder aber die spezifizierte Kausalstruktur enthält Fehler. Letzteres kann bedeuten, dass Parameter fälschlicherweise auf Null fixiert wurden oder aber die Struktur an sich ist problematisch, was in der Regel bei fehlspezifizierten Messmodellen der Fall ist. Im Folgenden werden wir Formen der Fehlspezifikation und ihre Implikationen für die Schätzung der Parameter kennenlernen. Dies soll nur exemplarisch erfolgen und dem/der Leser/in ein Gefühl zu vermitteln, wie der ML-Algorithmus im Falle einer Fehlspezifikation durch Verzerrung von Modellparametern versucht, dennoch eine Passung der S und $\Sigma(\theta)$ zu erreichen und trotzdem scheitert. Aus der Inspektion der Abweichungen zwischen beiden Matrizen lassen sich im Idealfall Spekulationen anstellen, wo das Modell falsch sein könnte.

5.1.1 Fälschlicherweise ausgelassene Parameter

Der erste Fall liegt vor, wenn im Modell Parameter fehlen (d.h. auf 0 fixiert sind), die zur Vorhersage der empirischen Kovarianzen notwendig sind (s. Abb. 22). Beispielsweise fehlt ein direkter Effekt (β_2) in einem als vollständigem Mediatormodell konzipierten Modell (links) oder eine Doppelladung (λ_{31}) in einem Faktormodell (rechts). In diesem Fall werden abhängig von der Richtung der „wahren" Effekte die Kovarianzen über- oder unterschätzt. Folge ist eine Diskrepanz zwischen S und $\Sigma(\theta)$, die sich in einem misfit äußert.

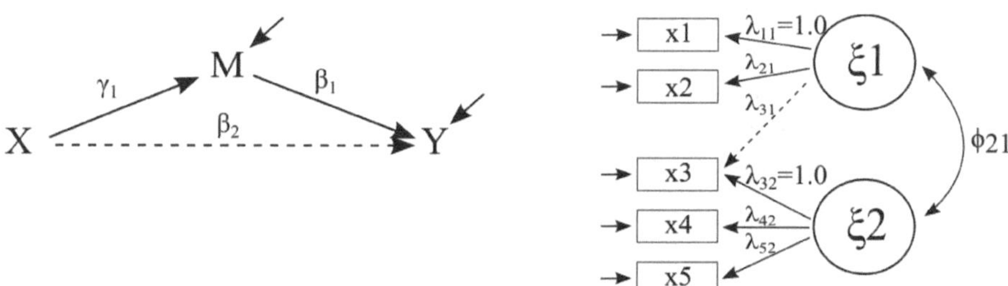

Abb. 22: Fehlspezifikation durch ausgelassene Effekte

So ergibt sich in dem genannten Mediatormodell die Kovarianz zwischen X und Y aus $Cov(X,Y) = Var(X) \times \gamma_1 \times \beta_1 + Var(X) \times \beta_2$. Fehlt der Effekt β_2 im getesteten Modell, ist die implizite oder vorhergesagte Kovarianz entweder geringer als die empirische – dies ist der Fall wenn alle indirekten und direkten Effekt positiv sind – oder höher, wenn sie unterschiedlicher Richtung sind.

Jenseits dieses misfits werden aber die geschätzten Koeffizienten γ_1 und β_1 verzerrt, weil der ML-Algorithmus versucht, durch Über- oder Unterschätzen dieser Koeffizienten S und $\Sigma(\theta)$ doch noch anzunähern. Sind also γ_1, β_1 und β_2 positiv, würden entweder γ_1 oder β_1 (oder beide) überschätzt werden, um noch die empirische Kovarianz zwischen X und Y zu „erreichen".

Ein zweites Beispiel für eine Fehlspezifikation durch ausgelassene Effekte ist das Auslassen von Doppelladungen in Faktormodellen wie es in Abb. 22 (rechts) illustriert ist. Doppelladungen können dann auftreten, wenn ein Indikator zwar ein valider Indikator der latenten Variable von Interesse ist, aber etwas in seiner Formulierung beinhaltet, was von einer anderen latenten Variable im Modell beeinflusst wird. Nehmen wir z.B. den schon häufig in diesem Buch verwendeten Indikator w01 („Meine beruflichen Anforderungen behindern mein Privat und Familienleben"), der latenten Work-Family-Conflict-Variablen. Stellen wir uns nun vor, dass die Item-Formulierung noch stärker auf die Stress-Verursachung von work-family conflict abzielen würde (z.B. „mein beruflicher Stress behindert mein Privat- und Familienleben"). Werden die Work-Family-Conflict-Indikatoren nun in ein Modell eingebettet, in dem es eine latente Stress-Variable gibt, ist es sehr wahrscheinlich, dass w01 ebenfalls auf der Stress-Variablen lädt. Entgegen häufiger Auffassungen ist das auch nicht schlimm und verringert seine Nutzbarkeit als Indikator von work-family conflict nicht. Allerdings sind solche Doppelbeeinflussungen von Indikatoren kein gutes Signal für die Einschätzung der Güte dieses Items als Messinstrument, da es ungünstig ist, wenn ein Messinstrument zwei Dinge gleichzeitig misst (z.B. was bedeutet ein „trifft überhaupt nicht zu" als Antwort auf dieses Item?). Für das Modell wird dies aber nur dann zum Problem, wenn diese Doppelladung fälschlicherweise ausgelassen wird. Was passiert nun in diesem Fall? Sehen wir uns dazu wieder das in Abb. 22 rechts abgebildete Modell an. Hier lädt x3 sowohl auf $\xi 1$ als auch auf $\xi 2$.

Zunächst wird die implizite Kovarianz zwischen x1 und x3 und zwischen x2 und x3 geringer sein, als die empirische, weil der Pfad $\lambda_{11}\phi_{21}\lambda_{32}$ (bzw. $\lambda_{21}\phi_{21}\lambda_{32}$) nicht ausreichen wird, um die empirische Kovarianz zu reproduzieren. Der Algorithmus versucht, das zu kompensieren, indem er die Kovarianz der Faktoren (ϕ_{21}) *über*schätzt. Die Alternative wäre, die Ladungen zu überschätzen, doch dies geht größtenteils nicht, da sowohl λ_{11} als auch λ_{32} auf 1 fixiert sind. Die Überschätzung von ϕ_{21} hat jedoch zur Konsequenz, das die Kovarianzen zwischen beiden Indikatoren von X1 und x4 und x5 *über*schätzt werden (z.B. ergibt sich

Cov(x1, x4) aus $\lambda_{11}\phi_{21}\lambda_{42}$). Dies kompensiert der Algorithmus nun durch *Unterschätzung* der Ladungen von x4 und x5.

Dieses Beispiel soll zum einen zeigen, wie umfassend Konsequenzen der Fehlspezifikation eines einzigen Parameters sein können und dass diese häufig nicht trivial sind und eine strenge Testung und Diagnostik erfordern (vgl. Abschnitt 2.4). Zum anderen zeigt das Beispiel, wie die Pfadregeln helfen können, diese Konsequenzen zu verstehen. Dies ist in der Tat eine komplexe Angelegenheit, aber mit etwas Übung kann man lernen, bei einem fehlspezifizierten Modell solche Fehler zu entdecken. Insbesondere in Kombination mit einer erneuten Inspektion der Item-Formulierungen können solche Probleme aufdeckbar sein. So werden mit etwas Erfahrung problematische Indikatoren, die eine Doppelladung wahrscheinlich machen, auffallen (wobei es sicher vorteilhaft ist, diese vor dem ersten Modelltest zu identifizieren).

Ein drittes Beispiel für fälschlicherweise ausgelassene Parameter sind unzulässig fixierte **Fehlerkovarianzen** von abhängigen Variablen. Dies kommt im Wesentlichen in zwei Fällen vor (s. Abb. 23): a) die beiden abhängigen Variablen werden von einer oder mehreren ausgeschlossenen gemeinsamen Ursachen beeinflusst (linker Teil der Abbildung), b) die kausale Richtung in einem hinteren Teil des Modells wird falsch spezifiziert.

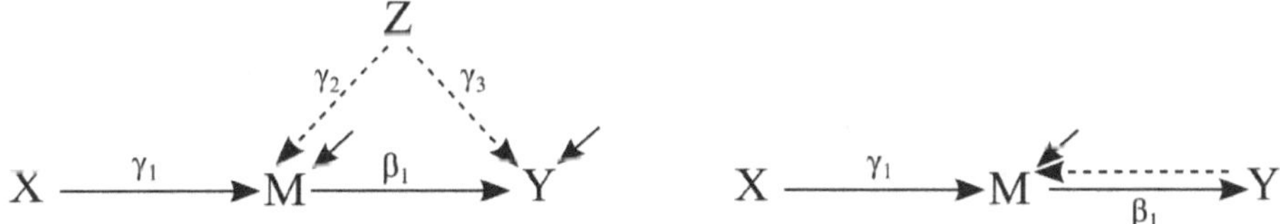

Abb. 23: Fehlspezifikationen, die zu Fehlerkovarianzen führen

Fall „a" kann jedes Modell mit mehr als einer abhängigen Variablen betreffen. Abb. 23 (links) zeigt ein Mediatormodell als Beispiel. Werden in einem solchen Modell M und Y von im Modell nicht berücksichtigten, weiteren gemeinsamen Ursachen Z beeinflusst (wobei Z eine oder mehrere Variablen sein können), wird der Effekt β_1 verzerrt (der Effekt γ_1 bleibt davon unberührt): Nach den Pfadregeln ergibt sich die Korrelation zwischen M und Y aus $\beta_1 + \gamma_2\gamma_3$. Durch Auslassen von Z wird $\gamma_2\gamma_3$ daher zu β_1 addiert (womit je nach Vorzeichen aller Effekte β_1 über- oder unterschätzt wird). Der Effekt von X auf M bleibt deshalb davon unberührt, weil Z und X keinerlei Kovarianz aufweisen.

Solch eine Struktur führt dazu, dass das Fixieren der Fehlerkovarianz von M und Y u.U. zu einem misfit führt, weil die Kovarianz zwischen X und Y nicht ausreichend reproduziert werden kann. Dies erscheint zunächst unplausibel; die d-separation-Regeln können aber helfen, dies zu verstehen: Im Modell ist M ein collider zwischen X und Z. Durch Konstanthalten von M (was die Spezifizierung im Modell ja impliziert) wird der Pfad zwischen X und Z geöffnet. Da Z einen Effekt auf Y hat, korreliert so X und Y – auch über die Einbeziehung von M hinaus. Hier neigt man häufig vorschnell dazu, diese verbleibende Beziehung als Grund für einen direkten Effekt von X zu interpretieren und diesen freizusetzen. Dies führt dann aber dazu, dass man sowohl einen nicht-vorhandenen Effekt schätzt, als auch, dass der Effekt von M auf Y verzerrt wird. Dieses Problem kann gelöst werden, indem die Fehlerkovarianz zwischen M und Y freigesetzt wird – vorausgesetzt, X hat einen *substantiellen* Effekt auf M, ist unabhängig von Z und hat *keinen* direkten Effekt auf Y (letzteres ist dadurch bedingt, dass solch ein Modell nicht identifiziert wäre). Diese Bedingungen können aber durch Hinzufügen einer weiteren unabhängigen Variablen erfüllt werden. Dies wird im nächsten Kapitel ausführlicher behandelt.

Ein weiteres Beispiel dieses Falls wäre eine Common-Cause-Struktur, in der eine oder mehrere unabhängige Variablen Effekte auf zwei abhängige Variablen haben. Werden diese beiden Variablen zusätzlich von ausgeschlossenen gemeinsamen Ursachen beeinflusst, resultiert dies in einem misfit des Modells, weil die Kovarianz zwischen beiden abhängigen Variablen nicht ausreichend durch die im Modell befindlichen Variablen erklärt werden kann. Auch hier sollten Fehlerkovarianzen spezifiziert werden – es sei denn, man hat die Annahme, dass sich alle gemeinsamen Ursachen der beiden abhängigen Variablen im Modell befinden.

Fehler in der kausalen Richtung (Fall „b“) liegen vor, wenn ein Effekt von M auf Y hypothetisiert wird, aber in Wirklichkeit Y einen Effekt auf M hat („**reverse causation**“) oder beides korrekt ist („**Simultanität**“). Der rechte Teil von Abb. 23 zeigt diesen Fall. Testet man nun ein simples Mediatormodell, wird β_1 verzerrt. Unter Umständen ist $\beta_1=0$ – d.h. M hat überhaupt keinen Einfluss auf Y und man erhält einen Scheineffekt. Wenn man das Modell testet, fittet es nicht, aus denselben Gründen, die auch für das in Abb. 23 links illustrierte Modell angeführt wurden: Wieder ist M ein collider und seine Konstanthaltung erzeugt eine artifizielle bedingte Abhängigkeit zwischen X und Y, die nicht durch den indirekten Effekt von X adressiert werden kann. Auch hier kann dieses Problem durch eine Schätzung der Fehlerkovarianz behoben werden: Das Modell fittet und der wahre Effekt von β_1 wird ausgegeben. Hier gelten dieselben Bedingungen wie im vorherigen Fall: X muss substantiell mit M kovariieren und keine Beziehung mit Y aufweisen.

Da Fehler in der kausalen Richtung nur eine Form des allgemeineren Problems der **Endogenität** sind (wie wir ausführlich in Kapitel 7 sehen werden) und dieses über eine Schätzung der Fehlerkovarianz behoben werden kann (sofern die geäußerten Annahmen über die Rolle von X valide sind), ist eine Schätzung der Fehlerkovarianz, wenn möglich, zu empfehlen (Antonakis, Bendahan, Jacquart, & Lalive, 2010).

5.1.2 Fehler in der Struktur

Diese Form der Fehlspezifikation ist fundamentaler als die bislang diskutierten, da hier die Struktur insgesamt fehlspezifiziert ist. Dies betrifft in erster Linie Messmodelle der latenten Variablen, bei denen die Anzahl der unterstellten latenten Variablen nicht korrekt ist. Nehmen wir ein einfaches Beispiel. Abb. 24 zeigt links ein Zweifaktorenmodell. Die Implikationen dieses Modells sind, dass sowohl x1 hoch mit x2 korrelieren sollte sowie x3 mit x4. Die Höhe der „Kreuzkorrelationen" (z.B. zwischen x1 und x3) hängt von der Kovarianz der Faktoren ϕ_{21} ab, welche die beiden jeweiligen Items verbindet. Der rechte Teil der Abbildung zeigt nun das Modell des/der Forscher/in. Statt eine differenzierte Faktorstruktur zu modellieren, wird angenommen, dass ein einziger Faktor hinter den vier Indikatoren steckt.

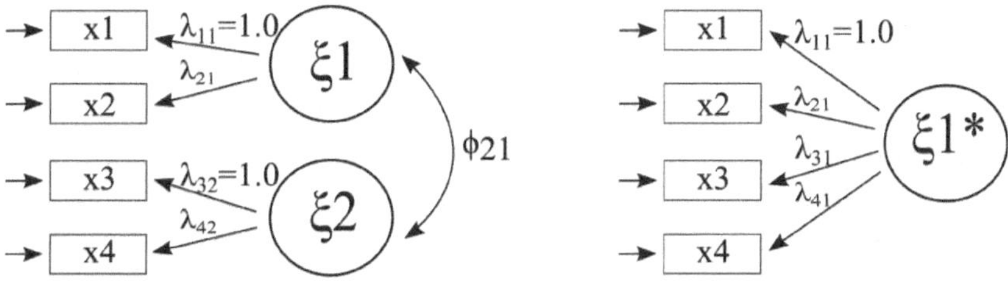

Abb. 24: Fehlspezifikation von Faktormodellen

Wie leicht zu erkennen, ist dies eine fundamentale Form der Fehlspezifikation, weil die latente Variable (deshalb die Kennzeichnung mit dem „*") schlicht nicht existiert. Ist das Messmodell Teil eines umfangreicheren Modells, in dem diese latente Variable mit anderen verknüpft ist, bedeutet dies folglich, dass auch alle Struktureffekte und andere Formen der Verknüpfung keinen Sinn ergeben.

Es sollte daher das wichtigste Ziel sein, bei der Modellierung von latenten Variablen mittels multiplen Indikatoren ein fittendes Modell zu erreichen. Wie bereits vorher diskutiert, hat die Verwendung multipler Indikatoren Vorteile – allerdings sollte der Wunsch danach nicht dazu führen, dass man Hinweise gegen das Modell ignoriert, oder einen misfit aus pragmatischen Gründen hinnimmt. Dies kann im Extremfall bedeuten, dass die Anzahl der Indikatoren radikal reduziert werden muss, denn kein Nachteil wiegt so schwer, wie eine falsche Modellierung einer latenten Variablen.

Was ist nun der Grund für den misfit solcher Modelle und wie versucht der Algorithmus, dennoch eine Annäherung von S und $\Sigma(\theta)$ zu erreichen? Faktormodelle mit mehreren Faktoren implizieren Cluster von inter-korrelierenden Items. D.h., dass in dem Modell im linken Teil der Abbildung x1 und x2 höher miteinander korrelieren als mit x3 und x4 (dasselbe gilt für die Korrelationen zwischen x3 und x4 und mit x1 und x2). Dies nivelliert sich zwar mit steigender Inter-Faktor-Kovarianz (Cov(ξ1,ξ2) bzw. ϕ_{21}), erreicht aber nie die Kovarianzstruktur, die aus einem Ein-Faktor-Modell folgen würde. Grund ist, dass beide Mengen von Indikatoren eine gemeinsame Ursache haben (ξ1 bzw. ξ2), die mit der jeweils anderen Ursache zwar korreliert ist, aber nicht identisch ist. Aus Sicht der Pfadregeln ergibt sich die Kovarianz zwischen x1 und x2 aus λ_{11}Var(ξ1)λ_{21} während die Kovarianz zwischen x1 und x3 aus λ_{21}Cov(ξ1,ξ2)λ_{32} folgt. Da die Kovarianz zweier Variablen (hier Cov(ξ1,ξ2)) niemals höher sein kann, als die höchste Varianz (also entweder Var(ξ1) oder Var(ξ2)), müssen folglich die Indikatoren einer latenten Variablen miteinander höher korrelieren als mit denen einer anderen latenten Variable. Ist beispielsweise Var(ξ1) die höhere der beiden Varianzen, sind x1 und x2 höher miteinander korreliert als mit x3 oder x4. Diese Muster von Korrelationen kann das Ein-Faktor-Modell auf der rechten Seite der Abbildung nicht adressieren. Hier werden alle Indikatoren durch dieselbe gemeinsame Ursache beeinflusst, woraus ein eher homogenes Muster an Kovarianzen folgt.

In dem hier illustrierten Fall fälschlicherweise eine einzige gemeinsame Ursache zu modellieren, würde dazu führen, dass die Kovarianz zwischen x1 und x2 sowie zwischen x3 und x4 *unterschätzt* würde – diese kovariieren in Wirklichkeit höher aufgrund ihrer Beeinflussung durch einen eigenen gemeinsamen Faktor. Jetzt wäre es prinzipiell für den Algorithmus möglich, diese Spezifität der Kovarianzen durch Manipulation der Ladungshöhen zu adressieren. Dies ist aus folgenden Gründen nur zum Teil möglich:

- Die Ladung von x1 ist auf 1.0 fixiert und somit nicht veränderbar
- Die Ladung von x2 zu erhöhen, würde Cov(x1,x2) zwar erhöhen – dies würde aber auch eine höhere Kovarianz mit x3 und x4 bedeuten, da die Ladung von x2 (λ_{21}) auch in den Gleichungen zur Bestimmung dieser Kovarianzen vorkommt (z.B. ist Cov(x2,x3) = λ_{21}Var(ξ1$*$)λ_{31}). Da diese aber durch die Modellierung einer gemeinsamen Ursache sowieso schon erhöht sind, würde damit der fit noch verschlechtert.
- Im Gegensatz zu x1 und x2 sind die Ladungen von x3 und x4 frei manipulierbar. Hier muss der Algorithmus einen Kompromiss finden zwischen dem Annähern der impliziten Kovarianz zwischen x3 und x4 an die empirische und dem ebenfalls aus der Schätzung dieser beiden Ladungen resultierenden Kovarianzen von x3 und x4 mit x1 und x2. Folge ist, dass nichts von beidem optimal erreicht werden kann: Sowie Cov(x1,x2) als auch Cov(x3,x4) bleiben zu niedrig während die Kovarianzen der „über-Kreuz-liegenden" Indikatoren (z.B. Cov(x1,x3) zu hoch bleiben.

Über den daraus resultierenden misfit des Modells und die Diagnose dieser Über- und Unterschätzungen sowohl der Kovarianzen als auch Parameter kann man in der Praxis Hypothesen über alternative Strukturen entwickeln. Hilfreich dazu sind neben deskriptiven Statistiken (wie den Korrelationen) und eine erneute Inspektion der Indikator-Formulierungen zwei Informationen, die fast jedes SEM-Programm liefert: Die standardisierten Residuen und (ferner) die Modifikationsindizes. Bevor diese erläutert werden, soll auf einen nicht selten auftretenden Umstand hingewiesen werden:

Es kann in der Praxis durchaus vorkommen, dass eine gewisse Menge an Indikatoren den Ein-Faktor-Test erfolgreich übersteht (d.h. das Modell fittet), das Modell aber in dem Moment, in dem eine oder mehrere weitere Variablen hinzukommen, plötzlich doch scheitert. Grund ist, dass durch das Hinzufügen dieser neuen Variable(n) neue Restriktionen impliziert sind, die entlarven, dass das Ein-Faktor-Modell doch eine Fehlspezifikation ist. So folgen im Wesentlichen weitere „proportionality constraints" aus dem erweiterten Modell. Zum Beispiel impliziert dieses Modell, dass die Kovarianz zwischen der hinzugefügten Variable „V" und jedem der *k* Indikatoren der latenten Variable ξ1 aus λ_{k1}Cov(ξ1, V)β_{11} ergibt, wobei β_{11} der Effekt von ξ1 auf V ist. Daraus folgt, dass diese Kovarianzen proportional zu den jeweiligen λ_{k1} sein müssen (da Cov(ξ1, V)β_{11} in allen Gleichungen enthalten ist).

Dieser Umstand hat zwei Implikationen: Erstens zeigt dies, dass Modelltests allgemein kontextabhängig sind und nie final verifiziert sind. Dies ist primär bedeutend für „validierte Skalen", die in einer Faktorenanalyse getestet wurden und nie davor gefeit sind, in zukünftigen Studien bzgl. Dimensionalität widerlegt zu werden, wenn sich mit dem Kontext die zusätzlich einbezogenen Variablen ändern. Zweitens zeigt dies, dass jede einzelne Variable Implikationen für die Testbarkeit des Modells und bestimmter Teile (z.B. der Messmodelle) hat. Dabei impliziert der Test immer höhere Anforderungen, je restriktiver und sparsamer das Modell ist. Beide Punkte – d.h. die strengen und somit häufig unrealistischen Annahmen solcher Modelle und die ausreichende Testbarkeit von Modellen mit weniger Indikatoren werden häufig unterschätzt.

5.2 Modifikationsindizes und standardisierte Residuen

Wie im vorherigen Abschnitt ausführlich diskutiert, impliziert ein misfit, dass das Modell von Interesse nicht in der Lage ist, die empirischen Kovarianzen zu reproduzieren. Dies impliziert eine Abweichung zwischen S und $\Sigma(\theta)$, die über diejenige hinausgeht, die man aufgrund des Stichprobenfehlers erwarten würde. Wie jedes andere SEM-Programm, bietet auch lavaan nun mit den Modifikationsindizes und den standardisierten Residuen Informationen, um einen Einblick in mögliche lokale Probleme des Modells zu gewinnen.

Die **Modifikationsindizes** geben für jeden fixierten Parameter des Modells einen Chi-Quadrat-Test mit einem Freiheitsgrad an, der widerspiegelt, ob die Fixierung dieses Parameters überzufällig zum misfit beiträgt. Dies gilt für die auf null fixierten Parameter genauso wie für Parameter, die auf einen anderen Wert fixiert wurden. Ebenso werden Modifikationsindizes für jeden Parameter ausgegeben, der mit einem anderen gleichgesetzt wurde, wie dies z.B. bei Gruppenvergleichen üblich ist (vgl. Abschnitt 6.1.2).

Modifikationsindizes sollten allerdings für übliche SEM-Anwendungen eher vorsichtig (wenn überhaupt) verwendet werden: Da sie nur fixierte Parameter adressieren, können sie nur Fehlspezifikationen aufdecken, die in fälschlicherweise ausgelassenen Parametern bestehen, wie sie in Abschnitt 5.1.1 diskutiert wurden (also z.B. fälschlicherweise fixierte Effekte, Ladungen, und Fehlerkovarianzen). Dies führt nur dann zu einer Verbesserung des Modells, wenn die bestehende Struktur an sich korrekt ist. Ist dies nicht der Fall, wird durch das freie Schätzen eines Parameters mit einem hohen Modifikationsindex der fit zwischen S und $\Sigma(\theta)$ zwar erhöht, das Modell bleibt aber u.U. fundamental fehlspezifiziert oder wird sogar verschlechtert (MacCallum, 1986; MacCallum, Roznowski, & Necowitz, 1992). Das „Befolgen" von Modifikationsindizes wird umso problematischer, je atheoretisch hierbei fixierte Parameter mit hohen Modifikationsindizes freigesetzt werden. Auf der anderen Seite muss sich ein Anwender, der die Freisetzung eines vormals fixierten Parameters als theoretisch plausibel einschätzt, fragen, warum er diesen Parameter dann nicht von vorne herein geschätzt hat.

Zusammen mit den Modifikationsindizes werden für jeden fixierten Parameter EPC-Werte („**expected parameter change**") ausgegeben, die darüber informieren, wie hoch der Parameter wäre, wenn er freigesetzt würde. Saris, Satorra, und van der Veld (2009) schlagen in diesem Zusammenhang vor, die statistische Power zur Aufdeckung eines fälschlicherweise fixierten Parameters zu berücksichtigen, um gerade bei Modellen mit einem großen N nicht irrelevante Parameter, für die zwar ein hoher Modifikationsindex, aber niedriger EPC angegeben ist, nachträglich zu schätzen. Die Nutzung dieser Strategie ändert allerdings nichts daran, dass a) sie nicht atheoretisch passieren sollte und b) nach wie vor voraussetzt, dass die Fehlspezifikation lediglich in ausgelassenen/fixierten Effekten besteht.

Die **standardisierten Residuen** resultieren aus der Differenz von $S - \Sigma(\theta)$. Sie werden als eine Matrix ausgegeben, deren Zellen jeweils die Differenz der betreffenden Zelle in S und jener in $\Sigma(\theta)$ sind. Somit geben sie für jede spezifische Varianz oder Kovarianz an, ob und wie gut diese durch das Modell adressiert bzw. reproduziert werden konnte. Die Residuen werden standardisiert, um bei Vorliegen von Modellvariablen mit unterschiedlichen Metriken dennoch eine Vergleichbarkeit zu gewährleisten. Dabei stellen die Einträge z-Werte dar; Werte $> |2|$ implizieren dabei signifikante Abweichungen von der H0, dass diese (zellspezifische) Abweichung nur durch Zufall entstanden ist und in der Population diese Differenz null beträgt.

Die standardisierten Residuen ermöglichen – in Kombination mit anderen Informationen (z.B. deskriptiven Statistiken, Item-Formulierungen, d-separation-Regeln und Pfadregeln) eine theoretisch-geleitete Re-Spezifikation des Modells. Entgegen den Modifikationsindizes implizieren sie keine direkten Vorschläge (z.B. dem Freisetzen von fixierten Parametern), sondern benötigen Hintergrundwissen, ein Loslösen von der bisherigen Theorie und etwas detektivisches Geschick, um zu einer Verbesserung des Modells zu gelangen. Dies bedeutet allerdings auch, dass man bei der Verbesserung des Modells scheitern kann.

Das Anfordern und Interpretieren beider diagnostischer Informationen wird nun mittels eines **simulierten Beispiels** illustriert. Wir greifen dabei auf das 2-Faktorenmodell zurück, das im vorherigen Abschnitt diskutiert wurde (s. Abb. 24). Für dieses Modell wurde nun eine Kovarianzmatrix simuliert, die wir mit S benennen. Diese kann mit der `matrix()`-Funktion nachgebildet werden:

```
S <- matrix(c(3.14, 1.85,1.24,1.07,
              1.85,2.70,1.18,0.96,
              1.24,1.18,3.24,2.05,
              1.07,0.96,2.05,2.68),4,4)
```

Die `matrix()` – Funktion schreibt alle Zahlen in der inneren Klammer spaltenweise in eine Matrix, deren Dimensionen am Ende der Funktion definiert wird (hier ist es eine 4x4-Matrix). Die Spalten und Zeilen benennen wir mit

```
rownames(S) <- c("x1","x2","x3","x4")
colnames(S) <- c("x1","x2","x3","x4")

S
      x1   x2   x3   x4
   x1 3.14 1.85 1.24 1.07
   x2 1.85 2.70 1.18 0.96
   x3 1.24 1.18 3.24 2.05
   x4 1.07 0.96 2.05 2.68
```

Diese Matrix wird nun die Grundlage für den Test eines (falschen) Ein-Faktor-Modells sein. Bereits die Inspektion dieser Matrix würde erste Zweifel an der Ein-Faktoren-Struktur aufkommen lassen, da die Kovarianzen zwischen x4 und sowohl x1 als auch x2 merklich von den anderen abweicht. Dies muss nicht unbedingt auf einen Fehler hinweisen, kann aber ein Indiz sein. Berechnen wir nun das Modell. Da die Daten simuliert und somit normalverteilt sind und auch keine fehlenden Werte vorliegen, benötigen wir diesmal weder FIML noch eine Korrektur (`mlm` oder `mlr`). Als Datengrundlage dienen nun aber nicht die Rohdaten, sondern die Kovarianzmatrix:

```
OneFactor <- '
  F1 =~ x1+x2+x3+x4
'

fit <- cfa(OneFactor, sample.cov=S, sample.nobs=300)
summary(fit, modindices=T, standardized=T)
```

Wie zu sehen, fordern wir nun mit `modindices=T` die Modifikationsindies an. Der Output zeigt einen signifikanten Chi-Quadratwert, der auf die Fehlspezifikation hinweist:

```
  Estimator                                          ML
  Minimum Function Test Statistic                95.450
  Degrees of freedom                                  2
  P-value (Chi-square)                            0.000
[...]

           Estimate  Std.err  Z-value  P(>|z|)   Std.lv  Std.all
Latent variables:
  F1 =~
    x1    1.000     0.939    0.531
    x2    0.928     0.135    6.863    0.000     0.871    0.531
    x3    1.593     0.188    8.473    0.000     1.496    0.832
    x4    1.386     0.164    8.471    0.000     1.302    0.796

[...]

Modification Indices:

   lhs op rhs     mi    epc sepc.lv sepc.all sepc.nox
1   F1 =~  x1     NA     NA      NA       NA       NA
2   F1 =~  x2  0.000  0.000   0.000    0.000    0.000
3   F1 =~  x3  0.000  0.000   0.000    0.000    0.000
4   F1 =~  x4  0.000  0.000   0.000    0.000    0.000
5   x1 ~~  x1  0.000  0.000   0.000    0.000    0.000
6   x1 ~~  x2 90.085  1.271   1.271    0.438    0.438
7   x1 ~~  x3 15.226 -0.669  -0.669   -0.211   -0.211
8   x1 ~~  x4 10.094 -0.474  -0.474   -0.164   -0.164
9   x2 ~~  x2  0.000  0.000   0.000    0.000    0.000
10  x2 ~~  x3 10.093 -0.506  -0.506   -0.172   -0.172
11  x2 ~~  x4 15.226 -0.540  -0.540   -0.202   -0.202
12  x3 ~~  x3  0.000  0.000   0.000    0.000    0.000
13  x3 ~~  x4 90.085  3.025   3.025    1.030    1.030
14  x4 ~~  x4  0.000  0.000   0.000    0.000    0.000
15  F1 ~~  F1  0.000  0.000   0.000    0.000    0.000
```

Im unteren Teil des Outputs werden nun die Modifikationsindices (`mi`), expected parameter change-Werte (`epc`) und verschiedene Arten standardisierter expected parameter change-Werte aufgelistet.

Wie zu sehen, haben die ersten fünf Zeilen einen Modifikationsindex von null oder sind nicht berechenbar, da diese im Modell frei geschätzt werden oder die Fixierung des Markers betreffen. Obwohl nun die Fehlspezifikation in der Gesamt-Struktur des Modells begründet liegt, werden substantielle Modifikationsindizes sowohl für die Messfehlerkovarianz zwischen x1 und x2 als auch zwischen x3 und x4 ausgeben – und zwar in identischer Höhe. Diese Modifikationsindizes adressieren die im vorherigen Abschnitt diskutierte Unterschätzung der empirischen Kovarianz zwischen den beiden jeweiligen Indikatoren. Nimmt man die Modifikationsindizes nun zum Anlass, eine der beiden **Messfehlerkovarianzen** freizusetzen, resultiert ein fittendes – aber immer noch fehlspezifiziertes Modell daraus. Grund ist, dass an der problematischen Ein-Faktoren-Struktur nichts geändert wurde, aber die Messfehlerkovarianz die verbleibende Kovarianz zwischen den jeweiligen Indikatoren adressiert. Dieses Beispiel zeigt zum einen, dass Fehlspezifikationen von Faktormodellen sich häufig in Modifikationsindizes ausdrücken, die Messfehlerkovarianzen betreffen und zum anderen, wie ein all zu automatisches Befolgen der Modifikationsindizes zwar den Modellfit verbessert, aber u.U. das Modell fehlspezifiziert lässt. Dies sollte jetzt nicht zur Schlussfolgerung führen, dass es unrealistisch ist, dass das angenommene Faktormodell zwar korrekt ist, es aber zusätzlich eine sinnvolle Messfehlerkovarianz gibt (z.B. weil derselbe Begriff in zwei Indikatoren vorkommt). Allerdings ist das Freisetzen von Messfehlerkovarianzen als post-hoc-Strategie zur „Behebung des misfits“ meist problematisch (MacCallum, 1986; MacCallum et al., 1992). Dazu kommt, dass selbst bei dem Freisetzen einer validen Messfehlerkovarianz unklar ist, was diese Messfehlerkovarianz inhaltlich bedeutet. Wenn der Grund wie o.g. das Auftreten desselben Begriffs in beiden Indikatoren ist, stellt sich dann ja doch die Frage, ob dies nicht gleich eine eigene latente Variable ist, die besser explizit als Teil des Modells modelliert werden sollte.

Die die **standardisieren Residuen** fordern wir an mit

```
resid(fit,"standardized")

$cov
   x1     x2     x3     x4
x1  0.000
x2  7.115  0.000
x3 -8.578 -5.406  0.000
x4 -4.467 -6.158  2.587  0.000
```

Bei der Interpretation der Vorzeichen sollte man sich immer die Reihenfolge der Matrizen vor Augen halten (d.h. S – Σ(θ)), die voneinander subtrahiert werden. Ist ein Residuum positiv, bedeutet dies folglich, dass die betreffende empirische Kovarianz höher ist, als die modell-implizite. Ist es negativ, zeigt es, dass die empirische Kovarianz niedriger ist. Die standardisierten Residuen spiegeln in unserem Fall exakt das **Muster an Über- und Unterschätzungen** wieder, wie sie im vorherigen Abschnitt detailliert diskutiert wurden: Die Kovarianzen der Indikatoren, die dieselbe latente Variable reflektieren werden *unter*schätzt (S - Σ(θ) ist positiv, also ist die empirische Kovarianz höher). So ist beispielsweise die Kovarianz zwischen x1 und x2 höher als die durch das Modell implizierte. Bei der Kovarianz zwischen x3 und x4 ist dies nicht so drastisch, was durch die Überschätzung Ladungen (standardisiert .832 und .796) erreicht wird. Die Kovarianzen zwischen Indikatoren, die verschiedene latente Variablen reflektieren, werden *über*schätzt. So ist beispielsweise die Kovarianz zwischen x1 und x3 nicht so hoch, wie durch das Modell, dass ja beide als Folgen einer gemeinsamen Ursache vorsieht), impliziert.

Wie zu sehen, zeigen die standardisieren Residuen ein komplexeres und umfassenderes Bild von den spezifischen Problemen des Modells, die empirischen Kovarianzen zu reproduzieren. Sie können allerdings nur Anhaltspunkt und Basis für eine erneute Konzeptualisierung des Modells sein. In unserem Beispiel wären sie die Basis für ein erneutes Betrachten der Item-Formulierungen und Erwägen einer u.U. doch differenzierteren Faktorstruktur.

5.3 Illustration am Beispiel des Work-Family-Conflict-Modells

Wenden wir nun das bisher Erörterte auf das Work-family-conflict-Modell an. Dies hatte einen signifikanten Chi-Quadrat-Wert, wenngleich akzeptable Fit-Indizes. Die folgende Beschreibung soll daher sowohl illustrieren, wie in der Praxis die Kombination von standardisierten Residuen, Korrelationen und Item-Formulierungen funktionieren kann, als auch einen Beleg dafür liefern, dass ein signifikanter Chi-Quadrat-Test trotz passabler Fit-Indizes ein Hinweis für eine fundamentale Fehlspezifikation sein kann. Schauen wir uns die relevanten Informationsquellen an. Die **standardisierten Residuen** bekommen wir wieder mit

```
resid(fitSEM, "standardized")  #Anm. fitSEM war das SEM-Objekt
                                 #(s. Abschnitt 4.10.2)
```

```
        w01    w02    w03    az01   az02   km01   km02   km03
w01   0.153
w02  -0.697  0.402
w03  -0.027     NA  0.014
az01 -2.916 -0.571  0.632 -0.225
az02 -1.695  0.566  1.004     NA     NA
km01  5.973  1.801 -0.071     NA     NA     NA
km02 -1.075 -0.710 -3.560  1.395  1.070 -1.338     NA
km03  0.123  0.278 -0.273  2.785  3.493 -0.908     NA     NA
```

Wie im vorherigen Abschnitt diskutiert, sind sowohl hohe Residuen an sich als auch Muster innerhalb einer Menge von Indikatoren häufig ein Zeichen dafür, dass die Restriktionen, die den entsprechenden Kovarianzen auferlegt sind, keine Reproduktion der empirischen Kovarianzen ermöglichen. Dies betrifft sowohl die Kovarianzen zwischen Indikatoren ein und derselben latenten Variable als auch Kovarianzen zwischen Indikatoren verschiedener latenter Variablen. So fällt hier bei km01 auf, dass dessen Kovarianz mit w01 deutlich unterschätzt wird, was am positiven Residuum von 5.973 sichtbar ist. Im Gegensatz dazu wird die Kovarianz zwischen km02 und w03 unterschätzt. Beide sind Indikatoren der jeweils selben latenten Variablen (d.h. work-family conflict und Kündigungsmotivation) – diese Muster an Abweichungen könnten ein Zeichen sein, dass eine von ihnen „Probleme“ hat. Gegen work-family conflict als Ursache spricht, dass die Residuen der Indikatoren untereinander unauffällig sind und auch die Kovarianzen zwischen den work-family-conflict-Items und denen von Arbeitszufriedenheit größtenteils keine hohen Werte zeigen (mit Ausnahme der -2.916).

Leider können für km01 nicht alle Residuen beurteilt werden, da lavaan hier „NA“ ausgibt. Dies kann passieren, wenn die Schätzung der Varianz des jeweiligen Residuums (durch deren Wurzel das Residuum zur Berechnung des Z-Wertes geteilt wird) negativ ist (Hausman, 1978). Allerdings fällt auf, dass die Kovarianz zwischen km03 und beiden Arbeitszufriedenheits-Items unterschätzt wird, während dies bei km02 nicht der Fall ist. Damit bleibt festzuhalten, dass die Kündigungsindikatoren als Messungen eines einzigen Faktors zu modellieren, ein Problem darstellen könnte. Als weitere Informationsquelle schauen wir uns dazu die Korrelationsmatrix an:

```
attach(wfcdata)
round(  cor(cbind(w01,w02,w03,az01,az02,km01,km02,km03),
            use="complete")  ,2)
detach(wfcdata)
```

Hierzu fügen wir den Datensatz durch `attach(wfcdata)` dem Suchpfad in R zu. Dadurch entfällt das Benennen des Datensatzes in der nachfolgenden Zeile (vor jedem Item). Den `cor()`-Befehl verschachteln wir in dem `round()`-Befehl, um die Dezimalstellen innerhalb der Korrelationsmatrix auf 2 zu beschränken. Zur besseren Sichtbarkeit dieser Verschachtelung wurden hier Leerzeichen eingefügt, die aber nicht nötig sind. Anschließend wird der Datensatz wieder aus dem Suchpfad entfernt.

```
      w01   w02   w03  az01  az02  km01  km02  km03
w01   1.00  0.67  0.65 -0.29 -0.28  0.27  0.15  0.15
w02   0.67  1.00  0.63 -0.24 -0.24  0.24  0.15  0.16
w03   0.65  0.63  1.00 -0.19 -0.21  0.17  0.06  0.13
az01 -0.29 -0.24 -0.19  1.00  0.80 -0.40 -0.29 -0.16
az02 -0.28 -0.24 -0.21  0.80  1.00 -0.48 -0.38 -0.21
km01  0.27  0.24  0.17 -0.40 -0.48  1.00  0.61  0.46
km02  0.15  0.15  0.06 -0.29 -0.38  0.61  1.00  0.56
km03  0.15  0.16  0.13 -0.16 -0.21  0.46  0.56  1.00
```

Um die Matrix inhaltlich zu interpretieren, sind die Formulierungen der Indikatoren sehr hilfreich – hier noch einmal die Formulierungen der drei Kündigungsmotivations-Indikatoren: *„Wie häufig kommt Ihnen der Gedanke, zu kündigen?“* (km01), *„Wie häufig haben Sie sich in letzter Zeit nach einem anderen Arbeitsplatz erkundigt (z.B. Stellenanzeigen gelesen, Bekannte gefragt etc.)“* (km02) und *„Wie wahrscheinlich ist es, dass Sie tatsächlich innerhalb des nächsten Jahres kündigen werden?“*

Bei der Betrachtung der Korrelationsmuster fällt auf, dass die drei Kündigungsmotivations-Indikatoren einen auffallenden Abfall der Korrelationen mit allen anderen Modellvariablen zeigen. So ein Abfall erinnert an eine Mediationsstruktur, bei der die Korrelation mit der ersten Variable (dem Mediator) höher ist, als die mit nachfolgenden Variablen. Als Kontrast dazu zeigen die beiden Arbeitszufriedenheits-Indikatoren (die fünfte und sechste Zeile) ein nahezu identisches Muster an Zusammenhängen mit allen anderen Variablen.

Zieht man jetzt noch die deutlich voneinander abweichenden Bedeutungen der drei Indikatoren hinzu (Gedanken daran, zu kündigen, Stellenanzeigen lesen und Kündigungsintention), ist eine plausible Alternative zum Ein-Faktoren-Modell, die drei Kündigungsmotivations-Indikatoren als Indikatoren völlig verschiedener empirischer Phänomene zu sehen – eben „Kündigungsgedanken“, „Arbeitsmarktanalyse“ und „Kündigungsintention“. Und nicht nur das: Die Muster – in den Korrelationen in Zusammenhang mit einem Revidieren der ursprünglichen (rudimentären) Kausaltheorie macht eine Mediator-Struktur als mögliche Erklärung dieser Muster plausibel, in der die Antezedenzen (work-family conflict und Arbeitszufriedenheit) Kündigungsgedanken auslösen und diese dann eine Marktanalyse bewirken (d.h. die Person ist motiviert, sich nach Alternativen umzuschauen). Wenn diese Arbeitsmarktanalyse dann entweder dazu führt, dass die Person ein Angebot findet ober ihr Vertrauen, ein passendes zu finden, gestärkt wird, erhöht dies die Intention, auch tatsächlich zu kündigen.

Bei der Modellierung des Alternativmodells besteht nun das Problem, dass die „neuen" latenten Variablen nur jeweils einen Indikator besitzen (vgl. Abschnitt 3.5). Der Messfehler kann also nicht geschätzt, sondern muss fixiert werden. Da eine Fixierung auf 0 in jedem Fall unplausibel ist, soll er hier auf konservative 10% der Indikator-Varianz fixiert werden. Diese erhalten wird mittels

```
var(wfcdata$km01, na.rm=T) *.10
var(wfcdata$km02, na.rm=T) *.10
var(wfcdata$km03, na.rm=T) *.10
```

Die entsprechenden Werte (.10, .10 und .09) benutzen wir dann zur Fixierung der Fehlervarianz. Die Modellspezifikation sieht jetzt folgendermaßen aus:

```
wfcSEM2 <- '
  WFC =~ w01+w02+w03
  AZF =~ az01 + az02
  Gedanken =~ km01
  Markt =~ km02
  Int =~ km03
  AZF ~ WFC
  Gedanken ~ AZF + WFC
  Markt ~ Gedanken
  Int ~ Markt
  km01 ~~.10*km01
  km02 ~~.10*km02
  km03 ~~.09*km03
  '
```

Die Fixierung der Ladungen muss nicht explizit angeordnet werden, da die entsprechenden latenten Variablen nur einen Indikator haben und deren Ladung immer auf 1 fixiert ist. Die letzten 3 Zeilen enthalten die fixierten Fehlervarianzen. Abb. 25 zeigt die Struktur des respezifizierten Modells:

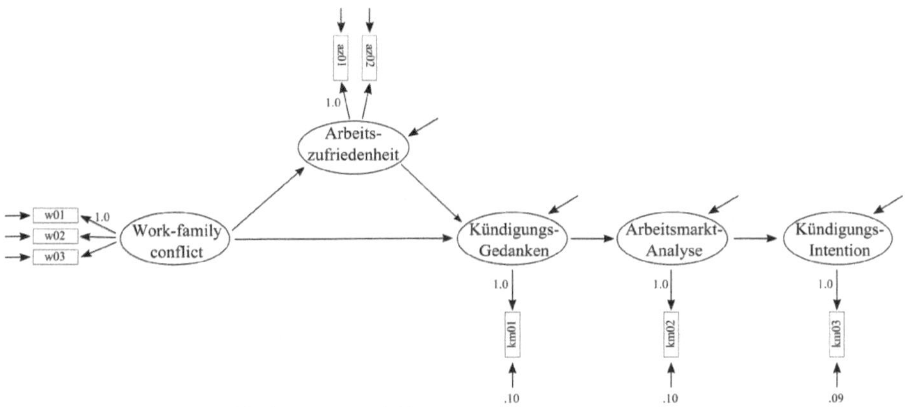

Abb. 25: Respezifiziertes Modell

Durch die Veränderung der Struktur hat sich der Modellfit deutlich verbessert ($\chi^2(18) = 24.74$, $p = .13$, RMSEA = .03, CFI = .99, SRMR = .03) und selbst der Chi-Quadrat-Test ist nun nicht mehr auffällig. Die Struktureffekte sehen plausibel aus:

```
               Estimate  Std.err  Z-value  P(>|z|)   Std.lv  Std.all
[...]
Regressions:
  AZF ~
    WFC        -0.313    0.064   -4.923    0.000   -0.328   -0.328
  Gedanken ~
    AZF        -0.542    0.084   -6.468    0.000   -0.468   -0.468
    WFC         0.162    0.069    2.345    0.019    0.146    0.146
  Markt ~
    Gedanken    0.723    0.050   14.533    0.000    0.692    0.692
  Int ~
    Markt       0.556    0.058    9.603    0.000    0.631    0.631
```

Insgesamt hat die Diagnostik zu einem alternativen Modell geführt, das mit den Daten übereinstimmt und eine plausible Alternative darstellt. Allerdings sollte dieses Modell als eine neue Hypothese angesehen werden, die einer Prüfung bedarf. Zwar ist die Respezifikation nicht allein aufgrund der Daten erfolgt, sondern theoretisch geleitet; allerdings wurde sie durch die Fehlanpassung des ersten Modells stimuliert und hat dadurch eben doch einen exploratorischen (genauer: abduktiven) Charakter. Wie eine Form der Prüfung dieser aussehen könnte, zeigt Kapitel 7. Vorher werden wir uns mit einer häufig anzutreffenden Form der Modellierung beschäftigen – nämlich wie man Modelle über Gruppen von Personen vergleicht und wie Moderatoren geprüft werden können.

6

Gruppenvergleiche und Tests von Moderatoren

Die bislang behandelten Anwendungen von SEM beinhalteten immer die Annahme der „**kausalen Homogenität**". Diese besagt, dass das Modell für eine Population gilt, in der die jeweiligen unabhängigen Variablen jeweils einen konstanten Effekt auf die abhängigen Variablen haben und es keine Subpopulationen gibt, in denen sich die Effekte unterscheiden. Oft besteht jedoch die Hypothese, dass sich ein kausaler Effekt über die Kategorien einer kategorialen Variablen oder über die Werte einer Dimension unterscheidet, d.h. moderiert wird. Solche **Moderatoreffekte** sind theoretisch wie praktisch bedeutsam, weil sie die Grenzen allgemeiner Haupteffekt-Hypothesen aufzeigen und die theoretische Vorstellung über die kausalen Effekte differenzieren. Moderatoren können **Gruppen** sein (z.B. Stichproben aus unterschiedlichen Kulturen, mit unterschiedlicher Bildung oder aus unterschiedlichen Berufen oder Branchen) oder **kontinuierliche Variablen**, die als Puffer oder Verstärker von Effekten wirken. Solche Variablen haben einen hohen Stellenwert für praktische Interventionen, da sie benutzt werden können, um z.B. negative Effekte mancher Variablen abzuschwächen oder positive zu verstärken. Dies kann v.a. in Kontexten eine wichtige Rolle spielen, in denen die unabhängigen Variablen selbst schlecht verändert werden können. Beispielsweise wird in der Stressforschung postuliert, dass der negative Effekt von Stressoren am Arbeitsplatz (z.B. Zeitdruck, Workload, Behinderungen) auf Wohlbefindensvariablen durch den Grad an Handlungsspielraum moderiert wird (Daniels & Guppy, 1994; Dwyer & Ganster, 1991; Ganster & Fusilier, 1989; Karasek, 1979, 1990). Handlungsspielraum bedeutet hier, dass die Person bestimmte Freiheiten hat, wann und wie sie ihre Aufgaben ausführt. Dadurch sollen sich Arbeitsstressoren nicht so schädlich auswirken, da die Person durch ihren Handlungsspielraum den Zeitpunkt, zu dem sie die Aufgabe erledigt und die Art und Weise, wie sie diese erledigt, frei wählen und auf ihre Leistungsfähigkeit abstimmen kann.

Als ein Beispiel für die Wirkungsweise von Moderatoren stelle man sich zwei Personen mit demselben Grad an Stressoren vor, die sich aber in dem Moderator Handlungsspielraum unterscheiden (z.B. Person A hat gar keine Spielräume, Person B völlige Autonomie). Falls Handlungsspielraum ein Moderator ist, sollte derselbe Grad an Stressoren sich bei Person A im Mittel (d.h. über alle Personen, die identisch zu Person A wären) negativer auf das Wohlbefinden auswirken als bei Person B.

Im Folgenden wird zunächst illustriert, wie kategoriale Moderatoren in SEM geprüft werden können. Dies funktioniert über **Gruppenvergleiche.** Anschließend wird gezeigt, wie man (semi)kontinuierliche Moderatoren behandelt. Dies funktioniert durch Einbeziehung von beobachteten oder latenten **Produktter-men**.

6.1 Gruppenvergleiche

Bei Gruppenvergleichen wird die Modellstruktur für die Gruppen (in weiterem Sinne: für die Kategorien der kategorialen Variable) separat spezifiziert. Dabei wird sowohl das Schätzen der Parameter als auch das Minimieren der Diskrepanz der betreffenden empirischen und modellimpliziten Kovarianzmatrizen (vgl. Abschnitt 2.4) nicht isoliert durchgeführt, sondern in einem **integralen Modell**. Folglich erhält man sowohl einen globalen Chi-Quadrattest als auch Tests für jedes Teilmodell.

In der Regel interessiert man sich bei Gruppenvergleichen für zwei Dinge: Zum einen soll untersucht werden, ob sich die Kausalstruktur über Gruppen unterscheidet; zum anderen interessieren **spezifische Parameter** (z.B. Struktureffekte oder Faktorladungen), für die entweder Hypothesen über die Ungleichheit über die Gruppen bestehen oder deren Gleichheit eine Voraussetzung für andere Analysen darstellt.

Gruppenvergleiche verlaufen über eine **Sequenz von Tests**, die das Modell von einer wenig restriktiven Struktur in eine restriktivere Struktur überführen. Die Restriktionen bestehen hierbei darin, dass man für eine bestimmte Art von Parametern (z.B. Faktorladungen) gruppenübergreifende **Gleichheitsrestriktionen** (equality constraints) definiert. Macht man das z.B. bei den Faktorladungen, führt dies dazu, dass das Programm bei der Variation der Parameter im Rahmen des Anpassungsprozesses der empirischen und modellimpliziten Kovarianzmatrix nicht mehr völlig frei ist. Stattdessen soll es einen Wert für die Parameter finden, der der Restriktion unterliegt, dass er in beiden Gruppen gleich sein muss.

Wenn der oder die Parameter in den Teil-Populationen tatsächlich denselben Wert haben sollten, wird diese Restriktion für das Programm kein Hindernis darstellen, die Parameter so zu schätzen, dass ein adäquater Fit daraus resultiert. Folglich wird der Chi-Quadrat-Wert in diesem Modell zwar in der Regel etwas höher sein als im Modell ohne die Gleichheitsrestriktion, aber der Unterschied wird nicht substantiell, d.h. nicht-signifikant sein. Wenn sich aber die Teil-Populationen in den Parametern unterscheiden, dann werden die Gleichheitsrestriktionen eine Verletzung der Gleichheits-Annahme sein. Folglich wird das Programm zwar die bestmöglichen Parameterschätzungen liefern, aber selbst dies wird nicht ausreichen, um einen adäquaten Fit zu bekommen. Zusätzlich wird der Abfall im Fit (bzw. der Anstieg des Chi-Quadrat-Wertes) deutlich und signifikant sein.

6.1.1 Spezifizieren eines Gruppenmodells

Als Beispiel soll untersucht werden, ob sich das in Abschnitt 5.3 modifizierte Work-Family-Conflict-Modell über Geschlechtskategorien hinweg strukturell unterscheidet. Bevor es allerdings an das Spezifizieren des Modells geht, müssen **fehlende Werte** in der Gruppierungsvariable eliminiert werden. Eine bessere Alternative wäre, die Daten für diese Variable durch moderne **Imputationsverfahren** (wie z.B. im mice-Paket vorhanden) zu ersetzen (Spieß, 1992). In unserem Fall fehlen allerdings lediglich Geschlechtsangaben für 3 Personen, so das durch deren Löschung kein Schaden zu erwarten ist. Diese Fälle eliminieren wir durch

```
wfcdata = wfcdata[!is.na(wfcdata$geschle),]
```

Diese Funktion ist über den konkreten Zweck (Löschung von fehlenden Werten) wichtig und soll zunächst erläutert werden. In ihr ist eine **Indizierung** von Elementen in einem Datensatz enthalten, die einen flexiblen Umgang mit R ermöglicht. So spricht man im Allgemeinen z.B. mit `Daten[2, ]` die zweite Zeile im Datensatz „Daten" an. Das Komma hinter der 2 ist wichtig, weil nach ihm keine Zahlen folgen, was bedeutet, dass alle Spalten ausgewählt werden. Alternativ würde `Daten[ ,3]` die dritte Spalte auswählen, weil die Zahl hinter dem Komma kommt und vor dem Komma die Freistelle. D.h., die eckige Klammer enthält zwei Stellen, vor und nach dem Komma, die die Möglichkeit für einen Zeilenindex (vor dem Komma) und einen Spaltenindex (nach dem Komma) bietet. Natürlich kann beides kombiniert (z.B. wählt `Daten[2,3]`von Variable 3 den zweiten Fall aus) und auf Bereiche von Variablen erweitert werden (z.B. wählt `Da-`

`ten[c(1-10), c(1-5)]` die ersten 10 Fälle für die Variablen 1-5 aus). In der o.g. Funktion werden nun diejenigen Fälle ausgewählt, die *nicht* („`!`") fehlende Werte („`is.na`") in der Variable Geschlecht („`wfcdata$geschle`") haben. Man beachte die Position des Kommas, was bedeutet, dass bei der Auswahl der Variablen keine Beschränkung vorgenommen wird, sondern alle Variablen einbezogen werden.

Um ein Gruppenmodell zu spezifizieren, muss lediglich zu der Syntax, durch die das Rechnen des Modells veranlasst wird, das `group="geschle"`-Argument hinzugefügt werden. Alles andere ist identisch zum Ein-Gruppen-Fall. Dies veranlasst, dass dieselbe Struktur in beiden Gruppen unterstellt wird, die Parameter aber (noch) frei variieren können.

```
wfcmodel2 <- '
 WFC =~ w01+w02+w03
 AZF =~ az01 + az02
 Phant =~ 1*km01
 Stell =~ 1*km02
 Int =~ 1*km03
 AZF ~ WFC
 Phant ~ AZF
 Stell ~ Phant
 Int ~ Stell
 km01 ~~.10*km01
 km02 ~~.10*km02
 km03 ~~.08*km03
'

fit <- sem(wfcmodel2,data=wfcdata, estimator="MLR",
       missing="FIML", group="geschle")

summary(fit,standardized=T)
```

Im Output erhalten wir zuerst den Fit des Gesamtmodells, dann die Chi-Quadrat-Werte für die beiden Teilmodelle:

```
  Number of observations per group
  1                                           137
  2                                           212

  Number of missing patterns per group
  1                                             7
  2                                             3
```

```
  Estimator                                         ML        Robust
  Minimum Function Test Statistic               52.439        46.604
  Degrees of freedom                                38            38
  P-value (Chi-square)                           0.060         0.160
  Scaling correction factor                                    1.125
    for the Yuan-Bentler correction

Chi-square for each group:

  1                                             18.638        16.564
  2                                             33.801        30.040
```

Wie zu sehen, hat das Modell einen nicht-signifikanten Chi-Quadrat-Wert. Folglich kann die Hypothese, dass die Struktur für beide Gruppen adäquat ist, beibehalten werden. Anschließend folgen die Parameterschätzungen getrennt für beide Gruppen. Dabei variieren die Parameter über die Gruppen noch. Da dieser Teil des Outputs keinerlei Neuerungen bietet, wird er hier ausgelassen.

6.1.2 Invarianz der Modellparameter

Häufig geht das Ziel eines Gruppenvergleichs über den einfachen Strukturvergleich hinaus. Stattdessen sollen z.B. Moderatoreffekte getestet werden, die darin bestehen, dass bestimmte Struktureffekte sich über Gruppen signifikant unterscheiden. Je nachdem, was das Ziel der Analyse ist, setzt ein Vergleich von Gruppen allerdings die Nicht-Unterschiedlichkeit (oder als Fachbegriff „Nicht-Invarianz") anderer Parameter voraus. So setzt z.B. ein Vergleich von Struktureffekten voraus, dass die Gruppen auch nicht-invariante Faktorladungen haben (sog. **metrische Invarianz**). Andernfalls würde man Effekte von Variablen vergleichen, die eine unterschiedliche Metrik und folglich logischerweise unterschiedliche Effekte oder im Extremfall eine völlig unterschiedliche Bedeutung haben. Einen Überblick über das Thema Invarianz geben Steenkamp und Baumgartner (1998), Vandenberg und Lance (2000) und Steinmetz, Schmidt, Tina-Booh und Wieczorek (2009).

Um ein Set von Parametern gleichzusetzen muss das Argument `group.equal=c("...")` in der `sem()` – oder `cfa()` -Funktion hinzugefügt werden. Innerhalb der Anführungszeichen wird dann der Parametertyp gekennzeichnet, der gleichgesetzt werden soll. Wenn dies mehrere Parametertypen sein sollen, werden sie, durch Kommata abgegrenzt, einfach hinzugefügt (z.B. `group.equal=c("loadings","regressions"))`. Im Folgenden werden die Formen der Invarianz und wofür sie notwendig sind kurz erläutert. Anschließend wird illustriert, wie die Sequenz der Testung einiger dieser Parameter in lavaan durchgeführt werden kann.

Konfigurale Invarianz. Dies ist die grundlegende Form der Invarianz. Sie betrifft die Gleichheit der kausalen Struktur in den Gruppen und wurde oben bereits getestet. Sie ist die Voraussetzung für jeden weiteren Vergleich der Gruppen. Abweichungen von dieser Form der Invarianz können allerdings triviale Unterschiede sein, die nicht zentral für die Art der Analyse sind. So kann es beispielsweise sein, dass es in einer Gruppe eine Doppelladung gibt, die es in der anderen Gruppe nicht gibt, ohne das davon andere Parameter oder die Bedeutung der Variablen betroffen ist. Gravierender sind dagegen Unterschiede, wie unterschiedliche Ladungsmuster, Anzahlen von latenten Variablen etc. Allerdings kann genau dies ein fruchtbares Ergebnis des Gruppenvergleichs sein, weil es auf fundamental unterschiedliche Kausalstrukturen, Bedeutungen und Anzahl latenter Variablen hinweist.

Metrische Invarianz. Diese Form der Invarianz bezieht sich auf die Faktorladungen. Sie bedeutet zum einen, dass davon ausgegangen werden kann, dass die latenten Variablen in den Gruppen dieselbe Bedeutung haben (Steinmetz et al., 2009), da die Ladungen als **Validitätskoeffizienten** angesehen werden können (sie sind die Brücke zwischen beobachtbaren und latenten Variablen). Metrische Invarianz belegt zum anderen die gleiche Metrik, die für jegliche quantitative Vergleiche nötig ist. Dazu ist die Voraussetzung, dass mindestens zwei Ladungen über die Gruppen gleich sind (Steenkamp & Baumgartner, 1998). Man spricht hier von **partieller Invarianz**; und sie ist eine abgemilderte Form der Restriktion gegenüber jener der **vollen Invarianz**, bei der alle Parameter eines Typs invariant sind. Genauso wie die konfigurale Invarianz ist die metrische Invarianz eine Voraussetzung für den quantitativen Vergleich weiterer Parameter (z.B. Struktureffekte). Um metrische Invarianz zu testen, fügt man dem Argument `group.equal=c("...")` den Begriff `"loadings"` hinzu.

Skalarinvarianz. Diese betrifft die Invarianz der Item-intercepts – d.h. der Konstanten in den Strukturgleichungen, die jedes Item mit ihrer zugrundeliegenden latenten Variablen verbinden (z.B. $w01 = \tau_1 + \lambda_1 WFC + \delta_1$, wobei τ der intercept ist, λ die Ladung und δ der Messfehler). Sie sind ausschließlich eine Voraussetzung für den Vergleich von Mittelwerten der latenten Variablen. Aus Platzgründen können diese in diesem Buch nicht behandelt werden. Steinmetz

(2010) gibt eine detaillierte Erörterung über die Bedeutung dieser intercepts und der Prozedur des Vergleichs latenter Mittelwerte. Um intercepts gleichzusetzen, fügt man den Begriff `"intercepts"` der o.g. Funktion hinzu.

Strukturelle Invarianz. Diese Form der Invarianz bezieht sich auf die Gleichheit der Struktureffekte der latenten Variablen. Wie oben angesprochen, erfordert diese einen erfolgreichen Test der konfiguralen und metrischen Invarianz. Sie wird durch den Begriff `"regressions"` induziert.

Sonstige Invarianzformen. Schließlich können Messfehler und mögliche Messfehlerkovarianzen über die Gruppen gleichgesetzt werden (`"residuals"` und `"residual.covariances"`), sowie Varianzen und Kovarianzen der latenten unabhängigen Variablen bzw. die Fehlervarianzen und deren Kovarianzen der abhängigen latenten Variablen (`"lv.variances", "lv.covariances"`). Sollte das Ziel ein Vergleich der (standardisierten) Korrelationen zwischen den latenten Variablen sein, müssen die Varianzen der latenten Variablen gleich sein.

6.1.3 Sequenz der Tests auf Invarianz und partielle Invarianz

Wie oben angesprochen, werden die erforderlichen Tests auf Invarianz innerhalb einer Abfolge von Modellen mit sukzessive steigenden Restriktionen getestet. Grundlage ist der einfache Strukturvergleich (konfigurale Invarianz). Darauf aufbauend werden sukzessive weitere Sets von Parametern gleichgesetzt. In jedem Schritt wird der Unterschied im Chi-Quadrat-Wert zum vorherigen Modell herangezogen, um zu evaluieren, ob es durch die jeweilige Gleichheitsrestriktion eine signifikante Verschlechterung des Modell gab. Es handelt sich hierbei um sogenannte **genestete Modelle**. Dies ist der Fall, wenn die Menge der geschätzten Parameter eines Modell eine Teilmenge der Parameter eines anderen ist. Vergleicht man genestete Modelle, kann zur Evaluation des Unterschieds der **Chi-Quadrat-Differenztest** angewendet werden. Dabei wird einfach die Differenz der beiden Chi-Quadratwerte errechnet und anhand einer Chi-Quadrat-Verteilung evaluiert, die durch die Differenz der Freiheitsgrade definiert ist:

$$\Delta\chi^2 = \chi^2_{\text{baseline Modell}} - \chi^2_{\text{restriktives Modell}}$$

und

$$\Delta df = df_{\text{baseline Modell}} - df_{\text{restriktives Modell}}$$

Eine Besonderheit liegt dann vor, wenn zur Schätzung `estimator="mlm"` oder `"mlr"` benutzt (wie in diesem Buch der Fall), wo diese Differenz korrigiert werden muss. Wir werden später eine handliche Funktion kreieren, mit der man das einfach durchführen kann.

Resultiert ein Schritt in einer signifikanten Verschlechterung des Modells, ist mindestens ein gleichgesetzter Parameter so unterschiedlich, dass ein zwangsweises Gleichsetzen eine Verletzung der Gleichheitsannahme darstellt. Allerdings weiß man nicht, welcher dies ist. In der Literatur wird daher empfohlen (Steenkamp & Baumgartner, 1998), die **Modifikationsindizes** für die gleichgesetzten Parameter und die **erwarteten Werte der Parameterschätzungen** bei einer Befreiung der Gleichheitsrestriktion (expected parameter change, EPC) zu analysieren. Beide werden in der `summary()`-Funktion durch das Argument „`modindices=TRUE`" angefordert. Die Modifikationsindizes werden für jeden fixierten Parameter ausgegeben und geben an, wie sehr der Chi-Quadrat-Wert sinken würde, wenn die betreffende Fixierung oder Gleichheitsrestriktion aufgehoben würde. Folglich sind die gleichsetzten Parameter mit den höchsten Modifikationsindizes diejenigen, die den größten Anteil am misfit haben.

Daneben sollte das Urteil über die Invarianz vs. Nicht-Invarianz nicht nur unter Gesichtspunkten der signifikanten Unterschiedlichkeit getroffen werden, sondern auch unter Gesichtspunkten der Stärke des Unterschieds – und darüber informieren die EPC-Werte. Es kann also durchaus sein, dass es einen signifikanten Unterschied gibt, der aber faktisch gering wäre. Dies wird v.a. in großen Stichproben der Fall sein, da hier die statistische Power hoch ist, sodass nicht-substantielle Unterschiede zwischen den Gruppen dennoch signifikant werden. Allerdings muss man aufpassen, sich nicht in Inkonsistenzen zu verfangen, wenn man zur Evaluation z.B. von Ladungsunterschieden die statistische Signifikanz herabwürdigt, sie aber zur Prüfung der späteren interessanten Unterschiede (z.B. in den Struktureffekten) als relevanten Evaluationsstandard heranzieht.

Entscheidet man sich für die Freisetzung einer Gleichheitsrestriktion, kann man dies durch Hinzufügen des Arguments "`group.partial="F=~x3"`" in die `sem()`- oder `cfa()`-Funktion bewirken, wobei „`F=~x3`" die Ladung ist, die freigesetzt werden soll (hier die Ladung von x3 auf der latenten Variablen F). Wie oben angesprochen, spricht man nun von **partieller Invarianz**. Ein vollständiges Beispiel für ein Modell mit metrischer partieller Invarianz wäre

```
sem(model, data=groupdata, group="group",
    group.equal=c("loadings"),"group.partial="F=~x3"")
```

Nun wird wiederum der Chi-Quadrat-Wert dieses modifizierten Modells herangezogen und mit dem letzten akzeptierten Modell verglichen. So würde das Modell mit partieller metrischer Invarianz mit dem Modell ohne jegliche Restriktionen (d.h. konfiguraler Invarianz) verglichen werden, da das Modell mit voller metrischer Invarianz ja vorher abgelehnt wurde. Ist der Unterschied immer noch signifikant, würde man wiederum Modifikationsindizes und EPC heranziehen und die entsprechenden Parameter freisetzen, bis es zu keinem Unterschied mehr kommt. Sind am Ende noch zwei Ladungen invariant (einer davon kann der Marker sein, dessen Ladung auf 1 fixiert ist), ist die Voraussetzung der partiellen Invarianz für weitere Analysen erfüllt. Es kann nun sein, dass die Modifikationsindices auf den Marker hinweisen, da die Fixierung auf 1 in beiden Gruppen ja eine Gleichheitsrestriktion impliziert. Dann kann man den Marker wechseln – entweder durch eine Umstellung der Reihenfolge in der Syntax (z.B. `WFC =~ w03 + w01 + w02`) oder durch Aufhebung der automatischen Fixierung des ersten Indikators und manuelle Fixierung (z.B. `WFC =~ NA*w01 + w02 + 1*w03`).

Wie bereits angesprochen, kann bei Verwendung von „`estimator=mlr`“ oder „`mlm`“ nicht die simple Differenz der korrigierten Chi-Quadrat-Werte berechnet werden. Dies lässt sich aber leicht beheben. Die Berechnung eines korrigierten Chi-Quadrat-Unterschiedes ist auf der Seite http://www.statmodel.com/chidiff.shtml erklärt. Wir benutzen diese Formeln, um eine eigene Funktion zu generieren, in die wir später als Input lediglich die beiden Objekt-Bezeichnungen der zu vergleichenden Modelle eingeben müssen und den Differenztest durchführen können:

```
CorrDiff <- function(fit.baseline, fit.restricted) {
  Corrc2Base <- fitMeasures(fit.baseline)[[5]]
  MLc2Base <- fitMeasures(fit.baseline)[[2]]
  dfBase <- fitMeasures(fit.baseline)[[3]]
  Corrc2Rest <- fitMeasures(fit.restricted)[[5]]
  MLc2Rest <- fitMeasures(fit.restricted)[[2]]
  dfRest <- fitMeasures(fit.restricted)[[3]]
  c0 <- MLc2Rest / Corrc2Rest
  c1 <- MLc2Base / Corrc2Base
  cd <- (dfRest*c0-dfBase*c1)/(dfRest-dfBase)
  CorrDiff <- (MLc2Rest-MLc2Base)/cd
  dfDiff <- dfRest - dfBase
  pDiff <- 1-pchisq(CorrDiff, dfDiff)
  result <- list(Corr.chisq.difference = CorrDiff,
           df.difference=dfDiff, p.value=pDiff)
```

```
return(result)
}
```

Die Funktion sieht recht unhandlich aus und Leser/innen mit Programmierkenntnissen mögen mir ihre eher primitive Natur bitte verzeihen. Dennoch ist sie hilfreich. Es sollte bedacht werden, dass sie eigentlich nur einmal abgetippt und ausgeführt werden muss. Dann kann sie mit dem gesamten image (siehe Abschnitt 4.5) des laufenden Projekts gespeichert werden („`save.image(...)`") und kann zukünftig mit „`load()`" geladen und für andere Projekte benutzt werden, um Satorra-Bentler (oder Yuan-Bentler) korrigierte Chi-Quadrat-Differenztests durchzuführen.

Die Funktion extrahiert aus beiden zu vergleichenden Modellen den korrigierten sowie den unkorrigierten Chi-Quadrat-Wert und die Freiheitsgrade und weist sie bestimmten Objekten zu. Beispielsweise extrahiert „`fitMeasures(fit.baseline)[[5]]`" den korrigierten Chi-Quadrat-Wert des Baseline – Modells und weist diesem dem Objekt „`Corrc2Base`" zu. Mit diesen Objekten werden anschließend die auf der o.g. Website diskutieren Korrekturwerte (z.B. c0, c1, cd) berechnet, mit denen die Chi-Quadrat-Differenz wiederum korrigiert wird.

Diese Funktion kann zukünftig für alle Vergleiche genesteter Modelle, die mit MLM oder MLR geschätzt wurden, benutzt werden, indem einfach in „`CorrDiff()`" die Objektnamen der zu vergleichenden Modelle eingegeben werden (z.B. „`fit.metric.inv`"), um die korrigierte Chi-Quadrat-Differenz zu benutzen. Wie diese Objekte heißen, ist dabei unerheblich. Es ist nur wichtig, dass das Baseline-Modell mit der geringeren Anzahl von Restriktionen zuerst kommt. Vertauscht man dies allerdings versehentlich, wird die Funktion lediglich „NA" (not available) ausgeben. Die Ergebnisse eines jeden Vergleiches innerhalb dieser Sequenz können dabei selbst wieder als Objekt gespeichert werden (z.B. „`MetricIN <- CorrDiff(...)`").

6.1.4 Beispiel

In diesem Abschnitt soll die gesamte Sequenz der Tests einmal vollständig durchlaufen werden. Einige der Stufen wurden aus didaktischen Gründen bereits vorher diskutiert, aber der Vollständigkeit halber wiederholt. Wir testen das modifizierte Work-Family-Conflict-Modell und vergleichen Männer und Frauen.

Als Erstes müssen die fehlenden Werte in der Geschlechtsvariablen entfernt werden (vgl. Abschnitt 6.1.1):

```
wfcdata = wfcdata[! is.na(wfcdata$geschle),]
```

Als nächstes wird die Modellstruktur spezifiziert werden (was in Abschnitt 6.1.1 bereits getan wurde). Wir rechnen das Baseline-Modell und testen damit auf konfigurale Invarianz; gleichzeitig erhalten wir dadurch die ersten für den Vergleich nötigen Werte (den korrigierten Chi-Quadrat-Wert, den „normal theory" ML Chi-Quadrat-Wert und die Freiheitsgrade des Modells). Die Ergebnisse des Modells werden im Objekt „`fit.base`" gespeichert.

```
fit.base <- sem(wfcmodel2,data=wfcdata, estimator="MLR",
                missing="FIML", group="geschle")

summary(fit.base)
```

Den Output inspizieren wir sowohl bzgl. des Modellfits, der Parameter und sonstiger Auffälligkeiten, die auf einen misfit hinweisen könnten (z.B. standardisierte Residuen). In unserem Fall kann das Modell akzeptiert werden. Als nächsten Schritt testen wir auf metrische Invarianz und setzen die Ladungen über beide Gruppen gleich. Dieses Modell nennen wird „`fit.metric.inv`":

```
fit.metric.inv <- sem(wfcmodel2,data=wfcdata,estimator="MLR",
                  missing="FIML", group="geschle",
                  group.equal="loadings")

summary(fit.metric.inv)
```

Auch dieses Modell hat einen guten Fit und wie der Output zeigt, sind die Ladungen von work-family conflict und Arbeitszufriedenheit nun über die Gruppen gleich gesetzt worden:

```
[...]
Group 1 [1]:
                   Estimate  Std.err  Z-value  P(>|z|)
Latent variables:
  WFC =~
    w01               1.000
    w02               0.997    0.066   15.150    0.000
    w03               0.971    0.065   14.936    0.000
  AZF =~
    az01              1.000
```

```
    az02                1.056    0.083   12.668    0.000

[...]

Group 2 [2]:
                   Estimate  Std.err  Z-value  P(>|z|)
Latent variables:
  WFC =~
    w01                 1.000
    w02                 0.997    0.066   15.150    0.000
    w03                 0.971    0.065   14.936    0.000
  AZF =~
    az01                1.000
    az02                1.056    0.083   12.668    0.000

[...]
```

Um beide Modelle zu vergleichen, nutzen wir die oben angelegte Funktion:

```
CorDiff(fit.base, fit.metric.IN)
```

Der Output dieser Funktion sieht folgendermaßen aus:

```
$SB.chisq.difference
[1] 1.417027

$df.difference
[1] 3

$p.value
[1] 0.7015484
```

Dies bedeutet, dass die korrigierte Chi-Quadrat-Differenz 1.417 ist. Evaluiert man diesen anhand einer Chi-Quadrat-Verteilung mit 3 Freiheitsgeraden, ergibt einen p-Wert von .70. Folglich kann die Hypothese, dass die Ladungen von Männern und Frauen in der Population gleich sind, beibehalten werden. Wenn dies nicht der Fall gewesen wäre, hätten die Modifikationsindizes mittels „`modindices=TRUE`“ in der `summary()`-Funktion inspiziert werden müssen und Kandidaten für unterschiedliche Ladungen mittels „`groups.partial=...`“ von der Gleichheitsrestriktion in der `sem()`-Funktion ausgeschlossen werden müssen. Dieses modifizierte Modell hätte dann in der Vergleichs-Funktion wiederum mit dem Baseline-Modell verglichen werden müssen (z.B. „`CorrDiff(fit.base,` fit.metric.Partinv"). In jedem Fall wird von nun an das letzte erfolgreiche Modell der neue Vergleichsmaßstab.

Da unser exemplarisches Ziel darin besteht, die Moderatorwirkung des Geschlechts auf die Struktureffekte der latenten Variablen zu testen, besteht der nächste und letzte Schritt darin, die Struktureffekte gleichzusetzen:

```
fit.reg.inv <- sem(wfcmodel2,data=wfcdata, estimator="MLR",
               missing="FIML", group="geschle",
               group.equal=c("loadings","regressions"))

summary(fit.reg.inv)
```

In diesem Fall muss der vorherigen Syntax einfach nur der Begriff „regressions" hinzugefügt werden. Alle Restriktionen des letzten erfolgreichen Modells bleiben erhalten. Vergleicht man dieses Modell nun mit dem Modell mit metrischer Invarianz, zeigt sich ebenfalls eine nicht-signifikante Verschlechterung:

```
CorDiff (fit.metric.inv, fit.reg.inv)

$SB.chisq.difference
[1] 8.393751

$df.difference
[1] 4

$p.value
[1] 0.07817402
```

In diesem Fall wäre die Hypothese der Moderatorwirkung des Geschlechts widerlegt. Eine Tabelle, die die Ergebnisse der Invarianztests wiedergibt, würde die korrigierten Chi-Quadratwerte, Freiheitsgrade der einzelnen Modelle und Outputs der `CorrDiff()`-Funktionen erhalten. Es sei darauf hingewiesen, dass in der Regel die Fitindizes (z.B. RMSEA, CFI und SRMR) für jedes Modell hinzugefügt werden. Diese müssten dann innerhalb der `summary()`-Funktion zusätzlich durch „`fit.measures=TRUE`" angefordert werden. Einige Autoren schlagen sogar vor, die Evaluation der Gleichheitsrestriktionen auf Basis der Differenzen der Fitindizes vorzunehmen. So schlagen Cheung und Rensvold (2002) aufgrund von Simulationsstudien vor, einen Unterschied im „comparative fit index" (CFI) von -.01 als Anzeichen von Nicht-Invarianz zu sehen. Kritisch an diesen Vorschlägen ist, dass diese primär empirisch getrieben sind (eben aufgrund der Simulationen, damit begrenzt auf die Kontexte, die in diesen Studien getestet wurden) und eigentlich keine statistische Grundlage haben.

Um die Tabelle herzustellen, kann man entweder die verschiedenen Test-Statistiken aus dem Output heranziehen, oder man extrahiert sie gezielt mit denjenigen Funktionen, die bereits in der `CorrDiff()`-Funktion verwendet wurden. Die Informationen, die in Tabelle 7 enthalten sind lassen sich beispielsweise durch den Block folgender Syntax-Befehle herstellen:

```
fitMeasures(fit.base)[[5]] #Korrigierter Chi-Quadrat-Wert
fitMeasures(fit.base)[[3]] #Freiheitsgrade

fitMeasures(fit.metric.inv)[[5]]
fitMeasures(fit.metric.inv)[[3]]
CorrDiff(fit.base,fit.metric.inv)

fitMeasures(fit.reg.inv)[[5]]
fitMeasures(fit.reg.inv)[[3]]
CorrDiff(fit.metric.inv,fit.reg.inv)
```

Tabelle 7: Exemplarische Tabelle über die Ergebnisse der Invarianztests

Modell	Art der Invarianz	$\chi 2$ (df)[a]	$\Delta\chi^2$ (Δdf)[b]	Vergleich
A	Konfigurale Invarianz	46.60 (38)		
B	Metrische Invarianz	47.69 (41)	1.42 (3)	A
C	Strukturelle Invarianz	56.24 (45)	8.39 (4)	B

Anm. [a]Nach Yuan-Bentler korrigierter Chi-Quadrat-Wert, [b]Die Differenz ist die korrigierte (und nicht nominale) Differenz der beiden Chi-Quadrat-Werte

Wie zu sehen, sollte in einer Anmerkung erwähnt werden, dass es sich um eine korrigierte Version des Chi-Quadrat-Werts handelt (was auch im Methodenteil kurz erklärt werden sollte), sowie, dass die Differenz der Chi-Quadrat-Werte korrigiert ist und sich nicht einfach aus der Subtraktion der Werte in der Tabelle ergibt.

6.2 Moderatoranalysen mit (semi)kontinuierlichen Moderatoren

Im vorherigen Abschnitt wurde beschrieben, wie man mittels Gruppenvergleichen untersuchen kann, ob sich ein Aspekt eines Modells (d.h. die gesamte postulierte Kausalstruktur oder einzelne Parameter) signifikant unterscheiden. In solchen Fällen ist die Gruppierungsvariable der Moderator. Ist der Moderator nicht kategorial, sondern kontinuierlich, ist dies kein gangbarer Weg mehr, weil der Moderator sonst künstlich in eine Gruppierungsvariable zerteilt werden müsste. Obwohl dies jahrzehntelang Praxis war, raten eine Reihe von Autoren davon ab und empfehlen dagegen, die Moderatoreffekte mittels Produkttermen zu testen (Dimitruk, Schermelleh-Engel, Kelava, & Moosbrugger, 2007; MacCallum, Zhang, Preacher, & Rucker, 2002; Maxwell & Delaney, 1993). Es sei dabei aber angemerkt, dass sich Gruppenunterschiede in Struktureffekten auch mittels Produktterm modellieren lassen. Allerdings bietet ein Gruppenvergleich die Möglichkeit, alle Parameter eines Modells auf signifikante Unterschiede zwischen Gruppen zu testen. Wie im vorherigen Abschnitt diskutiert, ist beispielswiese die Invarianz von Ladungen eine Voraussetzungen für Vergleiche der Struktureffekte.

Handelt es sich bei dem Modell um ein **Pfadmodell** mit beobachteten Variablen (vgl. Abschnitt 4.8), ist das Vorgehen identisch zur klassischen **moderierten Regression** (Aguinis & Stone-Romero, 1997; Champoux & Peters, 1987; Echambadi & Hess, 2007; Frazier, Barron, & Tix, 2004; Irwin & McClelland, 2001). Dabei werden die unabhängige Variable X und die Moderatorvariable Z multipliziert und der daraus entstandene Produktterm als weitere Variable in das Modell genommen. Ist der Regressionseffekt des Produktterms nun signifikant, bedeutet dies, dass die Multiplikation einen Anteil an der Varianz erklärt, der nicht bereits in den Haupteffekten von X und Z enthalten war.

Komplexer wird es, wenn es sich bei X und Z um **latente Variablen** handelt, die selbst durch mehrere Indikatoren reflektiert werden. Die Verwendung von latenten Variablen hat wegen der Messfehlerreduktion insbesondere bei Tests von **Moderatoreffekten** besonders große Vorteile, weil die Reliabilität des Produktterms das Produkt der Reliabilitäten der Ursprungsvariablen ist (Busemeyer & Jones, 1983). Haben diese z.B. eine Reliabilität von .70, ist sie beim Produktterm nur noch .49. Dieser hohe Messfehler resultiert in einem enormen Verlust an statistischer Power, was das Auffinden von Moderatoreffekten erschwert (Frazier et al., 2004). Während aber nun bei der Verwendung einer beobachteten X- und Z-Variable einfach nur diese beiden Variablen multipliziert werden, müssen bei der Verwendung latenter Produktterme die Indikatoren der latenten X- und Z-Variablen multipliziert werden. Daraus ergibt sich das Problem, dass die Messfehler der Indikatoren und Produktvariablen-Indikatoren nicht nur miteinanderkorrelieren, sondern bestimmten mathematischen Konsequenzen folgen (siehe

Steinmetz, Davidov, & Schmidt, 2011, für eine nähere Erläuterung). Diese müssen nun bei der Spezifikation des Messmodells berücksichtigt, d.h. fixiert werden. Beispielsweise zeigen frühe Arbeiten zu latenten Interaktionsmodellen (Jöreskog & Yang, 1996; Kenny & Judd, 1984), dass der Messfehler eines Produkttermindikators $x_i z_i$ eine komplexe Funktion ist, bestehend aus den Varianzen der latenten X- und Z-Variable, Messfehlern der Indikatoren x_i und z_i, aus denen er kreiert wurde, und seiner Ladung auf der latenten Produkttermvariablen. Anstatt diese Messfehlervarianz also frei schätzen zu können, muss er als diese Funktion fixiert werden. Wegen dieser Fixierungen („constraints“) heißt dieser Ansatz auch „**constrained approach**“ und war für lange Zeit für Forscher/innen kaum durchführbar, weil diese Fixierungen sehr komplex sind.

Nachdem in den Jahren darauf zwei starke Vereinfachungen dieses Ansatzes („unconstrained approaches“) veröffentlicht worden waren (Algina & Moulder, 2001; Marsh, Wen, & Hau, 2004) erschien 2006 mit dem „**residual centering approach**“ (Little, Bovaird, & Widaman, 2006) ein Ansatz, der am leichtesten durchführbar ist.

Im Folgenden soll zunächst ein Pfadmodell mit einer beobachteten Produkttermvariable vorgestellt werden, um die Bedeutung der Parameter in einem solchen Modell zu illustrieren und häufig anzutreffende Fehlannahmen zu adressieren. Anschließend wird der residual approach von Little et al. demonstriert.

6.2.1 Moderatoreffekte in Pfadmodellen

Die Beziehung von Interesse soll in diesem Beispiel die Beziehung zwischen work-family conflict und Arbeitszufriedenheit sein. Es wird nun angenommen, dass diese Beziehung vom Ausmaß negativer Affektivität moderiert wird. Negative Affektivität ist eine Persönlichkeitseigenschaft, die eine negativ getönte Wahrnehmung und Bewertung der eigenen Person und der Umwelt impliziert. Folglich sind solche Menschen wenig stressresistent, leicht zu verärgern und labil in ihren Stimmungen. Der Moderatoreffekt sollte jetzt dahingehend bestehen, dass mit steigender negativer Affektivität der negative Effekt von work-family conflict auf Arbeitszufriedenheit zunehmen sollte, in dem z.B. erlebte Konfliktsituationen aufgrund der stärkeren negativeren Bewertung sich stärker auf die Arbeits(un)zufriedenheit auswirken. Zur Messung von negativer Affektivität werden 3 Items benutzt:

- n01: Ich leide unter Nervosität.
- n02: In meinen Stimmungen gibt es häufig ein Auf und Ab.

- n03: Ich gerate leicht aus der Fassung, wenn es kritisch wird.

Analog zur Bildung von Skalenwerten, wie wir sie in Abschnitt 4.8 kennengelernt haben, bilden wir durch Berechnung des Mittelwertes dieser drei Items einen Skalenwert für negative Affektivität („nafS")

```
wfcdata$nafS <- apply(wfcdata[,c("na01","na02","na03")],1,mean)
```

Es hat sich bei moderierten Regressionsanalysen zudem als praktisch herausgestellt, die Variablen (in unserem Fall wfcS und nafS) zu zentrieren oder zu standardisieren. Dies hat den Vorteil, dass eine spätere grafische Veranschaulichung und die Interpretation der Parameter des Modells erleichtert werden. Es wurde lange angenommen, dass die Zentrierung die Multikollinearität, die durch die Einbeziehung eines Produktterms zwangsläufig erzeugt wird, reduziert. Wie Echambadi und Hess (2007) in ihrem Artikel belegen, kann eine Zentrierung dies nicht bewirken.

Bei der Zentrierung von Variablen wird von jedem Wert, den eine Person in der Variablen hat, der Mittelwert der Variablen über alle Personen abgezogen. Dies hat zur Folge, dass der Mittelwert dieser Variablen null ist. Wir legen also eine neue Variable im Datensatz an (der Zusatz „c" soll für „centered" stehen):

```
wfcdata$wfcSc <- wfcdata$wfcS - mean(wfcdata$wfcS,na.rm=TRUE)
wfcdata$nafSc <- wfcdata$nafS - mean(wfcdata$nafS,na.rm=TRUE)
```

Anschließend werden beide zentrierten Skalenwerte miteinander multipliziert, um den Produktterm zu erhalten:

```
wfcdata$prtrm <- wfcdata$wfcSc*wfcdata$nafSc
```

Schließlich wird das Pfadmodell spezifiziert, in dem alle 3 Variablen (wfcSc, nafSc und der Produktterm) einen Effekt auf die abhängige Variable haben:

```
intmod <- '
  azfS ~ wfcSc + nafSc + prtrm
  '

fit <- sem(intmod, data=wfcdata, missing="FIML", estimator="MLR")

summary(fit)
```

Die Ergebnisse zeigen, dass der Moderatoreffekt signifikant ist:

```
                   Estimate  Std.err  Z-value  P(>|z|)
Regressions:
  azfS ~
    wfcSc            -0.243    0.050   -4.913    0.000
    nafSc            -0.094    0.067   -1.414    0.157
    prtrm            -0.159    0.055   -2.886    0.004

Intercepts:
    azfS              0.680    0.046   14.724    0.000

Variances:
    azfS              0.723    0.061
```

Aber auch die Koeffizienten für wfcS und nafS (**„First-Order-Effekte“**) haben eine Bedeutung, die man kennen sollte, damit man v.a. Unterschiede zwischen einem Modell ohne und einem mit Produktterm interpretieren kann: Dadurch, dass ein Moderator vorliegt, gibt es ja keinen Haupteffekt mehr – stattdessen ändert sich der Effekt der unabhängigen Variable mit jeder Stufe des Moderators. Damit sind die First-Order-Effekte in einem Modell mit Produktterm „**konditionale Effekte**“, d.h. diejenigen Effekte der jeweiligen Variable an der Stelle der anderen Variable, wo diese den Wert 0 hat. Zum Beispiel ist der First-Order Effekt von wfcS $\gamma = -.243$ der Effekt von work-family conflict für Personen mit einer negativen Affektivität von 0. Hier kommt nun der Nutzen der oben besprochenen Zentrierung zum Tragen, denn durch diese ist 0 der Mittelwert der Variablen. Folglich ist der First-Order Effekt von work-family conflict der Effekt für Leute mit durchschnittlicher negativer Affektivität. Ohne Zentrierung wäre der Effekt schlecht interpretierbar, weil der Wert 0 dann „keiner“ negativen Affektivität entspräche. In solchen Fällen, wo der Wert 0 theoretisch sinnvoll ist, bietet es sich an, die unzentrierten Werte zu nehmen. Dies ist v.a. dann der Fall, wenn der Moderator eine binäre **dummy-Variable** ist, die die Werte 0 und 1 besitzt. So könnte das Geschlecht ein Moderator mit den Werten 0 = männlich und 1 = weiblich sein. Folglich wäre der First-Order-Effekt des Prädiktors derjenige Effekt in der männlichen Stichprobe.

Nun soll der Moderatoreffekt grafisch veranschaulicht werden. Auf der Seite http://www.uoguelph.ca/~scolwell/simplecontr.html findet sich ein hilfreicher R-Syntaxeditor, in dem man lediglich die Parameterschätzungen aus dem

Output eintragen muss und dann eine Syntax ausgegeben bekommt, die man zur Bildung der Grafik einfach in die R-Konsole kopiert (s. Abb. 26).

	Mean	Std. Dev.
1. Enter these values for X:	0	0.9322645
2. Enter these values for Z:	0	0.7121247

3. Enter the name of the Y-Axis: Arbeitszufriedenheit
4. Enter the name of the X-Axis: Work-family conflict
5. Enter the main title for the graph:
6. Enter the title for the legend:

(1) Generate R Code
(2) Select Code to Copy
Reset Form

Enter only the Unstandardized Beta Coefficients
(include the negative signs where applicable)

b_0	$b_1(X)$	$b_2(Z)$	$b_3(XZ)$
.68	-0.243	-0.094	-0.159

Abb. 26: Syntax-Editor zur Bildung der R-Syntax einer Interaktions-Grafik

In diesem Formular müssen zunächst die Mittelwerte und Standardabweichungen der First-Order-Effekt-Variablen eingetragen werden. Durch die Zentrierung sind die Mittelwerte 0; die Standardabweichungen bekommen wir durch

```
sd(wfcdata$wfcSc,na.rm=TRUE)
sd(wfcdata$nafSc,na.rm=TRUE)
```

Darunter kann man noch die Variablennamen eintragen. In der unteren Reihe werden schließlich die Koeffizienten aus dem lavaan-output eingetragen. Der Wert der Konstanten (b0) findet man unter „intercepts“ im lavaan-output. Klickt man nun auf „Generate R code“, wird in einem darunter liegenden Feld die R-Syntax dargestellt. Ein Klick auf „Select Code to copy“ markiert sie, „STRG-C“ kopiert sie in die Zwischenablage, von wo sie mit „STRG-V“ in die R-Konsole kopiert werden kann. Im Grafik-Fenster (s. Abb. 27) wird dann die Interaktion dargestellt:

Die beiden Geraden sind zwei exemplarische Effekte von work-family conflict auf Arbeitszufriedenheit; eine bei „niedriger“ negativer Affektivität (die gepunktete Gerade) und eine bei „hoher“ negativer Affektivität (die gestrichelte Gerade). Gewöhnlich werden als „niedrig“ eine Standardabweichung unter dem Mittelwert von negativer Affektivität dargestellt und als „hoch“ eine Standardabweichung über dem Mittelwert. Man sollte sich aber alle möglichen Geraden zwischen den beiden abgebildeten als eine kontinuierliche Fläche vorstellen, um ein Gefühl für die Art der Moderation zu bekommen. Den First-Order Effekt von work-family conflict kann man sich als zwischen beiden abgebildeten Gerade liegend vorstellen (eben da wo der wegen der Zentrierung der Mittelwert von negativer Affektivität liegt). Wie zu sehen, ist die Beziehung zwischen work-family conflict und Arbeitszufriedenheit für Personen mit niedriger negativer Affektivität schwach – u.U. sogar nicht-signifikant, während sie deutlich stärker für Personen mit hoher negativer Affektivität ist. Dies entspricht den theoretischen Annahmen.

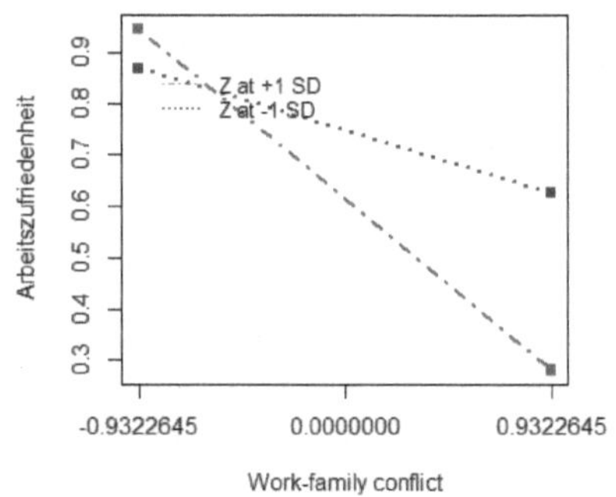

Abb. 27: Interaktionseffekt von work-family conflict und negativer Affektivität

Neben der Frage nach dem Vorhandensein eines Moderator- oder Interaktionseffekts besteht häufig das Interesse darin, die Signifikanz von Effekten *an bestimmten Stellen* des Moderators zu testen. Beispielsweise könnten beim Betrachten der oberen Abbildung Zweifel aufkommen, ob die gepunktete Linie – und damit der Zusammenhang zwischen work-family conflict und Arbeitszufriedenheit *bei niedriger negativer Affektivität* – überhaupt signifikant ist. Eine solche Analyse heißt **simple-slope-Analyse** (Aiken & West, 1991; Jaccard, Turrisi, & Wan, 1990). Diese besteht aus zwei Schritten: (1) Berechnung des Effekts von Interesse (d.h. der simple slope) und (2) Berechnung des Standardfehlers dieses Effekts.

1. Schritt: Berechnung der simple slope

Zur Berechnung der simple slope benötigen wir ihre Strukturgleichung – in unserem Fall die Gleichung für den Effekt von work-family conflict *an der Stelle unterdurchschnittlich negativer Affektivität*. Als „unterdurchschnittlich" wählen wir wie üblich eine Standardabweichung unterhalb des Mittelwerts von 0 aus (in diesem Fall -.71, siehe Abb. 26) – das heißt, wir suchen denjenigen Effekt von work-family conflict bei einer negativen Affektivität von -.71. Die Strukturgleichung für den Effekt an dieser Stelle ergibt sich nach Jaccard et al. (1990) durch

$$\gamma_{(NAF=-.71)} = \gamma 1 + \gamma 3 \times Z \qquad (9)$$

wobei $\gamma 1$ der First-Order-Effekt von work-family conflict ist, $\gamma 3$ der Effekt des Produktterms und Z die Stelle auf der negative-Affektivität-Dimension, für die wir den Effekt berechnen wollen. Dies ergibt

```
-.243 -.159*-.71
```

Als Ergebnis erhalten wir $\gamma_{(NAF=-.71)} = -.13$. Work-family-conflict hat also einen leicht negativen Effekt auf Arbeitszufriedenheit bei unterdurchschnittlich negativ affektiven Menschen.

2. Schritt: Berechnung des Standardfehlers der simple slope

Um die Signifikanz dieses Effekts zu beurteilen, benötigen wir den Standardfehler dieses Effekts. Dieser ergibt sich durch:

$$SE(\gamma 1_Z) = \sqrt{(var(\gamma 1) + Z^2 \times var(\gamma 3) + 2Z \times cov(\gamma 1, \gamma 3)} \qquad (10)$$

Außergewöhnlich an dieser Gleichung ist, dass hier die Varianzen und Kovarianzen der Parameter $\gamma 1$ und $\gamma 3$ benötigt werden – also Schätzungen der Variation dieser geschätzten Parameter, wenn die Stichprobenziehung und anschließende Analyse unendlich oft wiederholt würde. Die Kovarianz beider Effekte gibt an, ob deren Schätzungen systematisch zusammenhängen. Diese Informationen können wir leicht aus dem oben angelegten fit-Objekt extrahieren:

```
vcov(fit)
```

Als Ergebnis wird die Kovarianzmatrix der Parameter ausgegeben:

```
             azfS~wS azfS~nS azfS~p azS~~S azfS~1
azfS~wfcSc    0.002
azfS~nafSc    0.000   0.004
azfS~prtrm -0.001    0.000   0.003
azfS~~azfS   0.000   0.000   0.001  0.004
azfS~1       0.000   0.000   0.000 -0.001  0.002
```

Wie zu sehen, kürzt lavaan die Variablen ab. Die Varianz von $\gamma 1$ (work-family conflict → Arbeitszufriedenheit) ist .002 (dies ist das Quadrat des Standardfehlers); Var($\gamma 3$) ist .003 und die cov($\gamma 1$, $\gamma 3$) ist -.001. Der Standardfehler ergibt sich nach Gleichung 10 dann durch

```
sqrt(.002 + (-.71)^2*.003 + 2*(-.71)*-.001)
```

Daraus ergibt sich ein Standardfehler von .07. Um den statistischen Prüfwert auf der Z-Verteilung zu erhalten, teilt man den Effekt durch seinen Standardfehler (-.13 / .07). Dies führt zu z = -1.85, was nicht-signifikant ist. Dies bedeutet, dass work-family conflict für Menschen mit unterdurchschnittlicher negativer Affektivität keinen überzufälligen Effekt auf die Arbeitszufriedenheit hat. Die gepunktete Gerade in Abb. 27 weicht also nicht überzufällig von der Horizontale ab.

6.2.2 Interaktionseffekte in Modellen mit latenten Variablen

Der residual centering-Ansatz von Little et al. (2006) besteht aus drei Schritten. Im ersten Schritt werden einzelne Produktterme für jede Kombination der jeweils 3 Indikatoren (w01-w03 und na01 - na03) gebildet; anschließend wird jeder dieser Produktterme in einer Regressionsanalyse auf *alle* **first-order-Effekt-Indikatoren** (d.h. w01 etc.) regressiert und die Residuen werden im Datensatz als Variablen gespeichert. Diese werden schließlich im Modell als Indikatoren der latenten Produkttermvariable verwendet. Die R-Syntax all dieser Prozeduren wirkt auf den ersten Blick umfangreich und kompliziert; allerdings sind die einzelnen Elemente dabei nur Wiederholungen mit leichten Anpassungen. Mit einer erforderlichen Systematik ist die Umsetzung dieses Ansatzes nicht schwer.

1. Schritt: Multiplikation aller first-order-Variablen

```
wfcdata$pt11 <- wfcdata$w01 * wfcdata$na01
wfcdata$pt12 <- wfcdata$w01 * wfcdata$na02
wfcdata$pt13 <- wfcdata$w01 * wfcdata$na03
wfcdata$pt21 <- wfcdata$w02 * wfcdata$na01
wfcdata$pt22 <- wfcdata$w02 * wfcdata$na02
wfcdata$pt23 <- wfcdata$w02 * wfcdata$na03
wfcdata$pt31 <- wfcdata$w03 * wfcdata$na01
wfcdata$pt32 <- wfcdata$w03 * wfcdata$na02
wfcdata$pt33 <- wfcdata$w03 * wfcdata$na03
```

Der erste Index jedes Produktterms steht dabei jeweils für den verwendeten first-order-Indikator.

2. *Schritt: Regression jedes Produktterms auf alle first-order-Effekt-Indikatoren und Speicherung im Datensatz*

```
wfcdata$res11 <- resid(lm(pt11 ~ w01+w02+w03 + na01+na02+na03,
 + data=wfcdata,na.action = na.exclude))
wfcdata$res12 <- resid(lm(pt12 ~ w01+w02+w03 + na01+na02+na03,
+ data=wfcdata,na.action = na.exclude))
wfcdata$res13 <- resid(lm(pt13 ~ w01+w02+w03 + na01+na02+na03,
+ data=wfcdata,na.action = na.exclude))
wfcdata$res21 <- resid(lm(pt21 ~ w01+w02+w03 + na01+na02+na03,
+ data=wfcdata,na.action = na.exclude))
wfcdata$res22 <- resid(lm(pt22 ~ w01+w02+w03 + na01+na02+na03,
+ data=wfcdata,na.action = na.exclude))
wfcdata$res23 <- resid(lm(pt23 ~ w01+w02+w03 + na01+na02+na03,
+ data=wfcdata,na.action = na.exclude))
wfcdata$res31 <- resid(lm(pt31 ~ w01+w02+w03 + na01+na02+na03,
+ data=wfcdata,na.action = na.exclude))
wfcdata$res32 <- resid(lm(pt32 ~ w01+w02+w03 + na01+na02+na03,
+ data=wfcdata,na.action = na.exclude))
wfcdata$res33 <- resid(lm(pt33 ~ w01+w02+w03 + na01+na02+na03,
+ data=wfcdata,na.action = na.exclude))
```

Bei diesem Block muss lediglich das gebildete Residuum (z.B. „res11“) und der verwendete Produktterm (z.B. „pt11“) als abhängige Variable angepasst werden – der Rest kann kopiert werden. Zur Verdeutlichung wurden beide Elemente im o.g. Code unterstrichen. Die einzelnen Befehlszeilen wenden drei noch unbekannte Funktionen an. Zum einen wird mit `lm(Y ~ X)` eine Regression („linear model“) der abhängigem Variable Y auf die unabhängige Variable X durchgeführt, `"na.action=na.exclude"` bewirkt dabei, dass fehlende Werte auf Seiten der unabhängigen Variable auch in den Residuen erhalten bleiben (andernfalls hätten wir weniger Fälle in den Produkttermvariablen als in den anderen Modelvariablen). Schließlich ist die `lm()`-Funktion eingefasst in die `resid()`-Funktion, die die Residuen der Regression extrahiert.

Nach der Bildung der Produkttermvariablen sollten zwei Tests durchgeführt werden, die sicherstellen, dass keine Fehler aufgetreten sind. Zum einen sollten die Mittelwerte dieser Indikatoren alle exakt Null sein; zum anderen sollten die

Korrelationen mit den first-order-Effekt-Indikatoren Null und unter den Produktterm-Indikatoren substantiell sein. Ersteres überprüft man durch

```
mean(wfcdata$res11,na.rm=TRUE)
```

Für Letzteres berechnet man die Korrelationsmatrix aller Variablen:

```
modvar <- wfcdata[c("w01","w02","w03","na01","na02","na03",
+"res11","res12","res13","res21", Rest ausgelassen)]

round(cor(modvar,use="pair"),2)
```

Es gibt viele Wege, Korrelationsmatrizen zu generieren, die alle zugegebenermaßen nicht sehr komfortabel sind. Hier wurde erst ein Subdatensatz mit den gewünschten Variablen gebildet und auf diesen die `cor()` – Funktion, integriert in die `round()` – Funktion, angewendet.

3. Schritt: Spezifikation des Modells

Schließlich geht es an das Modell. Dieses hat zwei Besonderheiten: Erstens werden die Kovarianzen der latenten Produktterm-Variablen und den beiden first-order-Effekt-variablen work-family conflict und negative Affektivität auf 0 fixiert. Zweitens werden Messfehler-Kovarianzen zwischen denjenigen Produktterm-Indikatoren, in denen quasi dieselben Indikatoren stecken, freigesetzt: Da diese aus denselben Indikatoren bestehen, sollten sie folglich auch korrelieren (dürfen). Die unten abgebildete Syntax sieht zugegebenermaßen (wieder) aufwändig aus – aber mit ein bisschen Systematik ist das leicht machbar. Die Systematik wurde durch Blöcke, die mit Überschriften versehen sind, hervorgehoben:

```
intlat <- '
  NAF =~ na01+na02+na03
  WFC =~ w01+w02+w03
  AZF =~ az01 + az02
  PRTRM =~ res11+res12+res13+res21+res22+res23+res31+res32+res33
  PRTRM ~~0*WFC #Der Produktterm kovariiert nicht mit den
  PRTRM ~~0*NAF #first-order-Effekt-Variablen

  #Messfehler-Kovarianzen derjenigen Produktterm-Indikatoren, die
  #gemeinsame Elemente enthalten
  #a) Erste Komponente ist identisch (in res11, res12,
  #   res13)
```

```
  res11~~res12
  res12~~res13
  res11~~res13

  #res21, res22, res23
  res21~~res22
  res22~~res23
  res21~~res23

  #res31, res32, res33
  res31~~res32
  res32~~res33
  res31~~res33

  #b) Zweite Komponente ist identisch (in res11, res21, und
  #res31)
  res11~~res21
  res11~~res31
  res21~~res31

  #res12, res22, und res32
  res12~~res22
  res12~~res32
  res22~~res32

  #res13, res23, und res33
  res13~~res23
  res13~~res33
  res23~~res33

  #Strukturmodel
  AZF ~ WFC + NAF + PRTRM
    '

fit <- sem(intlat, data=wfcdata, missing="fiml",estimator="mlr")

summary(fit)
```

Im Output interessiert hier nur primär das Strukturmodell. Der Fit des Modells sollte in der Regel akzeptabel sein, wenn das Messmodell der latenten Variablen (ohne Produktterme) klar fittet. Wenn nicht, wird durch die Verwendung der Produktterme der vorhandene misfit ebenfalls stark gewichtet. Im nachfolgenden Output wurden der Sparsamkeit halber viele Zeilen gekürzt. Für die Moderatorwirkung zentral ist der Effekt des latenten Produktterms, der hier γ = -.30 ist.

Sein z-Wert beträgt 3.054. Somit ist dieser nur geringfügig höher als im Pfadmodelle (z = 2.89). Dies muss aber nicht immer so sein. Vor allem bei Indikatoren mit höherem Messfehler oder kleineren Stichproben kann der Unterschied zwischen einem Modell mit einem latenten Produktterm und einem Pfadmodell substantiell und entscheidend sein.

```
                   Estimate  Std.err  Z-value  P(>|z|)
Latent variables:
  NAF =~
    na01              1.000
    na02              0.950    0.159    5.991    0.000
    na03              0.739    0.123    5.986    0.000
  WFC =~
    w01               1.000
    w02               1.002    0.071   14.151    0.000
    w03               0.973    0.070   13.978    0.000
  AZF =~
    az01              1.000
    az02              0.833    0.132    6.302    0.000
  PRTRM =~
    res11             1.000
    res12             0.981    0.190    5.157    0.000
[...]

Regressions:
  AZF ~
    WFC              -0.333    0.062   -5.371    0.000
    NAF              -0.155    0.111   -1.397    0.163
    PRTRM            -0.300    0.098   -3.054    0.002

Covariances:
  WFC ~~
    PRTRM             0.000
  NAF ~~
    PRTRM             0.000
  res11 ~~
    res12             0.138    0.068    2.020    0.043
[...]

Variances:
    na01              0.559    0.083
    na02              0.544    0.077
    na03              0.547    0.060
    w01               0.308    0.048
```

```
    w02                   0.427    0.058
    w03                   0.485    0.056
    az01                  0.108    0.124
    az02                  0.239    0.093
    res11                 0.658    0.131
    res12                 0.711    0.136
[...]
    NAF                   0.399    0.088
    WFC                   0.739    0.080
    AZF                   0.716    0.131
    PRTRM                 0.454    0.207
```

Um den Interaktionseffekt grafisch zu illustrieren, benutzen wir wieder das in Abschnitt 6.2.1 beschriebene Web-Formular. Die Mittelwerte der latenten Variablen sind hier null, ihre Standardabweichung ergibt sich aus der Wurzel der Varianzen (.739 und .399). Die Form der Interaktion ist erwartungsgemäß identisch zu derjenigen, die das Pfadmodell ermittelte. Die **simple-slope-Analyse** für den Effekt von work-family conflict bei Personen mit unterdurchschnittlicher negativer Affektivität führen wir ebenfalls analog zu der im Pfadmodell durchgeführten durch:

(1) Wie im vorherigen Abschnitt gezeigt, berechnet sich die simple slope der latenten Work-Family-Conflict-Variablen bei negativer Affektivität = Z aus

$$\gamma_{(NAF = Z)} = \gamma 1 + \gamma 3 \times Z \quad \text{(Wiederholung Gleichung 9)}$$

Angewendet auf die geschätzten Parameter impliziert dies einen Effekt von

```
-.14 = -.33 -.30*-sqrt(.399)
```

Der Wert „.399“ ist die Varianz von negativer Affektivität (siehe SEM-Output) – und die Wurzel daraus ist demnach die gewünschte Standardabweichung. Er beträgt $\gamma_{(NAF = Z)}$ = -.14 und ist sehr ähnlich zu dem aus dem Pfadmodell.

(2) Der Standardfehler für die simple slope ergibt sich wieder aus

$$SE(\gamma 1_Z) = \sqrt{(var(\gamma 1) + Z^2 \times var(\gamma 3) + 2Z \times cov(\gamma 1, \gamma 3)}$$

Nun besteht das praktische Problem, dass `vcov(fit)` eine sehr große Matrix ergeben würde, da sie die Varianzen und Kovarianzen *aller* Modellparameter enthält. Wir können aber zur Extrahierung von var(γ1), var(γ3) und cov(γ1, γ3) einen direkteren Weg gehen, in dem wir direkt auf spezifische Elemente in der Matrix zugreifen:

```
var_g1 <- vcov(fit)["AZF~WFC","AZF~WFC"]
var_g3 <- vcov(fit)["AZF~PRTRM","AZF~PRTRM"]
cov_g13 <- vcov(fit)["AZF~WFC","AZF~PRTRM"]
```

Wie zu sehen, müssen lediglich die gewünschten Parameter adressiert werden. So adressiert „`AZF~WFC`" den Effekt von work-family conflict auf Arbeitszufriedenheit in der gewohnten Notation. Wenn wir die *Varianz* eines Parameters extrahieren möchten, kommt dieser Parameter zweimal in der eckigen Klammer vor, da die Varianz einer Variablen die *Kovarianz dieser Variablen* mit sich selbst ist. Möchten wir die Kovarianz zwischen zwei Parametern, kommen demnach beide gewünschten Parameter in die Klammer. Wie zu sehen, weisen wir das Ergebnis der Extraktion einem Objekt zu, welches wir dann in der Formel zur Berechnung des Standardfehlers nutzen können. Dies tun wir aus demselben Grund für die Standardabweichung von negativer Affektivität:

```
Z = -sqrt(.399)          #Wurzel aus der latenten Varianz (s. Output)
```

Schließlich fügen wir alle angelegten Objekte in die schon bekannte Formel ein:

```
sqrt(var_g1 + Z^2*var_g3  + 2*Z*cov_g13)
```

Als Ergebnis bekommen wir den Wert .099.

(3) Den z-Wert zur Beurteilung der Signifikanz der simple slope erhalten wir wieder, in dem wir die simple slope durch ihren Standardfehler teilen (-.14 / .099). Wir erhalten den z-Wert von -1.41. Das er geringer als 2 ist, zeigt, dass die simple slope nicht signifikant ist. Analog zu den Ergebnissen aus dem Pfadmodell hat work-family conflict für Personen mit niedriger negativer Affektivität keine Folgen für die Arbeitszufriedenheit.

Wie zu sehen, kommen das Pfadmodell und das Modell mit latentem Produktterm zu identischen Ergebnissen. Dies muss aber nicht immer so sein, und wird v.a. bei Indikatoren mit hohem Messfehler zu Unterschieden führen. Hier hat das Modell mit latenten Variablen eine höhere statistische Power, weil der Produktterm messfehlerfrei ist.

7 Endogenität und Instrumentalvariablen

In Kapitel 2 wurde die Philosophie hinter SEM dargelegt – nämlich, dass aus einer kausalen Struktur eine bestimmte Kovarianzstruktur der Variablen folgt und dass man auf diesem Wege Modellstrukturen testen kann, indem man prüft, ob die empirische Kovarianzstruktur mit der durch das Modell vorhergesagten Struktur übereinstimmt. Ein Grundproblem dabei ist, dass es alternative Modellstrukturen gibt, die die Kovarianzstruktur ähnlich gut, in identischer Weise oder gar besser vorhersagen können. Dies schränkt die kausale Evidenz eines Modells ein. Dazu kommt, dass unabhängige Variablen sowohl in SEM als auch in jedem Regressionsmodell **exogen** sein müssen, damit man den von ihnen ausgehenden Effekt als kausal interpretieren kann. Exogenität bedeutet hier, dass die unabhängigen Variablen nicht mit dem Fehlerterm der abhängigen Variable korrelieren dürfen (Eine genaue Erläuterung folgt später). Folglich besteht, selbst wenn das Modell den Test erfolgreich überstanden hat, immer die Möglichkeit, dass es ein anderes Modell gibt, welches das korrekte Modell ist, und man so die Effektschätzungen nie mit Sicherheit als kausale Effekte interpretieren kann. Zur Reduzierung beider Gefahren kann die Einbindung von Instrumentalvariablen sehr hilfreich sein. Diese können den Grad an Evidenz für ein Modell enorm erhöhen.

7.1 Der Feind jeder kausalen Interpretation: Endogenität

Experimente gelten allgemein als der Königsweg, um Kausalität nachzuweisen. In einem typischen Experiment werden Personen zufällig einer Treatment- und einer Kontrollgruppe zugewiesen; dadurch sollen sie sich in allen gemessenen und nicht gemessenen Merkmalen gleichen. Das ist vor allem essentiell für diejenigen Merkmale, die einen Einfluss auf die abhängige Variable des Experiments haben. Nehmen wir an, in dem Experiment soll die Wirkung eines Software-Trainings auf den Umgang mit dieser Software getestet werden. Hier sind eine Menge weiterer Einflussfaktoren auf den Umgang denkbar: Technische Affinität, Intelligenz, Erfahrung mit dem PC etc. Eine Möglichkeit wäre natürlich, all diese Faktoren zu messen und statistisch zu kontrollieren. Nur kann man eben nie sicher sein, dass man alle Einflussfaktoren kennt. Wenn man also keine „Randomisierung" durchführen würde, ist es denkbar, dass sich technisch-affine Personen wegen ihres Interesses an der Software-Schulung in die Trainingsgruppe **selbst-selektieren.** Zeigt diese Gruppe später nun eine bessere Leistung als die Kontrollgruppe, muss das nicht am Training liegen, sondern an ihrer Affinität – und damit zusammenhängend – höheren allgemeinen Kompetenz, Interesse, Offenheit etc. Dies ist das Problem, mit dem sog. Quasi-Experimente konfrontiert sind, bei denen eine randomisierte Zuteilung nicht möglich ist. Abb. 28 (links) zeigt ein Pfaddiagramm für diesen Fall.

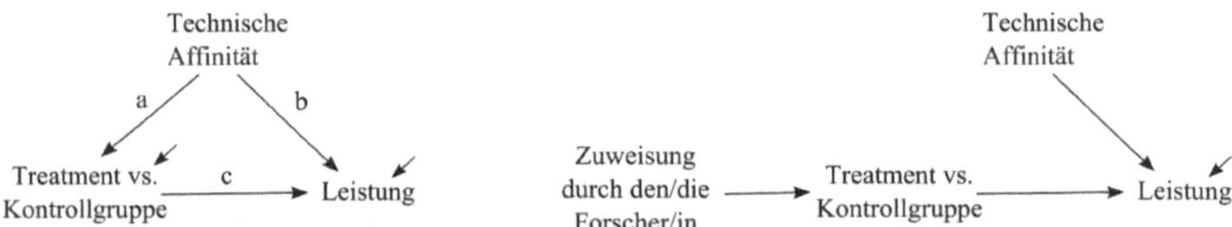

Abb. 28: Problem ausgeschlossener Drittvariablen bei nicht-randomisierter Gruppenzuteilung

Es ist ersichtlich, dass der Einfluss der Drittvariablen „Technische Affinität" den eigentlich interessierenden Effekt c um den Betrag a*b verzerrt. Man kann nicht einmal sicher sein, ob das Training überhaupt einen kausalen Effekt hat. Der Grund dafür ist die Endogenität der Treatment-Variable. Die Zufallszuweisung in einem randomisierten Experiment durchbricht den Pfad a mit der Folge, dass nicht mehr die technische Affinität (durch Selbst-Selektion) bestimmt, ob eine Person in die Trainings- oder Kontrollgruppe kommt, sondern der/die Forscher/in aufgrund einer Zufallsauswahl. Das rechte Pfaddiagramm in Abb. 28 zeigt diesen Fall. Es zeigt, dass die Treatment-Variable keinen Fehlerterm mehr hat, weil die Frage, ob eine Person zur Treatment- oder Kontrollgruppe gehört, vollständig durch die Zuweisung bestimmt wird. Die technische Affinität bestimmt weiterhin die Leistung, aber sie korreliert nicht mehr mit der Treatment-Variable: Die Treatment-Variable ist exogen.

In Feldstudien mit nicht randomisierten unabhängigen Variablen ist die Gefahr durch Endogenität allgegenwertig und der Einfluss nicht-gemessener Variablen ist nur ein Grund dafür. Bevor wir aber zu Lösungsansätzen kommen, soll erst die Frage beantwortet werden, was Endogenität genau bedeutet.

Formal bedeutet **Endogenität**, dass eine unabhängige Variable (gemäß üblicher Notationen nennen wir sie X) mit dem Fehlerterm der abhängigen Variable Y korreliert. Ist dies der Fall, wird der Effekt von X auf Y **verzerrt** (d.h. er ist nicht einmal im Mittel über alle möglichen Stichproben korrekt) und **inkonsistent** (d.h. er ist nicht einmal in einer unendlich großen Stichprobe korrekt); man kann weder die Effektstärke interpretieren, noch schlussfolgern, ob X überhaupt Y beeinflusst. Abb. 28 war ein Beispiel dafür. Wie schon häufiger in diesem Buch angemerkt, ist der Fehlerterm eine Zusammenfassung aller weiteren, nicht im Modell befindlichen Ursachen von Y. Alle Gründe, die zu einer Korrelation zwischen X und dem Fehlerterm von Y führen, lassen sich auf drei Kategorien von Gründen reduzieren:

(1) Ausgeschlossene Variablen
(2) Messfehler bei der Messung von X
(3) Fehler in der Modellierung der kausalen Richtung

7.1.1 Ausgeschlossene Variablen

Es gibt zwei Fälle ausgeschlossener Variablen, die zu Endogenität führen: a) ausgeschlossene Kovariate (sofern sie mit X korrelieren) und b) ausgeschlossene gemeinsame Ursachen von X und Y. Beide funktionieren über denselben Mechanismus: Wie oben angemerkt, ist der Fehlerterm von Y eine Zusammenfassung aller nicht erfassten weiteren Ursachen von Y. Korrelieren eine oder mehrere dieser Ursachen nun mit X, korreliert folglich auch der Fehlerterm mit X. Alternativ können die Pfadregeln benutzt werden, um den Einfluss ausgeschlossener Variablen zu verstehen. Demnach wird bei Ausschluss einer mit X korrelierenden weiteren Ursache dem geschätzten Effekt von X der **compound path**, welcher X über die ausgeschlossene Variable mit Y verbindet, hinzugefügt. Dieser besteht aus dem Effekt der ausgeschlossenen Variable multipliziert mit ihrer Korrelation mit X.

Im Fall **ausgeschlossener gemeinsamer Ursachen** (vgl. Abb. 28) ist der Mechanismus nahezu identisch. Hier wird dem Effekt von X das Produkt aus beiden Effekten der ausgeschlossenen Ursache hinzugefügt. Ausgeschlossene Ursachen können wie im oben behandelten Beispiel des Software-Trainings substantielle theoretisch bedeutsame Variablen sein – es kann sich bei ihnen aber auch um sog. Methodenfaktoren handeln. Diese implizieren bestimmte Stile einer Person,

Fragen in Fragebögen zu beantworten (z.B. **Aquieszenzbias**, **Tendenz zur Mitte**, **soziale Erwünschtheit**). Die Probleme, die durch diese Methodenfaktoren entstehen können, werden in der Literatur unter dem Begriff **„common method bias“** diskutiert (Podsakoff, MacKenzie, Lee, & Podsakoff, 2003).

Dies alles bedeutet, dass immer dann, wenn ein Modell nicht alle mit X korrelierenden weiteren Ursachen von Y oder gemeinsamen Ursachen enthält, wenig Hoffnung besteht, mit dem *geschätzten* Effekt auch den *tatsächlichen kausalen* Effekt von X zu erhalten. Hier hilft auch nicht der Test des Modells, da die verzerrenden Mechanismen sich nicht unbedingt auf den Fit auswirken müssen.

7.1.2 Messfehler

Eine weitere Ursache für Endogenität ist das Vorhandensein von Messfehlern in X (siehe Antonakis et al., 2010, für einen algebraischen Nachweis). Zwar können wir die Hoffnung haben, dass dies bei Verwendung von latenten Variablen und Modellierung von Messfehlern reduziert wird – allerdings muss dies nicht zwangsläufig so sein: Angenommen, das Messmodell ist korrekt spezifiziert und die Indikatoren reflektieren tatsächlich eine latente Variable. Wenn diese latente Variable aber nun nicht diejenige Entität reflektiert, die wir eigentlich messen wollen, sondern nur ein Proxy dieser Variable ist, erhält man trotz korrekter Spezifikation und Messfehlerkorrektur nicht die Effekte der eigentlichen Variable. Ein Beispiel dafür wäre, wenn wir statt der Arbeitszufriedenheits-Indikatoren Indikatoren im Modell hätten, die sich auf die emotionale Reaktion der Befragten auf ihre Arbeit beziehen (z.B. „Ich freue mich häufig über meine Arbeit“). Dies wären nicht Indikatoren der latenten Variable „Arbeitszufriedenheit“, sondern der ähnlichen Variable „positive Arbeitsemotionen“. Hier würde auch eine Modellierung einer latenten Variable nichts bringen, wenn die wahre Ursache Arbeitszufriedenheit und nicht positive Arbeitsemotionen ist. Wie schon oft in diesem Buch gesagt: der Modellfit belegt nicht, dass die latenten Variablen die Bedeutung haben, die wir unterstellen. Dies ist ein Grund, warum manche Autoren für eine theoriegeleitete Fixierung der Messfehler plädieren, statt einer freien Schätzung (Hayduk & Littvay, 2012). In jedem Fall sollte die Konzeptualisierung der latenten Variable und entsprechende Auswahl der Indikatoren präzise erfolgen (vgl. Kap. 3) und ein klarer Modellfit und theoretisch plausible Ladungen für die Bedeutung der latenten Variable sprechen, wie sie der kausalen Theorie des/der Forschenden entspricht.

7.1.3 Fehler in der Modellierung der kausalen Richtung

Fehler in der Modellierung der kausalen Richtung sind die gravierendste Form der Fehlspezifikation. Hier ist nicht X die Ursache von Y, sondern Y die Ursache von X („**reverse causation**"). Ein Spezialfall ist die **Simultanität** beider Variablen, bei der X zwar Y, Y aber auch X beeinflusst. Testet man in einem Modell nun einfach einen Effekt von X auf Y, bekommt man – je nach tatsächlichen Effekten – nach oben oder nach unten verzerrte Effekte.

7.2 Lösungsansätze für Endogenität: Instrumentalvariablen

Während Endogenität in den meisten Verhaltenswissenschaften (Psychologie, Sozialwissenschaft, Soziologie) zwar als integrales Konzept nahezu unbekannt ist, sind ihre drei Ursachen ein bekanntes Problem. Traditionelle Lösungsansätze sind

a) Kontrollvariablen einzubeziehen und die Messung von X und Y bei verschiedenen Personen (zur Lösung des omitted variable-Problems),

b) Messfehler zu reduzieren durch Aggregation multipler Items oder Modellierung latenter Variablen (zur Lösung der Messfehlerproblematik) und

c) Durchführung von Längsschnittstudien.

Die Ökonometrie dagegen (Kennedy, 2003) hat einen alternativen Lösungsansatz kultiviert, nämlich den Einsatz von **Instrumentalvariablen**. Diese können benutzt werden, um die Evidenz eines kausalen Effekts von X auf Y zu erhöhen, auch wenn X messfehlerbehaftet ist, es ausgeschlossene Variablen gibt und die Plausibilität für Simultanität gegeben ist. Eine Instrumentalvariable (oder kurz: Instrument W) ist eine Variable, die folgende Eigenschaften hat:

(1) Sie korreliert mit X, und dies möglichst hoch und nach Kontrolle aller anderen Prädiktoren (**„Relevanz-Kriterium"**). Diese Korrelation darf durch den Effekt von W auf X oder durch andere Faktoren zustande gekommen sein. Sie darf lediglich nicht aus einem Effekt von X auf W resultieren (zur Begründung siehe Punkt 3).

(2) Sie korreliert nicht mit Y, wenn X auspartialisiert/statistisch kontrolliert wird (**„Exklusions-Kriterium"**). Eingefügt in ein SEM impliziert diese Annahme, dass das Instrument keinen direkten Effekt auf Y hat.

(3) Sie korreliert nicht mit dem Fehlerterm von Y, d.h. sie muss exogen bezüglich Y sein (**„Exogenitätskriterium"**). Eine Korrelation mit dem Fehler wür-

de z.B. daraus resultieren, dass das Instrument mit ausgeschlossenen Prädiktoren von Y korreliert (die sich dann im Fehlerterm befänden) oder ebenfalls von einer gemeinsamen Ursache von Y beeinflusst wird. Geht die Korrelation zwischen X und W beispielsweise auf einen Effekt von X auf W zurück, korreliert W mit der ausgeschlossenen Ursache von X (und Y) und folglich mit dem Fehler von Y (gemäß Pfadregel 1).

Wie man leicht erahnen kann, ist die Identifikation von Instrumenten die zentrale Herausforderung. Sie erfordert bereits bei der Planung der Studie zusätzlichen Aufwand. Aber es lohnt sich. Die Identifikation von Instrumenten sollte vor allem vor dem Hintergrund zweier Fragen erfolgen: Als erstes stellt sich die Frage nach der Relevanz, d.h. was könnte eine Variable sein, die mit X möglichst hoch korreliert? Dies könnte z.B. eine direkte Ursache sein, am besten so eindeutig, klar und hoch korreliert mit X, dass es fast schon trivial ist.

In einer Studie über den Effekt von Stressoren im Dienstleistungsbereich auf Burnout suchten wir beispielsweise nach Instrumenten für „Kundenstressoren" und erwogen, was Gründe dafür sein könnten, dass Kunden unzufrieden sind und mit Ärger reagieren. Erfolgreich getestet wurden schließlich das Ausmaß an a) *Warteschlangen*, b) *defekten Produkten oder mangelhaftem Service* und c) *Gedränge im Laden*.

Ein Beispiel aus der Literatur ist die berühmte Kriminalitäts-Studie von Levitt (1997, 2002), in der die Frage untersucht wurde, ob die Polizeistärke in Städten eine Reduzierung der Kriminalität bewirkt. Problem hierbei war, dass die Kriminalitätsrate möglicherweise einen simultanen, aber positiven Effekt auf die Polizeistärke haben könnte. Dies würde wegen der gegenläufigen Effekte erklären, warum es keinen einfachen Zusammenhang gibt. Levitt identifizierte als Instrument die Frage, ob im Jahr der Studie in der Stadt eine Bürgermeisterwahl stattgefunden hatte, da Bürgermeister dazu tendierten, im Wahlkampf die Polizeistärke aufzustocken. Mit Hilfe dieses Instrumentes konnte er einen Effekt der Polizeistärke nachweisen.

Während eine Identifikation von Variablen, die mit X korrelieren sollten, relativ einfach sein dürfte, ist eine Beantwortung der zweiten Frage schwieriger, nämlich, ob sie keinen direkten Effekt auf Y haben, oder mit dem Fehlerterm von Y korrelieren (Angrist & Krueger, 2001). Beides ist dann der Fall, wenn sie einen direkten Effekt auf die Outcome-Variable haben oder mit ausgeschlossenen Variablen korrelieren (also mit Kovariaten von X oder von ausgeschlossenen Ursachen von Y beeinflusst werden). In der o.g. Studie über Kundenstress sollten beispielsweise die vermuteten Gründe für Kundenstress keinen direkten Effekt auf Burnout haben und auch nicht mit Kovariaten von Kundenstressoren korrelieren. Wenn eine Korrelation mit einer vermuteten Kovariate plausibel war, wurde diese mit erhoben und in das Modell integriert. So bestand z.B. die Möglichkeit, dass die Variable „defekte Produkte" (also eine negative Bewertung der eigenen Produkte und Dienstleistungen) von negativer Affektivität beeinflusst ist und negative Affektivität auch Burnout beeinflusst. Daher wurde negative Affektivität mit erhoben und die erwartete Struktur explizit modelliert. Diese bestand darin, negative Affektivität als „common cause" von „defekte Produkte" und Burnout zu modellieren.

7.3 Funktionsweise von Instrumentalvariablen

Warum sind Instrumentalvariablen so mächtig und wie können sie helfen, selbst mittels Querschnittdaten die Evidenz für kausale Interpretationen zu stärken? Der klassische Instrumentalvariablenschätzer in der Ökonometrie im Falle endogener Regressoren ist **two stage least squares** (TSLS, 2SLS). Dabei wird in einem ersten Schritt eine Regression von X auf W (das Instrument) – oder besser mehrere W's durchgeführt. Als Ergebnis liefert dieser Schritt erstens Informationen über die Relevanz der W's und zweitens durch die W's vorhergesagte X-Werte, d.h. die Werte, die auf der Regressionsgeraden liegen. Der zweite Teil der TSLS-Schätzung benutzt nun nur diese vorhergesagten Werte zur Schätzung des X-Y-Effekts. Die Logik dahinter ist folgende: Die vorhergesagten Werte sind eine gewichtete Summe der W's, die durch die Regressionsgleichung $\hat{X} = \beta_1 W_1 + \beta_2 W_2 + \ldots \beta_k W_k$ gebildet werden. Da die W's exogen sind, ist folglich auch jede ihrer Kombinationen exogen und der kontaminierte Teil der Varianz von X somit im Residuum dieser ersten Regressionsgleichung enthalten. Dies bedeutet, dass zur Vorhersage von Y nur *der Teil von X* verwendet wird, der nicht mit dem Fehler von Y korreliert. Dies führt zu unverzerrten und konsistenten Schätzern des Effekts von X auf Y, auch wenn nicht alle omitted variables berücksichtigt wurden, X messfehlerbehaftet ist oder Y zusätzlich X beeinflusst. Hat X überhaupt keinen Effekt auf Y, sondern liegt ein reiner reverse-causation-Prozess vor, wird diese Schätzung einen nicht-signifikanten Effekt des X - Y - Effekts bringen, wo eine naive OLS-Schätzung (ohne Instrumente) einen signifikanten und vielleicht substantiellen Einfluss suggerieren würde.

Anstatt nun eine TSLS-Regression zu rechnen, lassen sich äquivalente Ergebnisse in einem SEM erzielen, wenn denn die Instrumente die o.g. Eigenschaften besitzen (Foster & McLanahan, 1996; Gennetian, Magnuson, & Morris, 2008). Die Eigenschaften lassen sich allerdings ebenfalls in dem SEM testen. Abb. 29 zeigt die Grundstruktur des Modells, wie wir sie schon in Abschnitt 5.1 kennengelernt haben. Dabei werden die Instrumente einfach als Ursachen von X modelliert. Deren Effekte auf X und das R-Quadrat dienen zur Beurteilung des Relevanz-Kriteriums. Daneben erlaubt das Vorhandensein mehrerer Instrumente den sog. **Sargan-Tests** (Sargan, 1958) der beiden anderen Annahmen – nämlich, dass die Instrumente nicht mit dem Fehlerterm von Y korrelieren und auch keinen direkten Effekt haben dürfen. Das Analogon in einem SEM ist der Chi-Quadrat-Test der auf Null fixierten Effekte der W's auf Y. Der Fokus auf den statistischen Test macht auch bereits klar, dass der Instrumentalvariablen-Ansatz mit der üblichen Fokussierung auf Fitindizes schlecht durchführbar ist.

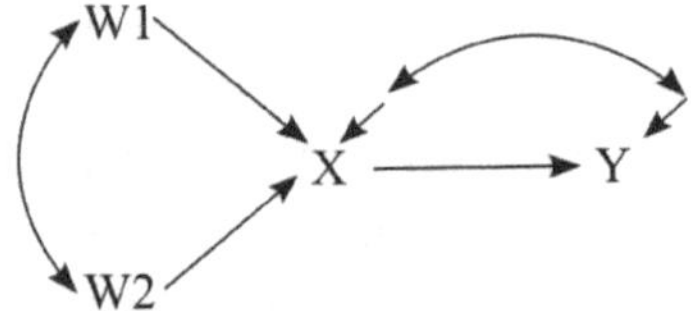

Abb. 29: Modell mit zwei Instrumentalvariablen W1 und W2

Was in einem Modell mit einem endogenen Regressor zentral ist, ist die **Schätzung der Fehlerkovarianz** von X und Y. Diese hat zwei Funktionen: (1) Zum einen stellt sie ein Analogon des sog. Hausman-(Durban-Wu)-Tests (Hausman, 1978) dar, der testet, ob X überhaupt endogen ist. Angewendet auf das SEM zeigt die Signifikanz der Fehlerkovarianz an, dass X endogen ist. Zum zweiten ist die Schätzung dieser Fehlerkovarianz im Fall von Endogenität nötig, um einen unverzerrten Effekt von X auf Y zu bekommen. D.h., das Vorhandensein exogener Instrumente hilft nicht für die Schätzung für den X-Y-Effekt, wenn die Fehlerkovarianz fixiert ist. Für die effiziente Schätzung der Fehlerkovarianz benötigt man relevante Instrumente. Sind Instrumente nur schwach mit X korreliert (sog. **„weak instruments“**-Situation), hat dies schwerwiegende Folgen:

a) Der Effekt von X wird vor allem in kleineren samples verzerrt; u.U. mehr, als wenn man das ganze Modell naiv und ohne Instrumente rechnen würde.
b) Der Standardfehler des Effekts von X steigt, was über die Verzerrung des Effekts hinaus dazu führt, dass er nicht signifikant wird.
c) Der Standardfehler der Fehlerkovarianz steigt ebenfalls deutlich an, was folglich die statistische Power des Hausman-Tests reduziert. Man bekommt so den Eindruck, dass X gar nicht endogen ist.

Aus dem Gesagten folgt, dass starke Instrumente (oder allgemeiner Prädiktoren der unabhängigen Variablen von Interesse) hilfreich sind, um die Fehlerkovarianz der nachgeschalteten Gleichung zu schätzen. Dies wiederum kann die negative Wirkung von endogenen unabhängigen Variablen eliminieren. Daher sollte eher eine Schätzung der Fehlerkovarianz angestrebt werden, anstatt diese wie üblich auf null zu fixieren. Hat man nur ein Instrument pro unabhängiger Variable, das aber plausibel ist, ist das eine Verbesserung gegenüber einem Modell ohne Instrument, hat aber die Limitation, dass man seine Exogenität nicht testen kann. Daher sind mehrere Instrumente hilfreich.

Schlägt der Test auf Exogenität (Sargan-Test) fehl, ist mindestens eines der Instrumente selbst endogen. Ist er erfolgreich, erhöht das (wie allgemein bei Tests der Fall) die Plausibilität für Exogenität, beweist sie aber nicht. So kann es durchaus sein, dass alle Variablen im Modell (d.h. die W's, X und Y) von einer gemeinsamen Ursache beeinflusst werden (z.B. die Verwendung einer gemeinsamen Methode führt zu **common method bias**). Hier kann es durchaus passieren, dass dies der Chi-Quadrat-Test nicht bemerkt – v.a. wenn es sich um einen sehr homogenen Einfluss auf alle Variablen handelt. Wenn man also eine Lösung des Common-Method-Bias-Problems durch die Einbeziehung von Instrumenten anstrebt, sollte man zur Messung der Instrumente eine andere Methode verwenden (z.B. anderes Skalenformat oder die Form der Befragung), um das Risiko für einen common method bias zu vermindern. Im besten Fall führt dies dazu, dass das Instrument überhaupt nicht von demjenigen Antwortstil beeinflusst wird, der X und Y beeinflusst. Im weniger idealen Fall impliziert dies zumindest differentielle Effekte, was die statistische Power des Tests erhöht.

Die bislang betrachtete Situation bezog sich auf einen einfachen Effekt von X auf Y. Häufig zielen Modellspezifikationen aber auf die Schätzung **indirekter Effekte** von X auf Y über den **Mediator** M ab. In einem solchen Modell ist nicht nur der Effekt von X auf den Mediator M möglicherweise von einer reziproken kausalen Richtung oder ausgeschlossenen Drittvariablen betroffen, sondern auch der Effekt von M auf Y. Daher ist Endogenität auch in **experimentellen Designs** eine Gefahr, in denen die Wirkung des Treatments auf das Outcome über einen Mediator geprüft werden soll. Denn obwohl X durch die Randomisierung exogen ist, kann die Beziehung zwischen M und Y durch eine ausgeschlossene gemeinsame Ursache beeinflusst sein oder aus einem simultanen Effekt resultieren. Beispielsweise könnte vermutet werden, dass der Effekt eines Trainings (X) auf die Leistung (Y) vermittelt wird über Selbstwirksamkeit (M). Selbst wenn die Trainings- und Kontrollgruppe **randomisiert** sind, ist es plausibel, dass Selbstwirksamkeit und die Leistung von ausgeschlossenen gemeinsamen Ursachen beeinflusst werden. Spezifiziert man ein Mediatormodell, wird der indirekte Effekt des Treatments verzerrt.

Die Lösung kann darin bestehen, X als Instrument für den M → Y - Effekt zu benutzen: Ist X exogen bzgl. Y und hat einen substantiellen Effekt auf M, reicht dies, um die Kovarianz der Fehler von M und Y zu schätzen. Nachteil ist natürlich, dass bei Vorhandensein von nur einer X-Variable deren Exogenität lediglich auf Plausibilität beruht und nicht getestet werden kann. Weiterhin setzt die Nutzung von X als Instrument voraus, dass es sich um eine volle Mediation handelt und X somit **exklusiv** ist (die oben genannten Regeln für Instrumente gelten dann für X genauso).

In Fällen, in denen X randomisiert wurde, sollte Bedingung 1 (Exogenität bzgl. M und Y) in der Regel erfüllt sein. Eine Ausnahme kann dadurch bedingt sein, das – z.B. bedingt durch eine geringe Stichprobengröße – die Treatment-Variable trotz Randomisierung zufällig doch mit ausgeschlossenen Drittvariablen korreliert. Kritisch ist in jedem Fall Bedingung 2 – d.h. die Notwendigkeit einer vollen Mediation (Exklusivität) und eine ausreichende Stärke des Effekts des Treatments (Relevanz). Ist eines von beidem nicht gegeben, sollte man ein oder mehrere Instrumente für den Mediator identifizieren, um den M-Y-Teil des Modells und damit den indirekten Effekt des Treatments konsistent schätzen zu können.

In Fällen aber, in denen X nicht randomisiert, sondern frei gemessen wurde, ist X unter Umständen selbst endogen, und muss ebenfalls instrumentiert werden. Durch die Instrumentierung wird $\hat{X}$ wiederum zum Instrument für den M - Y - Effekt. Auch hier bleibt die Voraussetzung bestehen, dass a) der Effekt von $\hat{X}$ auf M substantiell ist (Relevanz-Kriterium) und b) $\hat{X}$ keinen direkten Effekt auf Y hat (Exklusions-Kriterium). Andernfalls sollten auch hier für M eigene Instrumente einbezogen werden. Dies wird in unserem Anwendungsbeispiel im nächsten Abschnitt illustriert.

7.4 Betrachtung des Work-Family-Conflict-Modells und Fazit

Betrachten wir jetzt das in diesem Buch diskutierte Work-Family-Conflict-Modell und das modifizierte, sauber fittende Modell in Abschnitt 5.3, so muss eingestanden werden, dass die kausale Überzeugungskraft nicht sehr hoch ist. Allem voran fehlen Kovariate von work-family conflict – also weitere Einflussfaktoren von Arbeitszufriedenheit, die mit work-family conflict korrelieren könnten; daneben besteht die Möglichkeit reziproker Kausalität (Arbeitszufriedenheit beeinflusst in Wirklichkeit work-family conflict) oder ausgeschlossener gemeinsamer Ursachen, was eine Gefahr für die Validität des Modells darstellt. Die Schwierigkeit hierbei ist, dass nicht nur Instrumente für work-family conflict gefunden werden müssen, sondern auch für Arbeitszufriedenheit. Grund dafür ist der direkte Effekt von work-family conflict auf Kündigungsgedanken.

Im Folgenden soll Schritt für Schritt das Vorgehen illustriert werden – auch die Schwierigkeiten, die während des Prozesses der Stärkung des Modells aufgetreten sind. Diese bestanden darin, dass im ersten Schritt Instrumente ausgewählt wurden, die wohl nicht ausreichend genug mit den Zielvariablen korreliert waren. Also wurden weitere Instrumente ausgewählt. Das finale Modell ist in Abb. 30 abgebildet. Da die angedeuteten Probleme häufig vorkommen dürften, wird im Folgenden der Prozess der Modellentwicklung detailliert geschildert.

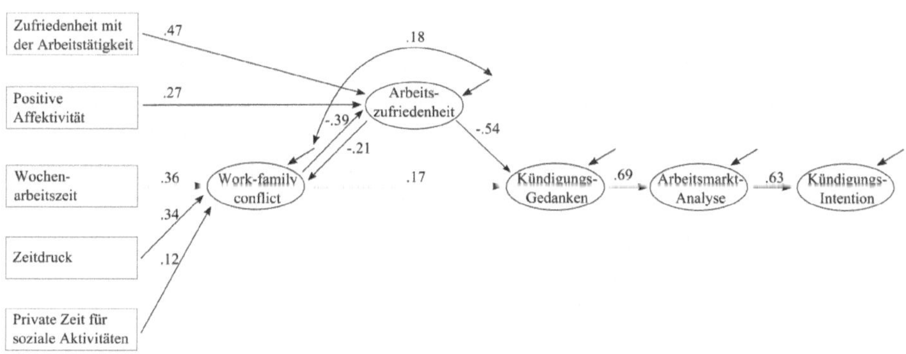

Abb. 30: Modell mit Instrumentalvariablen und reziproken Effekten

Wie im vorherigen Kapitel besprochen, war das Ziel zunächst, zwei Instrumente für jeweils work-family conflict und Arbeitszufriedenheit zu finden. Diese sollten möglichst hoch mit diesen beiden Variablen korrelieren und im Modell weder direkte Effekte auf alle anderen Variablen haben noch mit ausgeschlossenen Variablen korrelieren, die auf die anderen Variablen Effekte haben. Da die Identifikation von Instrumenten *theoretisch fundiert* sein sollte (Gennetian et al., 2008), lagen der Inspektion des vorhandenen Datensatzes die Fragen zugrunde, welche Konstrukte möglichst klare und enge Zusammenhänge mit diesen beiden Variablen aufweisen dürften. Eine theoretisch basierte Suche grenzt den Suchbereich ein und erhöht die Plausibilität der Rolle eines Instruments.

Für Arbeitszufriedenheit fiel die Wahl auf das Persönlichkeitsmerkmal *„positive Affektivität"* und die *„Zufriedenheit mit den Arbeitsaufgaben"*. Positive Affektivität kennzeichnet die stabile Disposition einer Person und hat einen gut belegten Einfluss auf die Arbeitszufriedenheit (Ilies & Judge, 2003). Zufriedenheit mit den Arbeitsaufgaben sollte ebenfalls einen Einfluss haben, da allgemeine Bewertungen der Arbeit auf der Bewertung der Tätigkeit als essentiellem Teil beruhen sollten. Insbesondere dieses Beispiel zeigt, dass häufig starke Instrumente solche Variablen sind, deren Zusammenhang so auf der Hand zu liegen scheint, dass er wissenschaftlich selbst kaum von Interesse ist. Zur Messung von positiver Aktivität wurde eine aus sechs Items bestehende Skala von Tellegen und Waller (1992) benutzt. Diese Items wurden zu einem manifesten **Variablenwert** zusammenaddiert. Da diese Variable hier eine reine Hilfsfunktion hatte, war auf die Modellierung der Messstruktur verzichtet worden. Die Zufriedenheit mit den Arbeitsaufgaben wurde mit der Frage „Wie zufrieden sind Sie im Allgemeinen mit ihrer Arbeitstätigkeit?" erfasst.

Für work-family conflict wurden (zunächst) die *Wochenarbeitszeit* und die Erfordernis, Zeit für soziale Aktivitäten im Privatleben aufwenden zu müssen (*„private Zeit für soziale Aktivitäten"*), einbezogen. Die Wochenarbeitszeit stellt einen wichtigen Einflussfaktor von work-family conflict dar und sollte weder auf Arbeitszufriedenheit noch auf Kündigungsgedanken direkt, sondern nur vermittelt über work-family conflict wirken. Diese vier Instrumente wurden in das SEM eingefügt – dabei wurden lediglich ihre Effekte auf ihre entsprechende abhängige Variable (d.h. Arbeitszufriedenheit oder work-family conflict) geschätzt.

Von Interesse waren hier zunächst a) die mögliche Veränderung des Modellfits (Sargan-Test) sowie b) die Struktureffekte der Instrumente und die Höhe der Varianzerklärung der Zielvariablen durch die Instrumente, weil diese die Relevanz der Instrumente kennzeichnen. Ein auftretender misfit des Modells könnte unter anderem nun darin begründet sein, dass eine oder mehrere Instrumente direkte Effekte auf die nachfolgenden Variablen haben oder, dass die Kovarianz der Fehler zwischen der Zielvariable und ihrer Folge signifikant ist. Somit sollte mittels Modifikationsindizes analysiert werden, ob für diese fixierten Parameter ein hoher Modifikationsindex angegeben ist. Dabei sollten allerdings auch theoretische Einschätzungen über die Plausibilität eines möglichen direkten Effekts im Vordergrund stehen.

Die Syntax dieses Modells ist identisch zu der des modifizierten Modells in 5.3 – nur das hier nun die Instrumente hinzukommen:

```
wfcSEM3 <- '
  WFC =~ w01+w02+w03
  AZF =~ az01 + az02
  Gedanken =~ km01
  Markt =~ km02
  Int =~ km03
  AZF ~ WFC
  Gedanken ~ AZF + WFC
  Markt ~ Gedanken
  Int ~ Markt
  km01 ~~.10*km01
  km02 ~~.10*km02
  km03 ~~.09*km03

#Instrumente
  WFC ~ WAZ + soz_akt
  AZF ~ pos_aff + azf_aktiv
'
```

Bei der Ausgabe des Outputs fordern wir nun die R-Quadrat-Werte an:

```
fit3 <- sem(wfcSEM,data=wfcdata, missing="FIML",
           estimator="mlr")
summary(fit3,standardized=T, modindices=T, rsquare=T)
```

Die Ergebnisse zeigen, dass das Modell weiterhin fittet ($\chi^2(46) = 54.34$, $p = .19$). Die Effekte der Instrumente sind substantiell und signifikant, die R-Quadrat-Werte sind für Arbeitszufriedenheit .44 und für work-family conflict .23. Nach Staiger und Stock (1997) spricht ein **F-Wert** des Teilmodells, in dem die Zielvariable auf ihre Instrumente regressiert wird, in Höhe von über 10 für eine ausreichende Relevanz der Instrumente. Den F-Wert berechnen wir aus dem R-Quadrat mittels $(R^2/k) / [(1-R^2) / (n-k-1)]$, wobei k die Anzahl der Instrumente und n die Stichprobengröße ist. In Fällen, in denen die Zielvariable sowohl von den Instrumenten als auch von anderen Modellvariablen beeinflusst wird, muss das partielle R-Quadrat (d.h. das Inkrement durch die Berücksichtigung der Instrumente) berechnet werden. Für Arbeitszufriedenheit erhalten wir einen F-Wert von $(.44/2) / [(1-.44) / (352-2-1)] = 137.11$ und für work-family conflict einen Wert von 52.12.

Es spricht also erst einmal nichts gegen die Validität der Instrumente. Lässt man jetzt allerdings die Kovarianz der Fehler zwischen work-family conflict (mit `AZF~~WFC`) schätzen, um zu testen, ob für work-family conflict überhaupt ein Endogenitätsproblem vorliegt (Hausman-Test), passiert etwas scheinbar Paradoxes: Während der Effekt von work-family conflict durch die Einbeziehung der Instrumente auf -.23 abfällt, steigt er nun wieder auf -.37; die Fehlerkovarianz ist aber nicht signifikant (z = 1.7, p = .09). Dieses Verhalten suggeriert hierbei trotz des ausreichenden F-Wertes die typische Reaktion auf eine „*weak-instrument*"-Situation, da bei schwachen Instrumenten sowohl die Standardfehler des Effekts als auch die Fehlerkovarianz steigen und die Tests dieser Parameter somit an statistischer power einbüßen. Dies kann soweit führen, dass der Effekt nicht mehr signifikant ist, obwohl der Hausman-Test eigentlich kein Zeichen von Endogenität feststellt.

Um dies zu vermeiden, wird mit der Variable „Zeitdruck" ein weiteres Instrument für work-family conflict eingefügt; ein Effekt von Arbeitsplatzbelastungen auf work-family conflict ist theoretisch plausibel (Greenhaus & Beutell, 1985) und empirisch gut belegt (Byron, 2005). Das Einfügen dieses weiteren Instruments verläuft erfolgreich: Das Modell fittet ($\chi^2(53) = 60.71$, p = .22); das R-Quadrat für work-family conflict steigt auf .35. Schätzt man nun die Fehlerkovarianz, ist diese nun signifikant (z = 2.22, p = .03) und der Fit verbessert sich weiter ($\chi^2(52) = 55.76$, p = .33). Dieses Ergebnis stützt die Vermutung, dass die vorherigen Instrumente für work-family conflict zu schwach waren und dass work-family conflict bezüglich seines Effekts auf Arbeitszufriedenheit tatsächlich endogen ist.

Um die Endogenitätsvermutungen für die nachfolgenden Effekte zu prüfen, werden sukzessive die Fehlerkovarianzen der nachfolgenden Variablenpaare geschätzt. Dies macht die vollständige Mediatorstruktur möglich. Da work-family conflict und Arbeitszufriedenheit nun „instrumentiert" sind, können sie bzgl. ihrer nachfolgenden Wirkungen so behandelt werden wie exogene Variablen. Dazu sind ihre Effekte auf Kündigungsgedanken und die nachfolgenden Effekte stark genug, um die Fehlerkovarianzen dieser Variablenpaare schätzen zu können. Tut man dies, verändert dies die Effekte nicht, und die Fehlerkovarianzen sind nicht-signifikant. Die Kausalitätsannahme erfährt somit eine Stärkung – und zwar deutlicher als im ursprünglichen Modell.

Da wir nun in der glücklichen Situation sind, sowohl für work-family conflict als auch für Arbeitszufriedenheit Instrumente zu haben, ergibt sich die Möglichkeit, simultane Effekte beider Variablen aufeinander zu prüfen. Dazu wird der Effekt von Arbeitszufriedenheit auf work-family conflict geschätzt (mit `WFC ~ AZF`). Die Syntax dieses finalen Modells lautet:

```
wfcSEM4 <- '
  WFC =~ w01+w02+w03
  AZF =~ az01 + az02
  Gedanken =~ km01
  Markt =~ km02
  Int =~ km03
  AZF ~ WFC
  Gedanken ~ AZF + WFC
  Markt ~ Gedanken
  Int ~ Markt
  km01 ~~.10*km01
  km02 ~~.10*km02
  km03 ~~.09*km03

#Instrumente
  WFC ~ WAZ + soz_akt + Zdruck + AZF
  AZF ~ pos_aff + azf_aktiv
'
```

Als Ergebnis steigt der Fit nochmals auf einen exzellenten Wert an ($\chi^2(51) = 48.85$, $p = .56$) und der Effekt ist signifikant. Daneben verändert sich keiner der anderen Parameter: Wie Abb. 30 zeigt, bleibt der Effekt von work-family conflict auf Arbeitszufriedenheit sowie die Fehlerkovarianz beider Variablen unverändert und signifikant.

Wie man in Abb. 30 sieht, hat sich das Modell von seinen Anfängen als Pfadmodell (Abschnitt 4.8) über die Modellierung der latenten Struktur (Abschnitt 4.10), der Respezifikation der Struktur der Kündigungsvariablen (Abschnitt 5.3) und nun durch die Einbeziehung der Instrumentalvariablen enorm verändert. Mit dieser Veränderung ist ein Lernprozess einhergegangen. Gegenstand dieses Lernprozesses war, dass Kündigungsmotivation – obwohl als intuitiv plausible gemeinsame Ursache dreier Aspekte von Kündigungen (Gedanken, Stellenanzeigen lesen und Kündigungsabsicht) konzeptualisiert – keine existierende Variable ist und die mit ihr verbundene Kausalstruktur wohl falsch ist. Basis dieser Respezifikation war der signifikante Chi-Quadrat-Test. Eine Berücksichtigung der gebräuchlichen Indizes hätte diese Erkenntnis nicht möglich gemacht.

Dazu haben wir durch die Einbeziehung von Instrumentalvariablen die Endogenität von work-family conflict nachgewiesen. Ihr Schaden wäre sicher begrenzt geblieben, da der Effekt im ursprünglichen Modell -.31 ist und nun auf -.39 ansteigt. Dies muss jedoch nicht immer so sein. Endogenität kann wahre Effekte völlig verschleiern oder nach unten oder oben verzerren. Zudem konnten wir eine reziproke Beziehung zwischen Arbeitszufriedenheit und work-family conflict testen, was durch die Einbeziehung von Instrumentalvariablen beider Variablen möglich war. Durch die Modellierung wurde die kausale Evidenz für dieses Modell gestärkt, auch wenn immer noch Möglichkeiten für alternative Strukturen vorhanden sein mögen. So kann der Sargan-Test, der die Exogenität der Instrumente testet, nie ausschließen, dass nicht doch ein Fehler vorliegt. Beispielsweise könnte ein uniformer Einfluss eines Methodenfaktors (Podsakoff et al., 2003) unerkannt bleiben, wenn er alle Modellvariablen (inkl. Instrumente) beeinflusst. Hier bleibt letztlich die theoretische Plausibilität als wichtiges Kriterium. Im Fall des hier getesteten Modells ist beispielsweise die Wochenarbeitszeit über eine Auflistung der typischen Anfangs- und Feierabendzeiten errechnet worden und kann somit unmöglich durch dieselben Beantwortungstendenzen beeinflusst sein wie die anderen Modellvariablen.

Die Durchführung des hier illustrierten Modells mitsamt Modellvariablen lief erfolgreich: Die Rolle der Instrumente und ihre Koeffizienten sind plausibel, sie hielten den verschiedenen Tests stand und der Chi-Quadrat-Test des gesamten Modells war trotz der substantiellen Stichprobengröße nicht-signifikant mit einem hohen p-Wert. Dies muss alles nicht immer so sein. In manchen Kontexten könnten Instrumente schwer zu finden sein und die Suche nach ihnen häufig umsonst erfolgen. Auch kann der Chi-Quadrat-Test auf Probleme des Modells hinweisen, aber kein diagnostischer Schritt führt zu einer fittenden und plausiblen Alternative. Und dennoch lohnt sich die Suche nach Instrumenten und ein Ernstnehmen angezeigter Probleme im Modell, denn das Wegerklären eines misfits oder vorzeitiges Akzeptieren eines vielleicht problematischen Modells kann nicht im Sinne seriöser Wissenschaft sein.

8

Mythen und Fallstricke

In dem nun folgenden Kapitel soll kurz auf bestimmte Überzeugungen eingegangen werden, die häufig einer erfolgreichen Analyse von SEM entgegenstehen. Einige davon sind die Grundlage für Skepsis gegenüber SEM oder sogar für die dessen völlige Ablehnung; andere Rechtfertigung für die Herabwürdigung des Tests von Modellen. Diese werden nun kurz adressiert. Die Absicht hierbei ist weniger, eine ausführliche Abhandlung zu liefern als viel mehr, Denkanstöße zu liefern. Schließlich sei erwähnt, dass die folgenden Abschnitte modular gestaltet sind. Somit können sie unabhängig voneinander gelesen werden, weisen dadurch allerdings leichte Redundanzen auf.

8.1 SEM ist (k)eine Methode

SEM wird von vielen Forscher/innen in erster Linie als Methode angesehen, die man anwendet, so wie man Faktorenanalyse, Mittelwertsvergleiche, Regressionsanalyse oder die cluster-Analyse anwendet. Erarbeitet sich ein Forscher eine profunde Expertise in SEM, bekommt er (oder sie) unter Kollegen schnell den Status des Methodikers. Der Kern dieser Vorstellung ist sicher wahr – SEM impliziert die Notwendigkeit, sich methodisches und statistisches Wissen anzueignen, da dieses eine bedeutende Rolle bei der Schätzung und Interpretation der Ergebnisse spielt. Dies sollte allerdings nicht darüber hinwegtäuschen, dass SEM in erster Linie ein Werkzeug des Theoretikers ist, weil in der Spezifikation eines SEM ein Grad an Konkretheit und Präzision erforderlich ist, wie er in keiner anderen Methode verlangt wird (z.B. Varianzanalyse, Regressionsanalyse). Während die Theorie- und Hypothesenentwicklung meist auf der (oft sehr vagen) Ebene von „Konstrukten" und ihren „Beziehungen" verbleibt, zwingt die Spezifikation eines SEM a) zur Konzeptualisierung eines Konstrukts in Termini spezifischer singulärer Entitäten und b) aller vermuteten kausalen Effekte und Restriktionen. SEM ist also ein allgemeines Rahmenwerk zu Übersetzung allgemeiner kausaler Theorien in ein empirisches Modell. Es geht daher über eine bestimmte Analyseform hinaus und integriert alle Methoden zur Spezifikation kausaler Effekte (Pearl, 2010).

8.2 Kausale Interpretation von SEM

Während in der Anfangsphase SEM noch (korrekt) als Kausalmodelle bezeichnet wurden, ist die Skepsis gegenüber dem kausalen Gehalt in den letzten Jahrzehnten enorm gestiegen. Grund ist, dass Parameterschätzungen in den Anfängen von SEM zu leichtfertig kausal interpretiert wurden. Als Folge bewegten sich die wissenschaftlichen Disziplinen, in denen SEM angewendet wird, in das Gegenteil. Kausale Sprache wird völlig vermieden, stattdessen werden Begriffe wie „Zusammenhänge" (zwischen Variablen), „Beziehungen" oder „Assoziationen" genutzt. SEM als Kausalmodellierung zu bezeichnen, war viele Jahrzehnte verpönt (bzw. ist es immer noch).

Im Gegensatz zu dieser wenig hilfreichen oder sogar schädlichen Vermeidung kausalen Denkens (siehe dazu auch 8.7) sollte nach der Lektüre dieses Buches klar geworden sein, dass die theoretische Grundlage eines Modells kausal ist und auch sein muss. Die Pfeile drücken kausale Annahmen aus und die „Lücken" in einem Modell (d.h. die Restriktionen) ebenfalls. Es ist ein allgemeines wissenschaftstheoretisches Problem, dass im Falle einer Übereinstimmung der Daten mit einer Hypothese dennoch nie mit Sicherheit von der Korrektheit der Hypothese ausgegangen werden kann (andernfalls begeht man die „fallacy of affirming the consequent"). D.h. selbst wenn das Modell fittet, kann es dennoch falsch sein. Die einzige Lösung ist: a) die kausalen Annahmen zu testen und so den Grad der Evidenz im Falle eines fittenden Modells zu erhöhen und b) die Parameterschätzungen im Sinne einer „Wenn-Dann-Beziehung" zu interpretieren (wenn das Modell korrekt ist, dann gibt der Parameter den kausalen Effekt an). Die Rigidität der Testung und die Anstrengungen, die Endogenitäts-Gefahr zu reduzieren (vgl. Kap. 7) stärkt die „Wenn"-Komponente dieser Beziehung.

8.3 All models are wrong (?)

Häufig besteht die Auffassung, dass ein Modell gar nicht korrekt sein *kann* (bzw. dass Modelle an sich falsch sind) und dass es deshalb auch keinen Sinn mache, statistische Tests zur Überprüfung seiner Korrektheit einzusetzen. Dabei wird sich auf ein berühmtes Zitat des Statistikers George Box berufen („*All mo-*

dels are wrong, but some are useful“). Im Folgenden soll kurz argumentiert werden, dass dies für den Bereich von SEM ein Mythos ist. Modelle mögen bezüglich mancher Aspekte grundsätzlich falsch sein, allerdings hat das keine Implikationen für die Essenz von Strukturgleichungsmodellen. Diese besteht darin, a) Variablen zu modellieren, die tatsächlich existierenden Phänomenen entsprechen und keine statistischen Artefakte sind und b) kausale Annahmen über vorhandene und nicht vorhandene kausale Effekte zu bilden. Es gibt keinen Grund zur Annahme, dass *alle* Modelle bezüglich dieser beiden Aspekte (immer) falsch sind.

Zunächst aber schauen wir uns das Zitat von George Box genauer an. Was bei der Nutzung dieses Zitats häufig übersehen wird, ist, dass es aus einem völlig anderen Kontext stammt. Kontext des Zitats war die Modellierung von Polynomen in einer komplexen Gleichung (z.B. $Y = a + b1X + b2Z + b3XZ + b4X^2 + b5Z^2$) zur Darstellung einer „response surface“, wie sie z.B. bei Kongruenz-Analysen angewendet wird. Bei solchen Analysen besteht die Hypothese, dass der Effekt der Interaktion zweier Variablen von ihrer Passung abhängt (Edwards, 1994; Edwards & Cable, 2009; Edwards & Parry, 1993). Das genaue Zitat lautet diesbezüglich „...*The fact that the polynomial is an approximation does not necessarily detract from its usefulness because all models are approximations. Essentially, all models are wrong, but some are useful. However, the approximate nature of the model must always be borne in mind.*" (Box & Draper, 1987, S. 424). Dies zeigt, dass es um die Angemessenheit des Polynoms für den wahren Koeffizienten ging und nicht um Modelle im Allgemeinen (und schon gar nicht um Kausalmodelle und deren Implikationen). Analog könnte man fragen, ob eine Regressionsgerade in der Population wirklich exakt eine Gerade ist, oder ob sie nicht die eine oder andere Abweichung von der Gerade aufweist.

Worin besteht nun aber der Zweck der Nutzung dieses (verkürzten) Zitats im Kontext von Kausalmodellen? Wenn alle Modelle grundsätzlich falsch sind: warum entwickeln wir dann Modelle und erheben Daten, um sie zu evaluieren? Und wenn manche dieser (falschen) Modelle nützlich sind, wie erkennen wir diese Modelle? Welche Form von „Inkorrektheit“ eines Modells minimiert den Nutzen des Modells und welcher lässt ihn unangetastet? Und was ist überhaupt *der* Nutzen eines Modells? Ist es ein Nutzen für die Vorhersagbarkeit? Für das Verständnis kausaler Effekte? Für Möglichkeiten einer Intervention?

In den meisten Fällen wird das Zitat als Argument für die Ablehnung des Chi-Quadrat-Tests und für die Benutzung der Fitindizes verwendet. Die Grundlogik ist dabei, dass – da ja alle Modelle falsch sind – der Chi-Quadrat-Test nie nicht-signifikant sein kann. Häufig wird dabei nicht spezifiziert, welcher Aspekt eines Modells immer falsch ist und ob die Inkorrektheit des entsprechenden Aspekts überhaupt Implikationen für die Essenz des Modells hat, nämlich den Test kausaler Annahmen. Wie unten gezeigt wird, kann es durchaus Aspekte geben, für die die Aussage der grundsätzlichen Inkorrektheit zutrifft – dadurch ist aber nicht impliziert, dass Aussagen über kausale Effekte und Restriktionen grundsätzlich falsch sein müssen. Daher lohnt es sich, zunächst zu diskutieren, für welche Aspekte es tatsächlich plausibel ist, dass wohl alle Modelle falsch sind. Ist dies erfolgt, kann die Frage beantwortet werden, welche Implikationen dies für den Test von Modellen hat.

Für folgende Aspekte ist es zumindest plausibel, anzunehmen dass sie immer „falsch“ sind (ob daraus folgt, dass alle Modelle zwangsläufig falsch sein *müssen*, sei dem/der Leser/in überlassen):

a) **Das Modell ist lediglich eine Abstraktion der Realität.** Häufig wird aus der Tatsache, dass das Modell nur eine Abstraktion und Vereinfachung der realen Welt sei und nicht *die* Welt (vgl. auch Abschnitt 8.9) geschlussfolgert, dass dadurch alle Modelle falsch sind (d.h. das Modell ist schon deshalb falsch, weil es nur ein Modell ist). Hierbei wird häufig die Analogie mit einer Stadtkarte hergestellt, die natürlich nicht „die Stadt“ ist. Wenn allerdings die grundlegenden Annahmen, die eine Stadtkarte erfüllen muss, darin bestehen, dass alle Straßen abgebildet sind sowie Abzweigungen und Kreuzungen korrekt eingezeichnet und benannt wurden, gibt es keinen Grund anzunehmen, dass eine Straßenkarte (bzgl. dieser Annahmen) immer falsch sein sollte. Das gleiche gilt für Kausalmodelle. Sie mögen nicht *die* Realität sein, aber sie sind eine Zusammenstellung von Annahmen, für die gilt, dass sie entweder richtig oder falsch sind.

b) **Messfehler.** Ein weiteres Argument für die grundsätzliche Inkorrektheit von Modellen ist häufig, dass die Variablen in einem Modell nie perfekt gemessen werden können. Dies ist allerdings für ein SEM kein Problem, da Messfehler explizit modelliert und so die Effekte von Variablen geschätzt werden können, die von Messfehlern befreit sind. Außerdem sind Messfehler in abhängigen Variablen unproblematisch, weil diese keinen verzerrenden Effekt auf die Schätzung von Struktureffekten haben.

c) **Unvollständigkeit.** Hierunter fällt das Argument, dass ein Modell immer nur einen Ausschnitt der Realität ist und die abhängigen Variablen von einer Vielzahl weiterer, ausgeschlossener Variablen beeinflusst werden. Dies ist wahr, macht aber ein Modell nicht grundsätzlich inkorrekt: Ausgeschlossene Einflussgrößen der abhängigen Variablen sind Teil ihres **Fehlerterms**, was wie oben angemerkt, zu keiner Verzerrung der geschätzten Effekte führt. Zum Problem werden sie dann, wenn sie *mehrere Variablen* im Modell beeinflussen *Kovariate der unabhängigen Variablen* sind (siehe Kap. 5 über Formen der Fehlspezifikation und Kap. 7 über Endogenität). Dies ist eine plausible Gefahr und führt dazu, dass Modelle nie abschließend verifiziert werden können. Einen **„omitted variable bias“** grundsätzlich allen Modellen zu unterstellen, ist jedoch illegitim. Außerdem haben wir in Kapitel 7 mit der Instrumentalvariablenschätzung eine Methode kennengelernt, wie das Risiko des verzerrenden Einflusses möglicher ausgeschlossener Variablen reduziert werden kann.

d) **Parametrisierung.** Die Effektschätzungen in einem Modell sind – sofern die Bedingung der Exogenität erfüllt ist – statistische Repräsentationen eines kausalen Effekts. Sie sind aber geknüpft an die Messung der Variablen, d.h. aus einer anderen **Metrik** folgt ein anderer Effekt. Wie in Abschnitt 3.4 diskutiert, muss einer latenten Variablen eine Metrik zugewiesen werden, die sie in der Natur nicht hat. Dies macht im Grunde jedes Modell artifiziell. Aber auch dieser Aspekt minimiert nicht den Sinn eines Tests der Kausalstruktur, der sich auf diese spezifizierten Effekte als auch die Annahmen von Restriktionen (z.B. als nicht existierend angenommene Effekte) beziehen. Dies bedeutet, dass ein kausaler Effekt zu einer Beziehung zwischen Variablen führt und nicht-vorhandene Effekte unabhängig von der Metrik. Gleichermaßen wird die Annahme eines nicht-vorhandenen Effektes dann halten, wenn sie korrekt ist, egal welche Metrik zugrunde gelegt wurde.

e) **Annahmen über die Daten sind falsch.** Modelle machen Annahmen über die Daten. So besteht eine Annahme innerhalb eines SEM in der **Multinormalverteilung** der Modellvariablen. Analog liegen einem Regressionsmodell die Annahmen zugrunde, dass die Residuen normalverteilt sind und die gleiche Varianz besitzen. Jede noch so kleine Abweichung von der theoretisch erwarteten Verteilung macht ein Modell daher „falsch“. Aber auch hier gilt, a) dass diese Annahmen nicht grundsätzlich verletzt sein müssen, b) die Abweichungen getestet und der Modelltest gegenüber Abweichungen korrigiert werden kann und c) Abweichungen den Test auf die Korrektheit der kausalen Struktur nur selten so sehr beschädigen, dass er unbrauchbar wird. So wird von Gegnern des Chi-Quadrat-Tests oft die Tatsache, dass der Chi-Quadrat-Test eine Multinormalverteilung voraussetzt, und diese nie exakt vorliegt, als Grund für die Ablehnung des Tests ins Feld geführt. Allerdings wird hier die Wirkung der Abweichung auf den Test überschätzt – v.a. im Vergleich mit der Wirkung einer kausalen Fehlspezifikation, die meist noch wirksam entdeckt werden kann. Darüber hinaus wird übersehen, dass Schätzer existieren (z.B. die **Satorra-Bentler-Korrektur** oder **Yuan-Bentler-Korrektur**), mit denen Abweichungen korrigiert werden können.

f) **Das Modell erlaubt keine zu 100% exakten Vorhersagen.** Wie in Abschnitt 8.7 noch ausführlicher diskutiert werden wird, ist ein wissenschaftliches wie praktisches Ziel bei der Überprüfung von Modellen, diejenigen Phänomene, die durch die abhängigen Variablen repräsentiert sind, zu erklären und vorherzusagen. So ist beispielsweise ein Ziel der medizinischen Forschung, Risikofaktoren für Krankheiten aufzudecken, um die Genese der Krankheit zu verstehen aber auch, um Risiken abzuschätzen, die sich durch eine bestimmte Ausprägung in den bekannten Risikofaktoren ergeben. Analog wird in der beruflichen Eignungsdiagnostik versucht, mit bestimmten Testverfahren den beruflichen Erfolg vorherzusagen. Allerdings kann man selbst durch Einbeziehung aller derzeit bekannten Prädiktoren in der Regel nur einen moderaten wenn nicht kleinen Teil der Varianz der meisten abhängigen Variablen erklären. Wenn ein Modell also nur dann korrekt sein sollte, wenn es die Welt zu 100% exakt erklärt, sind gängige Modelle in der Regel falsch. Analog zu den bislang besprochenen Aspekten minimiert das den Sinn des Tests von Modellen nicht, da die unbekannten Einflüsse einer Variablen im Rahmen des Fehlerterms ein Teil des Modells sind.

Diese beispielhafte Auflistung zeigt, dass es auf denjenigen der oben diskutierten Aspekte eines Modells ankommt, auf dem der Fokus zur Beurteilung seines Wahrheitsgehaltes liegt. Hinsichtlich der diskutierten Aspekte ist es tatsächlich plausibel, dass Modelle häufig bezüglich irgendwelcher (oder all dieser) Aspekte „falsch“ sind. Daraus folgt dennoch nicht, dass dies für *alle* Modelle und *immer* gilt. Aber selbst für den extremen Fall, dass Modelle grundsätzlich bezüglich eines Aspekts immer falsch sind, sind die Schlüsse, die wir daraus ziehen, abhängig davon, warum wir das Modell überhaupt entwickeln.

Denn auch wenn z.B. Aspekte wie die Genauigkeit der Schätzungen oder Präzision der Vorhersagen wichtig sind, ist die *Korrektheit der kausalen Annahmen* zentraler Aspekt eines SEM. Dies setzt zunächst voraus, dass die Variablen in dem Modell existieren und die theoretische Bedeutung haben, die wir ihnen zuschreiben. Hierbei kann man den Fehler begehen, statistische Artefakte zu modellieren, wie dies zum Beispiel bei fehlspezifizierten **Common-Factor-Modellen** der Fall sein kann (vgl. Abschnitt 5.1.2) oder einer latenten Variable eine Bedeutung zuzuweisen, die sie nicht hat (vgl. Abschnitt 8.9). Es gibt aber keinen Grund, anzunehmen, dass diese Fehler grundsätzlich und immer passieren.

Neben der Repräsentation zentraler Variablen ist der Kern eines Kausalmodells das System der Annahmen über die kausalen Effekte dieser Variablen, wie es in den geschätzten und fixierten Effekten zum Tragen kommt. Bzgl. dieses Aspekts zu unterstellen, dass alle Modelle immer falsch sind, käme der Überzeugung gleich, dass Variablen entweder keine Effekte haben oder dass jede Variable jede andere Variable verursacht und somit die Restriktionen falsch sind. Eine zentrale Annahme hinter einem Modell ist daher, dass manche Variablen andere Variablen beeinflussen und andere nicht. D.h. selbst wenn wir bescheiden konstatieren, dass Messungen mit Messfehlern behaftet sind, Modelle unvollständig und Parameterschätzungen allenfalls Approximationen der wahren Effekte sind, macht ein Modell kausale Aussagen, die wahr oder falsch sind und somit überprüfbar sind. Aber sie sind nicht grundsätzlich falsch.

Nun können allerdings Aspekte wie Messfehler, Unvollständigkeit und die Verletzungen der Verteilungsannahmen dazu führen, dass Parameterschätzungen verzerrt sind. Daraus folgt nicht, dass ein Modell bzgl. der kausalen Annahmen *immer* falsch ist, sondern dass sie eine Gefahr für die Validität darstellen und adressiert werden sollten – z.B. durch die Verwendung messfehlerbereinigter latenter Variablen, die Einbeziehung von Instrumentalvariablen oder den Gebrauch von robusten Schätzern.

8.4 Probleme von fittenden Modellen

Wie unter Punkt 8.1 erwähnt, weiß man auch im Falle eines fittenden Modells nicht, ob es korrekt ist. Es könnte ein noch besser fittendes Modell geben, oder das korrekte Modell hat einen ähnlichen oder sogar äquivalenten Fit (Tomarken & Waller, 2003, 2005). Daher sollte die traditionelle Signifikanzschwelle von $p < .05$ nicht als „Schallmauer" interpretiert werden. Ein Modell mit einem p-Wert von .06 ist geringfügig über dieser Schwelle und stellt trotz der nicht-signifikanten Abweichung von S und $\Sigma(0)$ nur eine geringfügig stärkere Evidenz für die kausalen Annahmen dar, als eines unter der Schwelle. Daher sollte in jedem Fall überprüft werden, ob das Modell Hinweise für eine Fehlspezifikation gibt, z.B. über die **standardisierten Residuen**, oder die Sinnhaftigkeit der Parameterschätzungen.

Daneben hat ein fittendes Modell umso weniger dieser Probleme, je überzeugender die Versuche sind, Endogenitätsrisiken zu verringern. So ist das Ziel von Modellen mit **Instrumentalvariablen** ja eben, alternative Modelle auszuschließen. Hier besteht allerdings das Problem, dass Modelle häufig komplex sind und so kaum für jede Gleichung Instrumente hinzugefügt werden können. Es bleibt also bei zwei schon mehrfach geäußerten Implikationen: a) das Modell rigide zu testen, da eine Akzeptanz laxer Kriterien die Anzahl akzeptabler Alternativmodellen erhöht und b) im Rahmen der unter Punkt 8.1 geäußerten Wenn-Dann-Beziehung zu interpretieren.

Ein spezielles Problem eines fittenden Modells besteht dann, wenn die Stichprobengröße gering ist und somit der Chi-Quadrat-Tests eine geringe statistische Power besitzt, ein falsches Modell zurückzuweisen. In diesem Fall ist ein nichtsignifikanter Chi-Quadrat-Wert keine besonders starke Evidenz. Da dies häufig als Argument gegen die Verwendung von SEM geäußert wird und als Grund, um zu (aus kausaler Sicht) schwächeren Analyseverfahren wie der OLS-Regressionsanalyse oder **Partial-least-squares** (**PLS**) zu wechseln, soll die Rolle kleiner Stichproben im nächsten Abschnitt gesondert diskutiert werden.

8.5 Die Stichprobengröße ist zu gering für ein Modell

Eine Annahme bei der Durchführung von SEM ist eine „ausreichend“ große Stichprobe. Maximum-likelihood-Parameterschätzungen sind asymptotisch unverzerrt – d.h. sie sind in kleinen Stichproben verzerrt und das Ausmaß der Verzerrung sinkt kontinuierlich mit steigender Stichprobengröße. Daneben hat der Chi-Quadrat-Test eine geringe statistische Power in kleinen Stichproben mit der Konsequenz, dass ein fehlspezifiziertes Modell akzeptiert wird. Im schlimmsten Fall konvergiert das Modell nicht oder es gibt völlig unsinnige Parameterschätzungen (Boomsma & Hoogland, 2001). Eine kleine Stichrobe kann aber kompensiert werden durch die Anzahl der Indikatoren pro latenter Variable und hohe Faktorladungen. Boomsma und Hoogland geben eine Stichprobengröße von 200 als ausreichend an. Da in den Verhaltenswissenschaften Stichprobengrößen oft kleiner sind, sehen Nutzer/innen dies oft als Grund und Legitimation, SEM ganz zu vermeiden und zu (OLS) Regressionsanalysen oder PLS überzuwechseln, in denen kein Test kausaler Restriktionen möglich ist (Antonakis et al., 2010; Rouse & Corbitt, 2008; Scholderer & Balderjahn, 2006).

Es gibt aber die Möglichkeit, mittels der **swain-Korrektur** (Herzog & Boomsma, 2009) negative Einflüsse der Stichprobengröße auf die Fitstatistiken zu korrigieren. Es gibt unter http://www.ppsw.rug.nl/~boomsma/swain.pdf eine von Herzog und Boomsma verfasste R-Funktion, in die man die Fitstatistiken und die Stichprobengröße eingibt und korrigierte Versionen des Chi-Quadrat-Wertes und der Fit-Indizes bekommt. Eine geringe Stichprobengröße als Grund zu sehen, kein SEM durchführen zu können, ist also daher das Ausschütten des Babys mit dem Badewasser: Der Test der kausalen Struktur hat Priorität, da bei einer Fehlspezifikation alle Parameterschätzungen im Extremfall nicht interpretierbar sind.

Eine sinnvolle Vorgehensweise in solchen Situationen ist die Durchführung einer **Monte-Carlo-Simulation** (Muthén & Muthén, 2002). Hierbei wird das Modell als Populationsmodell unterstellt und analysiert, welcher Bereich von Fitstatistiken und Parameterschätzungen in der konkreten Situation (d.h. mit der gegebenen Stichprobengröße) auftreten würde. Dies erlaubt den Einfluss der Verzerrung abzuschätzen. Monte-Carlo Simulationen sind sehr einfach mit dem `simsem()`-Paket möglich.

8.6 Rolle von Replikationen und Kreuzvalidierungen

Replikationen sind *Wiederholungen einer Studie* zur Überprüfung der Konsistenz der in einer Studie gefundenen Ergebnisse. Das Ausmaß der Nähe zur ursprünglichen Studie kann dabei variieren (von einer exakten Wiederholung der Studie mit einer Stichprobe aus der gleichen Population, gleichen Messungen, gleichem Design etc. bis zu einer Abweichung von diesen Aspekten). Die exakte Wiederholung dient dabei der Überprüfung, ob das Ergebnis auf idiosynkratrische Eigenschaften genau dieser Stichprobe zurückzuführen ist, was durch eine Überprüfung in einer neuen Stichprobe aufgedeckt würde. Die lediglich ähnliche Wiederholung ist im Grunde keine wirkliche Replikation, dient aber der Überprüfung der Robustheit der Ergebnisse über variierende Studien-Charakteristika. Es gäbe z.B. Grund zur Sorge, wenn ein Experiment mit aufsehenerregenden Ergebnissen an die Durchführung bestimmter Forscher/innen gebunden wäre und fehlschlägt, wenn ein/e andere/r Forscher/in es zu wiederholen versucht – selbst unter ansonsten identischen Bedingungen. Replikationen haben daher sinnvolle Funktionen und werden leider zu selten durchgeführt.

Bezogen auf Kausalstrukturen werden Replikationen oft empfohlen, um die Validität des kausalen Modells zu überprüfen. Ein Beispiel ist die **Exploration** einer Faktorenstruktur mittels einer **exploratorischen Faktorenanalyse**. Hier besteht oft die Ansicht, dass eine neue Studie, in der die explorierte Struktur dann entweder durch eine erneute exploratorische Faktorenanalyse wieder zu Tage tritt oder mittels einer konfirmatorischen Faktorenanalyse getestet wird, die Evidenz dafür erhöht, dass die explorierte Struktur korrekt ist. Analog soll bei **respezifizierten Modellen**, die z.B. mittels der **Modifikationsindizes** so verbessert wurden, dass sie einen ausreichenden Fit aufweisen (vgl. Abschnitt 5.2), mittels einer Replikation die Validität der Respezifikation überprüft werden. Die schlechte Nachricht hierbei ist, dass dies keine Möglichkeit zur Überprüfung ist und simple Wiederholungen der Studie mit erfolgreicher Testung des respezifizierten Modells keine Evidenz für die Korrektheit der Struktur darstellen.

Eine exakte Replikation mit einer Stichprobe aus derselben Population würde erlauben zu überprüfen, ob die Ergebnisse in der initialen Stichprobe (z.B. exploratorische Faktorenanalyse) auf idiosynkratische Merkmale dieser Stichprobe zurückzuführen sind oder über Stichproben hinweg stabil sind. Dieser Aspekt adressiert die zu erwartende Stichprobenvariation und damit den zufälligen **Stichprobenfehler**. Letzteres kann natürlich der Fall sein und wird bei einer Replikation auffallen. Das größere und wahrscheinlich häufiger auftretende Problem besteht dann, wenn man eine falsche Struktur an die Daten anpasst. Diese Struktur wird in neuen Studien mit hoher Wahrscheinlichkeit wieder diese Anpassung zeigen, weil sie den über Stichprobenziehungen hinweg systematischen und damit stabilen Teil der Daten betrifft. Somit ist eine erneute Anpassung kein Beleg dafür, dass die Struktur korrekt ist. Um eine explorierte oder respezifizierte Struktur zu testen, müssen stattdessen *zusätzliche Restriktionen* in einer neuen Studie implementiert werden. Die weitergehenden Forschungsfragen sollten sich demnach darauf beziehen, welche Zusammenhänge, Effekte und Nicht-Effekte sich zeigen müssten, wenn das Modell korrekt ist.

Ein Modell *unter veränderten Bedingungen* (erneut) zu testen ist hilfreich – dies überprüft die Robustheit der Struktur über die entsprechenden Bedingungen hinweg (z.B. Überprüfung eines an Studierenden getesteten Modells in einer Erwerbstätigen-Stichprobe) – ist aber weniger eine Überprüfung der ersten Analyse sondern betrifft den Aspekt der **Generalisierbarkeit** des Modells.

Sogenannte **Kreuzvalidierungen** bestehen schließlich darin, die Stichprobe in zwei zufällige Hälften zu trennen. Da die beiden Hälften dadurch bereits äquivalent sind, werden damit Unterschiede in jeglichen statistischen Analysen auf (unwahrscheinliche) Ungleichheiten bei der Trennung zurückzuführen sein.

Insgesamt bleibt festzuhalten, dass Replikationen sehr sinnvoll für eine Reihe wichtiger Fragen sind, welche die Robustheit der Ergebnisse (z.B. Parameterschätzungen oder Outcomes eines Experiments) gegenüber Merkmalen der Studie oder Einschätzungen der Stichprobenfehleranfälligkeit betreffen und die bis zur Absicherung gegen Fehler bei der Erhebung, Dateneingabe und Auswertung oder sogar absichtlichen Fälschungen reichen. Für die Überprüfung der Korrektheit der kausalen Spezifikation tragen sie – im Sinne einer reinen Wiederholung der Studie und Testung des Modells – wenig bei. Hier besteht die bessere Vorgehensweise darin, das Modell zu erweitern oder dahingehend zu verändern, dass dies neue Restriktionen für das Modell impliziert.

8.7 Die Verwendung anti-kausalen Jargons

In der Literatur – v.a. der psychologischen – haben sich seit vielen Jahrzehnten bestimmte Begriffe etabliert, die dadurch gekennzeichnet sind, dass sie a) **ontologisch** (d.h. seins-philosophisch) sehr vage sind und b) eine Vermeidung von kausalen Konzepten darstellen. Dies ist v.a. bei der Verwendung von **nicht-experimentellen** Daten der Fall und mag oft unabsichtlich geschehen oder resultiert entweder aus Sorge vor einer allzu naiven kausalen Interpretation nicht-experimenteller Analysen oder explizit aus den **positivistischen** historischen Wurzeln der entsprechenden Disziplin. So mag eine positivistische und damit anti-realistische Orientierung dazu führen, dass manche Forscher/innen die empirische Seite theoretischer Konstrukte weniger präzise konzeptualisieren – oder gar ein Unbehagen verspüren, überhaupt latente Variablen und Kausalität als wissenschaftliche Begriffe zu akzeptieren. So hat beispielsweise Karl Pearson, der nicht nur Statistiker und der Erfinder des **Produkt-Moment-Korrelations-Koeffizienten** war, sondern auch Wissenschaftsphilosoph mit positivistischer Orientierung, den Kausalitätsbegriff in Gänze abgelehnt. Vor diesem Hintergrund war der Korrelationskoeffizient nicht nur ein hilfreiches statistisches Konzept, sondern diente schlicht dazu, kausales Denken aus der Wissenschaft zu verbannen (Mulaik, 2009).

In der Folge haben sich bestimmte Begriffe so sehr in der natürlichen wissenschaftlichen Sprache etabliert, dass das kausale Denken selbst davon betroffen ist. So sehr diese Begriffe in vielen Situationen angebracht und hilfreich sind, so sehr schaden sie, wenn sie kausale Konzepte ersetzen. Dies ist v.a. im Rahmen von SEM der Fall, die einen klaren **Entitäten-Realismus** (d.h. die Vorstellung kausal wirksamer Phänomene) und kausale Konzepte erfordern. Die Folge ist eine nebulöse und unpräzise Sprache, die speziell im Fall der Spezifikation eines SEM häufig in eine fehlerhafte Übersetzung theoretischer Vorstellungen in das Modell mündet. Im Folgenden sollen diese Begriffe kurz diskutiert werden. Das

Ziel dieser kurzen Diskussion soll schlicht sein, sich dieser Tatsache bei ihrer Verwendung bewusst zu sein.

Der Begriff der **Varianzaufklärung** beispielsweise beschreibt, wie viel der Varianz einer abhängigen Variablen durch die einbezogenen Variablen erklärt werden kann. Dies entspricht damit der Frage nach dem Wissenstand zum Verständnis der Ursachen eines Phänomens und ist damit eine fundamentale Frage in der Wissenschaft. Die Varianzaufklärung ist aber kein Kriterium zur Beurteilung der Korrektheit der kausalen Spezifikation eines Modells. Modelle können korrekt spezifiziert sein und dennoch eine geringe Varianzaufklärung haben, oder sie können – trotz Fehlspezifikation – eine hohe Varianzaufklärung haben.

Hinzu kommt, dass das Ziel der Betrachtung der Varianzaufklärung häufig darin besteht, mit einem Set an Variablen eine andere Variable möglichst gut vorherzusagen oder (statistisch) zu erklären. Darunter fällt dann aber auch, eine Ursache durch seine Wirkung oder eine Variable mit Kovariaten ihrer eigentlichen Ursachen vorherzusagen. In all diesen Fällen bekommt man u.U. eine substantielle Varianzaufklärung – ohne aber irgendwas über das Funktionieren der Welt verstanden zu haben. Es bleibt somit festzuhalten, dass eine Erhöhung der Varianzaufklärung ein wichtiges und interessantes Ziel ist; eine sinnvolle theoretische Einschätzung der Varianzaufklärung allerdings ein korrektes Modell voraussetzt. Zum Problem wird das Ziel der Erhöhung der Varianzaufklärung, wenn diese das Streben nach korrekter kausaler Spezifikation ersetzt.

Die Begriffe **prädiktive Validität** und **inkrementelle Validität** sind Begriffe, die v.a. in Situationen, in denen es um Nützlichkeit von Prädiktoren geht (z.B. in der Eignungsdiagnostik), verwendet werden. Der Begriff prädiktive Validität wird in Situationen benutzt, in denen ein Prädiktor eine abhängige Variable vorherzusagen vermag und ist ersichtlich an einem substantiellen Regressionskoeffizienten oder **R-Quadrat**. Zum Beispiel ist Intelligenz ein starker Prädiktor des Berufserfolgs (Schmidt & Hunter, 2004). Inkrementelle Validität bezeichnet die Tatsache, dass ein zusätzlicher Prädiktor einen Zusatznutzen gegenüber einem ersten Prädiktor hat – d.h. der zusätzliche Prädiktor erlaubt eine Vorhersage bei statistischer Kontrolle des ersten. So ist beispielsweise die Anwendung von Assessment Centern ebenfalls prädiktiv valide für den Berufserfolg. Inkrementell valide wären Assessment Center dann, wenn sie bei Kontrolle von Intelligenz einen signifikanten Regressionskoeffizienten aufweisen.

Beide Begriffe sind mit zwei Problemen verbunden. Erstens haben sie wenig mit **Validität** zu tun, bei der es ja um die Frage geht, ob eine Messung das misst, was sie messen soll (siehe die exzellente Kritik an gängigen Validitätskonzepten von Borsboom, Mellenbergh, & van Heerden, 2004), sondern eher mit „Nützlichkeit für die Vorhersage“. Zweitens mögen sie das Kriterium erfolgreich vorhersagen und somit einen bestimmten praktischen Nutzen haben (z.B. für Auswahlverfahren) – zur Beantwortung zentraler wissenschaftlicher Fragen, ob „prädiktiv valide“ Prädiktoren auch kausale Einflüsse haben, tragen sie wenig bei. Ähnlich wie im Fall der Varianzaufklärung ist die Verwendung dieser Begriffe dann wissenschaftlich schädlich, wenn sie kausale Denken ersetzen.

Zwei weitere Formulierungen beziehen sich auf Zusammenhänge von Variablen. Diese werden häufig als **„gemeinsame Varianz“** (common variance, shared variance) oder **„Überlappung“** bezeichnet. Obwohl beide Begriffe so populär sind, sollten sie schlicht aus dem Sprachgebrauch gestrichen werden, weil sie ein problematisches Verständnis von Kovarianzen oder gar Variablen selbst reflektieren. Beide dürften auf der Verwendung von sogenannten **Ballentines** (Kennedy, 2002, 2003) bzw. **Venn-Diagrammen** zurückzuführen sein. Ballentines haben durchaus ihren Sinn; sie dienen zur Veranschaulichung von Zusammenhängen zwischen Variablen, den Konsequenzen der Kontrolle von Variablen oder deren Auslassung, oder Multikollinearität (s. Abb. 31). Zum Problem werden Ballentines dann, wenn sie das grundsätzliche Denken über Variablen und deren Zusammenhänge prägen.

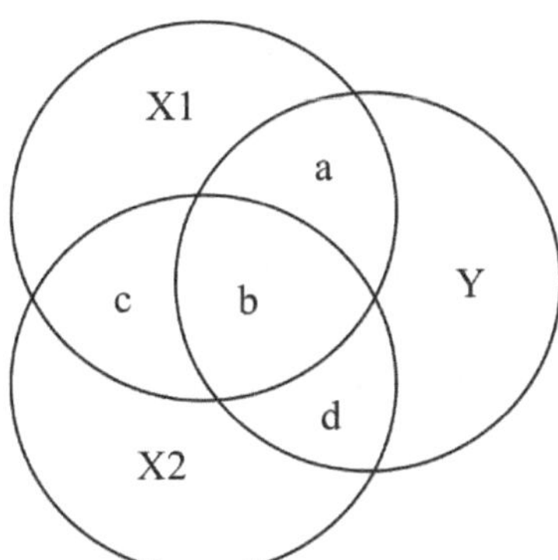

Abb. 31: Ballentine oder Venn-Diagramm zur Veranschaulichung der Zusammenhänge zwischen Variablen

Grundlage einer Ballentine ist eine Kreisfläche, die die Variation einer Variablen symbolisiert. Zum Beispiel illustriert der Kreis „Y“ in Abb. 31 die Varianz von Y. Wird Y auf X1 regressiert (oder umgekehrt), resultiert ein Regressionskoeffizient daraus, der die Kovarianz beider Variablen reflektiert. Diese „gemeinsame Varianz“ ist in der Abb. durch die Überlappung beider Kreisflächen (Schnittmenge a und b) illustriert. Mit dem Ausmaß der Überlappung ist auch die Menge an Information verbunden, die X1 für die Vorhersage von Y enthält – diese impliziert wiederum eine höhere „Sicherheit“ in der Vorhersage von Y, die sich in geringeren Standardfehlern ausdrückt. Wird ein weiterer Prädiktor X2 einbezogen, der mit X1 kovariiert (Schnittmenge c und b) wird dieser bei der Vorhersage von Y berücksichtigt. Die Menge an Informationen, die für die Schätzung der jeweiligen Regressionskoeffizienten zur Verfügung steht, ist durch die statistische Kontrolle des jeweils anderen Prädiktors geringer – Folge sind steigende Standardfehler. Je höher die Kovarianz zwischen X1 und X2, umso mehr sinkt die Information, die in jedem der Prädiktoren (a und d respektive) enthalten ist und umso stärker steigen die Standardfehler (**Multikollinearität**, siehe Abschnitt 8.10). Wird X2 fälschlich in der Analyse ignoriert, folgt ein verzerrter Koeffizient von X1 daraus, weil die Schnittmenge „b“ fälschlicherweise X1 zugeschlagen wird, wenn auch mit geringerem Standardfehler.

Ballentines sind Werkzeuge, um solche Gesetzmäßigkeiten zu illustrieren, und in der methodischen Lehre hilfreich. Sie sind aber wie alle Metaphern irreführend, wenn sie als ein generelles Modell verinnerlicht und automatisiert werden. Der Grund ist, dass sie das Denken von Variablen als zweidimensionale Flächen fördern. Variablen sind aber keine Flächen und ihre Variation ebenfalls nicht. Folglich ist es (außerhalb des oben skizzierten Anwendungsbereichs) nicht sinnvoll, von Kovarianzen als Schnittmenge, Überlappung oder gemeinsamer Varianz zu sprechen. Die Konzepte Varianz und Kovarianz sind statistisch eindeutig definiert; das Konzept „gemeinsame Varianz“ gibt es in der Statistik schlicht nicht. Auch dürften Ballentines die häufig anzutreffende Meinung verursacht haben, dass durch den Einbezug von Kontrollvariablen gemeinsame Varianz „verloren“ geht (in Abb. 31 die Schnittmenge b).

8.8 Breite von latenten Variablen

Bei der Thematisierung von (latenten) Variablen wird man immer wieder mit der Vorstellung von der „**Breite**“ latenter Variablen, bzw. ihrer „**Vielschichtigkeit**“, „**Mannigfaltigkeit**“ oder ihres „**Facettenreichtums**“ konfrontiert. In der Regel wird hier aber fälschlicherweise die latente Variable mit dem theoretischen Konstrukt gleichgesetzt. Dies ist dann problematisch, wenn – wie in Abschnitt 3.3 ausführlich diskutiert – das Konstrukt mehrdimensional ist und dadurch die Vorstellung der Vielschichtigkeit bedingt ist. Das fatale daran ist, dass diese

Multidimensionalität nicht konzeptualisiert wird, sondern fälschlicherweise eine Singularität oder Eindimensionalität des Phänomens unterstellt wird. Bei der Messung eines solchen Konstrukts versucht die/der Forschende dann, eine Vielzahl unterschiedlicher Indikatoren zu berücksichtigen – die Multidimensionalität wird damit unbewusst und quasi „durch die Hintertür" adressiert. Abschnitt 3.3 hat detailliert Alternativen aufgezeigt.

Es sei also nochmals festzuhalten, dass eine latente Variable oder Dimension per definitionem eindimensional ist und keine Breite, Vielschichtigkeit oder ein Facettenreichtum besitzt (siehe auch dazu Abschnitt 8.7 über die Vorstellung einer Variable als zweidimensionale Fläche). Sie bezeichnet ein empirisches Phänomen, von dem die/der Forschende ausgeht, dass es in Form einer Variablen darstellbar ist. Der Begriff „Variable" impliziert dabei, dass Personen in der Population hinsichtlich *eines* Merkmals (Häufigkeit, Stärke, Zustimmungsgrad etc.) variieren. Würden sie nämlich hinsichtlich mehrerer Merkmale variieren, wären es folglich mehrere Variablen.

Ein verwandtes Problem tritt auf, wenn der/die Forschende den singulären Aspekt latenter Variablen zwar noch erkennt, bei der Spezifikation von Second-Order-Faktormodellen diesen Aspekt dann wieder ignoriert. Folglich wird dem Second-Order-Faktor dann doch die besagte Breite unterstellt. Allerdings ändert sich auf der Sekundärfaktoren-Ebene diese Logik nicht: Hier ist der Second-Order-Faktor eine ebenso spezifische, eindimensionale latente Variable, von der angenommen wird, dass sie die Primärfaktoren verursacht und so für die Kovarianzen der Primärfaktoren verantwortlich ist.

8.9 Reifikationsfehler

Die Grundphilosophie hinter der Modellbildung, die in diesem Buch vertreten wurde, ist die einer möglichst **naturgetreuen Abbildung** der zentralen Variablen und ihrer kausalen Effekte. Auch wenn man im wissenschaftlichen Alltag niemals abschließend die Validität eines Modells beurteilen kann, sollte die Modellierung eines tatsächlichen Zustandes das anzustrebende Ideal sein (wenn man dieser Philosophie folgen mag).

Das sollte dennoch nicht zu dem Fehler führen, die Künstlichkeit eines Modells zu übersehen. Das Modell ist und bleibt ein Modell; eine Abstraktion und nicht die Wirklichkeit. Der sogenannte **Reifikationsfehler** (reification error) bezeichnet den Fehler, das Modell mit der Realität gleichzusetzen. Insbesondere besteht dabei die Gefahr, theoretische Konstrukte als reale **Dinge** zu verstehen, die einen objektiven – d.h. von einem betrachtenden Subjekt unabhängigen Existenzstatus – besitzen. Bartholomew (2004) diskutiert in seinem Buch als Beispiel die von manchen Autoren vorgebrachte Ansicht, Intelligenz eine reale Existenz zuzusprechen. Er konstatiert, dass Intelligenz in der Tat kein „Ding" sei, wie das der **Materialismus** als Voraussetzung für die Existenz theoretischer Entitäten vorsieht (Hacking, 1987). Allerdings könne man aber dennoch in den Fällen von Existenz sprechen, wenn die Entität wahrnehmbare kausale Effekte hat (S. 144-145). Dies lässt sich nach dem Prinzip „kein Effekt ohne Existenz" umschreiben bzw. „wenn etwas einen Effekt hat, muss es existieren". Hacking (1987) diskutiert diese Ansicht als **Kausalismus.**

Daraus folgt, dass insbesondere das *Einbetten des Konstrukts in eine kausale Struktur* die Realitätsangemessenheit erhöhen und die Gefahr eines Reifikationsfehlers vermindern kann. In diesem Sinne kommt dem Test von kausalen Restriktionen ein besonderer Wert bei der Überprüfung und Stärkung der Konstruktvalidität zu. Somit ist ein SEM der Entwicklung eines **nomologischen Netzes** (Whitely, 1983), bei dem Messungen des Konstruktes mit denen anderer Konstrukte lediglich korreliert werden, überlegen (siehe hierzu insbesondere die Kritik von Borsboom et al., 2004).

Ein Problem hinsichtlich des Reifikationsfehlers besteht allerdings bei der Modellierung von **Index-Variablen**, die aus Einzel-Indikatoren zusammenaddiert wurden (siehe dazu auch den nächsten Abschnitt). Da man selbst den Index durch das Zusammenaddieren ins Leben gerufen hat, hat dieser unabhängig von dieser Prozedur keine natürliche und eigenständige Existenz (Borsboom, 2008; Borsboom, Mellenbergh, & van Heerden, 2003).

Ein weiterer Fehler besteht darin, zwar ein Faktorenmodell zu unterstellen, den Faktor dann aber jedoch als eine *Folge* des Einflusses der Indikatoren (siehe die vertiefende Diskussion in Abschnitt 3.3) oder deren *Zusammenfassung* zu interpretieren. Ein Beispiel dafür wäre, dass Indikatoren in eine Faktorenanalyse einbezogen wurden, die sich auf das Ausführen von Tätigkeiten beziehen und der extrahierte Faktor dann als „Verhaltensfaktor" bezeichnet wird. Aus der Perspektive des Faktors als den diversen Verhaltensweisen *zugrundeliegende Ursache* ergibt sich dagegen eher die plausible Alternative, dass er eine Disposition (z.B. Kompetenz, Motivation, Werte) darstellt, die diese Verhaltensweisen beeinflusst. Auch hier liegt der Schlüssel zur Beantwortung dieser Fragen in der Einbettung des Faktormodells in ein komplexeres Kausalmodell, in dem über die getesteten Effekte von unabhängigen Variablen auf den Faktor oder von dem Faktor auf abhängige Variablen auf seine Bedeutung geschlossen werden kann.

Schließlich weisen Lee und Cadogan (2013) auf die Gefahren solcher Reifikationsfehler von Second-Order-Faktoren hin und kritisieren die häufige Modellierung solcher Strukturen. Nach Auffassung der Autoren ist das Modell einer solchen Struktur nur dann vertrauenswürdig, wenn die Primärfaktoren konzeptionell extrem ähnlich sind. Ansonsten bleibt völlig im Unklaren, welche Bedeutung der Second-Order-Faktor hat. So könnte er auch die Methodenkovarianz ausdrücken oder überhaupt keine existierende Entität repräsentieren.

8.10 Multikollinearität

Multikollinearität ist ein häufig auftretendes Problem, über dessen Konsequenzen und Lösungen häufig falsche Vorstellungen bestehen. Multikollinearität liegt dann vor, wenn ein Prädiktor in einer Regressionsanalyse (und auch in einem SEM) eine Funktion anderer Prädiktoren im Modell ist und somit das Modell keine Informationen über die Assoziation genau dieses Prädiktors mit der abhängigen Variable zur Verfügung hat, um den Regressionskoeffizienten zu schätzen. Hierzu sind die in Abschnitt 8.7 diskutierten **Ballentines** bzw. **Venn-Diagramme** (Abb. 31) hilfreich. Wäre z.B. Prädiktor X1 kollinear mit Prädiktor X2, würde die Fläche „a" kleiner, je höher die Kollinearität ist. Im Extremfall ließen sich Werte in X1 vollständig aus X2 vorhersagen, womit beide Flächen deckungsgleich wären und keine Flächen mehr übrigbleiben, die die Assoziation nur dieses Prädiktors mit Y abschätzen. Folge ist eine deutliche Erhöhung der **Standardfehler**, da die für die Schätzung eines jeweiligen Regressionskoeffizienten vorhandene Information umso geringer ist, je redundanter die Information in beiden Prädiktoren ist. Die Formulierung „ein Prädiktor ist eine Funktion" kann dabei bedeuten, dass die Korrelation dieses Prädiktors mit einem oder mehreren Prädiktoren sehr hoch ist, aber auch, dass er sich aus der *Kombination* der anderen Prädiktoren vorhersagen lässt. Dies kann dazu führen, dass ein Prädiktor sogar nur moderate Korrelationen mit den anderen Prädiktoren hat und sich dennoch durch sie gemeinsam sehr gut ergeben würde.

Wie oben beschrieben ist eine Folge der Kollinearität ein Ansteigen der Standardfehler. Gegenteilig zur weitläufigen Auffassung sind auch in Fällen deutlicher Kollinearität die Koeffizienten selbst unverzerrt (d.h. im Schnitt sind sie korrekt). Allerdings wird durch die Kollinearität die Variationsbreite der geschätzten Regressionskoeffizienten steigen – diese reflektiert der Standardfehler ja. Dies kann dazu führen, dass Koeffizienten mit negativem Vorzeichen geschätzt werden, wo eigentlich positive vorliegen. Folglich lässt sich Kollinearität am effektivsten mit der Erhöhung der Stichprobengröße bekämpfen, da diese die Variationsbreite verringert und so die Standardfehler senkt.

Sichtbar wird Kollinearität meist an den Korrelationen zwischen den Prädiktoren, aber dies muss aus den oben genannten Gründen nicht sein. Sie wird aber deutlich, wenn die Regressionskoeffizienten zwar eine substantielle Höhe haben, aber nicht-signifikant sind. Außerdem ist in diesen Fällen das Ausmaß der Varianzaufklärung der abhängigen Variablen (R-Quadrat) substantiell und signifikant, die Koeffizienten aber nicht.

Häufig nehmen Forscher/innen die Korrelation zwischen zwei Prädiktoren zum Anlass, die beiden Prädiktoren zusammenzuaddieren und den resultierenden Summenwert als neuen Prädiktor in das Modell einzufügen. Auch wenn dadurch die Probleme augenscheinlich gelöst werden, bleibt die Frage, was der Summenwert inhaltlich bedeutet. Diese Vorgehensweise ist nur sinnvoll, wenn beide Prädiktoren als Messungen einer zugrundeliegenden latenten Ursache aufgefasst werden können und diese der eigentliche Einflussfaktor ist. Ist dies aus theoretischer Sicht unwahrscheinlich, bekommt man einen **Index** (vgl. vorangehender Abschnitt). Im Rahmen der Zielstellung einer naturgetreuen Abbildung von Phänomenen und ihren Effekten ist ein Index unbefriedigend, weil dieser selbst generiert wurde und so außerhalb der Forschungssituation keine eigenständige Existenz und somit auch keine kausalen Effekte haben kann (vgl. Abschnitt 8.9).

Zusammenfassend beziehen sich die o.g. fehlerhaften Auffassungen auf folgende Aspekte:

- Mit steigendem Zusammenhang zwischen den Prädiktoren steigen kontinuierlich die Standardfehler. Es gibt dabei keinen Sprung. Manche Autoren/innen sprechen daher dann von Multikollinearität, wenn die Software nicht mehr in der Lage ist, überhaupt die Regressionseffekte zu schätzen (z.B. $r > .90$). Häufig wird aber schon ab einer Korrelation von $r > .60$ von Multikollinearität gesprochen. Diese Nennung eines solchen Wertes ist erstens beliebig; zweitens sollten Regressionskoeffizienten bei einer substantiellen Stichprobengröße ausreichend präzise geschätzt werden können, do dass die Nullhypothese, wenn sie falsch ist, zurückgewiesen werden kann.
- Die Kollinearität verzerrt nicht die Regressionskoeffizienten. Richtig ist: Regressionseffekte werden trotz Multikollinearität unverzerrt geschätzt; allerdings steigen die Standardfehler und damit sinkt die statistische Power.
- Multikollinearität betrifft nicht nur Regressionsmodelle – SEM sind davon genauso betroffen. Das Problem besteht darin, dass bei Multikollinearität ein

Mangel an Information herrscht, um einen Effekt zu schätzen. Dies betrifft alle Schätzer und damit auch diejenigen in einem SEM.

9

Zentrale Syntax-Codes auf einen Blick

Syntax zur Spezifikation von Parametern in Strukturmodellen

`~`	Pfad- bzw. Struktureffekt
`~~`	Varianz oder Kovarianz
`Y ~ 1.5*X`	Fixieren des Effektes von X auf Y auf den Wert 1.5
`Y1~~.5*Y2`	Fixieren der Fehlerkovarianz von Y1 und Y2
`Y1~~.7*Y1`	Fixieren der Fehlervarianz von Y1

Syntax zur Spezifikation von Parametern in Messmodellen mit latenten Variablen

`WFC =~ w01`	Ladung des Indikators w01 auf der latenten Variable WFC
`WFC ~~.8*WFC`	Die Varianz der latenten Variablen WFC wird auf den Wert .80 fixiert
`WFC ~~ 1.2*AZF`	Die Kovarianz zwischen den latenten Variablen WFC und AZF wird auf 1.2 fixiert
`w01~~.10*w01`	Die Messfehlervarianz des Indikators w01 wird auf .10 fixiert.
`w01~~az01`	Die Kovarianz der Messfehler von w01 und az01 wird frei geschätzt (diese ist als Grundeinstellung auf 0 fixiert). Ähnlich zu den Fixierungen in der vorherigen Zeile kann diese auch mit * auf einen bestimmten Wert fixiert werden.
`WFC =~ w01+a*w02+a*w03`	Die Ladungen von w02 und w03 werden mit einem Label „a“ versehen und so mittels einer Gleichheitsrestriktion gleichgesetzt. Sollten diese Ladungen in der Population ungleich sein, führt dies zu einem misfit des Modells.

Optionale Argumente der Schätz-Funktionen `sem()` oder `cfa()`

`sample.cov=<#>,` `sample.nobs=<#>`	Verwendung einer Kovarianzmatrix als Datengrundlage. Die #'s enthalten den Namen der Matrix und die Nennung der Stichprobengröße

`estimator =`
`"gls"`
`"uls"`
`"wls"`
`"ml"`
`"mlm"`
`"mlr"`
`...`
	Auswahl der Schätzer
	• Generalized least squares (GLS)
	• Unweighted least squares (ULS)
	• Weighted least squares (WLS) / Asymptotically distribution free (ADF)
	• Maximum likelihood (ML; Grundeinstellung)
	• Robust maximum likelihood (Satorra-Bentler Korrektur und robuste Standardfehler, MLM)
	• Robust maximum likelihood mit einer Yuan-Bentler Korrektur der Chi-Quadrate-Statistik und nach Huber-White geschätzten Standardfehlern (MLR)

`missing="fiml"`	Schätzung mittels full information maximum likelihood bei fehlenden Werten

`toot =`
`"Satorra-Bentler"`
`"Yuan-Bentler"`
`"boot"`
	Korrektur der Chi-Quadrat-Statistik
	• Satorra-Bentler-Korrektur
	• Yuan-Bentler-Korrektur
	• Bollen-Stine bootstrap

`se =`
`"robust.mlm"`
`"robust.mlr"`
`"boot"`
	Anforderung robuster Standardfehler (SE)
	• Konventionelle robuste SE (robust.mlm)
	• Huber-White SE (robust.mlr)
	• Bootstrapping der SE (boot)

Optionale Argumente der summary-Funktion

`rsquare=T`	Anzeigen des R-Quadrats für alle abhängigen Variablen
`fit.measures=T`	Anzeigen der Fitindizes
`standardized=T`	Anzeigen der standardisierten Koeffizienten
`modindices=T`	Anzeigen der Modifikationsindizes

10

Literatur

Abraham, W. Todd, & Russell, Daniel W. (2004). Missing data: a review of current methods and applications in epidemiological research. *Current Opinion in Psychiatry, 17*, 315-321.

Aguinis, Herman, & Stone-Romero, Eugene F. (1997). Methodological artifacts in moderated multiple regression and their effects on statistical power. *Journal of Applied Psychology, 82*(1), 192-206.

Aiken, Leona S. , & West, Stephan G. (1991). *Multiple regression*. Newbury Park, CA: Sage.

Algina, James, & Moulder, Bradley C. (2001). A note on estimating the Jöreskog-Yang model for latent variable interaction. *Structural Equation Modeling, 8*(1), 40-52.

Anderson, James C., & Gerbing, David W. (1988). Structural equation modeling in practice: A review and recommended two-step approach. *Psychological Bulletin, 103*, 411-423.

Angrist, Joshua D., & Krueger, Alan B. (2001). Instrumental variables and the search for identification: From supply and demand to natural experiments. *The Journal of Economic Perspectives, 15*(4), 69-85.

Antonakis, John, Bendahan, Samuel, Jacquart, Philippe, & Lalive, Rafael. (2010). On making causal claims: A review and recommendations. *The Leadership Quarterly, 21*, 1086-1120.

Barrett, Paul. (2007). Structural equation modelling: Adjudging model fit. *Personality and Individual Differences, 42*(5), 815-824.

Bartholomew, David J. (2004). *Measuring intelligence - Facts and fallacies*. Camebridge: University Press.

Beaujean, A. Alexander. (2014). *Latent variable modeling using R: A step-by-step guide*. New York: Routledge.

Bentler, Peter M. . (1990). Comparative fit indexes in structural models. *Psychological Bulletin, 107*, 238-246.

Bentler, Peter M. , & Bonet, D.G. (1980). Significance tests and goodness of fit in the analysis of covariance structures. *Psychological Bulletin, 88*(3), 388-606.

Bollen, Kenneth A. (1989). *Structural equations with latent variables*. New York: Wiley.

Bollen, Kenneth A., Glanville, Jennifer L. , & Stecklov, Guy. (2001). Socioeconomic status and class in studies of fertility and health in developing countries. *Annual Review of Sociology, 27*, 153-185.

Boomsma, Anne, & Hoogland, Jeffrey J. (2001). The robustness of LISREL modeling revisited. In R. Cudeck, S. du Toit & D. Sörbom (Eds.), *Structural equation models: Present and future. A festschrift in honor of Karl Jöreskog* (pp. 139-168). Chicago: Scientific Software International.

Borsboom, Denny. (2005). *Measuring the mind: Conceptual issues in contemporary psychometrics*. Cambridge: University Press.

Borsboom, Denny. (2006). The attack of the psychometricians. *Psychometrika, 71*(3), 425-440.

Borsboom, Denny. (2008). Psychometric perspectives on diagnostic systems. *Journal of Clinical Psychology, 64*(9), 1089-1108.

Borsboom, Denny, Mellenbergh, Gideon J., & van Heerden, Jaap. (2003). The theoretical status of latent variables. *Psychological Review, 110*(2), 203-219.

Borsboom, Denny, Mellenbergh, Gideon J., & van Heerden, Jaap. (2004). The concept of validity. *Psychological Review, 111*(4), 1061-1071.

Box, George E.P., & Draper, Norman R. (1987). Empirical Model-Building and Response Surfaces: Wiley.

Browne, Michael W., & Cudeck, Robert. (1993). Alternative ways of assessing model fit. In K. A. Bollen & J. S. Long (Eds.), *Testing structural equation models* (pp. 36-162). Newbury Park, CA: Sage.

Busemeyer, J. , & Jones, L. (1983). Analysis of multiplicative combination rules when the causal variables are measured with error. *Psychological Bulletin, 93*, 549-562.

Byron, Kristin. (2005). A meta-analytic review of work-family conflict and its antecedents. *Journal of Vocational Behavior, 67*, 169-198.

Champoux, Joseph E. , & Peters, William S. (1987). Form, effect size and power in moderated regression analysis. *Journal of Occupational Psychology, 60*, 243-255.

Cheung, Gordon W., & Rensvold, Roger B. (2002). Evaluating goodness-of-fit indexes for testing measurement invariance. *Structural Equation Modeling, 9*(2), 233-255.

Cohen, Jacob, Cohen, Patricia, West, Stephan G., & Aiken, Leona S. (2003). *Applied multiple regression/correlation analysis for the behavioral sciences*. Mahwah, NJ: Lawrence Erlbaum.

Daniels, K., & Guppy, A. (1994). Occupational stress, social support, job control and psychological well-being. *Human Relations, 47-1523-1544.*

Dimitruk, Polina, Schermelleh-Engel, Karin, Kelava, Augustin, & Moosbrugger, Helfried. (2007). Challenges in nonlinear structural equation modeling. *Methodology, 3*(3), 100-114.

Dwyer, D.J. , & Ganster, D.C. (1991). The effects of job demands and control on employee attendance and satisfaction. *Journal of Organizational Behavior, 12*, 595-608.

Echambadi, Raj, & Hess, James D. . (2007). Mean-centering does not alleviate collinearity problems in moderated multiple regression models. *Marketing Science, 26*(3), 438-445.

Edwards, Jeffrey R. (1994). The study of congruence in organizational behavior research: Critique and a proposed alternative. *Organizational Behavior and Human Decision Processes, 58*, 51-100.

Edwards, Jeffrey R. (2001). Multidimensional constructs in organizational behavior research: Towards an integrative and analytical framework. *Organizational Research Methods, 4*(2), 144-192.

Edwards, Jeffrey R. (2011). The fallacy of formative measurement. *Organizational Research Methods, 14*(2), 370-388.

Edwards, Jeffrey R., & Cable, Daniel M. (2009). The value of value congruence. *Journal of Applied Psychology, 94*(3), 654-677.

Edwards, Jeffrey R., & Parry, Mark E. (1993). On the use of polynomial regression equations as an alternative to difference scores in organizational research. *Academy of Management Journal, 36*(6), 1577-1613.

Enders, Craig K. (2001). A primer on maximum likelihood algorithms available for use with missing data. *Structural Equation Modeling, 8*(1), 128-141.

Enders, Craig K., & Bandalos, Deborah L. (2001). The relative performance of full information maximum likelihood estimation for missing data in structural equation models. *Structural Equation Modeling, 8*(3), 430-457.

Finney, Sara J., & DiStefano, Christine. (2006). Non-normal and categorical data in structural equation modeling. In G. R. Hancock & R. O. Mueller (Eds.), *Structural equation modeling: A second course* (pp. 269-314). Greenwich, CT: Information Age.

Flora, David B., & Curran, Patrick J. (2004). An empirical evaluation of alternative methods of estimation for confirmatory factor analysis with ordinal data. *Psychological Methods, 9*(4), 466-491.

Floyd, Frank J., & Widaman, Keith F. (1995). Factor analysis in the development and refinement of clinical assessment instruments. *Psychological Assessment, 7*(3), 286-299.

Forero, Carlos G., Maydeu-Olivares, Alberto, & Gallardo-Pujol, David. (2009). Factor analysis with ordinal indicators: A Monte Carlo study comparing DWLS and ULS estimation. *Structural equation modeling, 16*, 625-641.

Fornell, Claes, & Yi, Youjae. (1992a). Assumptions of the two-step approach to latent variable modeling. *Sociological Methods & Research, 20*(3), 291-320.

Fornell, Claes, & Yi, Youjae. (1992b). Assumptions of the two-step approach: Reply to Anderson and Gerbing. *Sociological Methods & Research, 20*(3), 334-339.

Foster, E. Michael, & McLanahan, Sara. (1996). An illustration of the use of instrumental variables: Do neighborhood conditions affect a young person's chance of finishing high school? *Psychological Methods, 1*(3), 249-260.

Frazier, Patricia A., Barron, Kenneth E., & Tix, Andrew P. (2004). Testing moderator and mediator effects in counseling psychology. *Journal of Counseling Psychology, 51*(1), 115-134.

Ganster, D.C. , & Fusilier, M.R. (1989). Control in the workplace. In C. L. Cooper & I. T. Robertson (Eds.), *International review of industrial and organizational psychology* (pp. 235-280). New York: John Wiley.

Gennetian, Lisa A., Magnuson, Katherine, & Morris, Pamela A. (2008). From statistical associations to causation: What developmentalists can learn from instrumental variables techniques coupled with experimental data. *Developmental Psychology, 44*(2), 381-394.

Gerbing, David W., & Anderson, James C. (1988). An updated paradigm for scale development incorporating unidimensionality and its assessment. *Journal of Marketing Research, 25*, 186-1982.

Gigerenzer, Gerd. (2004). Mindless statistics. *The Journal of Socio-Economics, 33*, 587-606.

Grace, James B., & Bollen, Kenneth A. (2005). Interpreting the results from multiple regression and structural equation models. *Bulletin of the Ecological Society of America, 86*(4), 283-295.

Graham, John W. (2009). Missing data analysis: Making it work in the real world. *Annual Review of Psychology, 60*, 549-576.

Greenhaus, Jeffrey H., & Beutell, Nicholas J. (1985). Sources of conflict between work and family roles. *Academy of Management Review, 10*, 76-88.

Greenland, Sander, Schlesselman, James J., & Criqui, Michael H. (1986). The fallacy of employing standardized regression coefficients and correlations as measures of effect. *American Journal of Epidemiology, 123*(2), 203-208.

Hacking, Ian. (1987). *Representing and intervening - Introductory topics in the philosophy of natural science*. Cambridge: University Press.

Haig, Brian D. (2005). An abductive theory of scientific method. *Psychological Methods, 10*(4), 371-388.

Haig, Brian D. (2008). Précis of 'an abductive theory of scientific method'. *Journal of Clinical Psychology, 64*(9), 1019-1022.

Hausman, J. A. (1978). Specification tests in econometrics. *Econometrica, 46*(6), 1251-1271.

Hayduk, Leslie A. (1987). *Structural equation modeling - Essentials and advances*. Baltimore and London: The Johns Hopkins University Press.

Hayduk, Leslie A. (2014). Seeing perfectly fitting factor models that are causally misspecified: Understanding that close-fitting models can be worse. *Educational and Psychological Measurement*, 1-22.

Hayduk, Leslie A., & Beckie, Theresa M. (1997). Measuring quality of life. *Social Indicators, 42*, 21-39.

Hayduk, Leslie A., Cummings, Greta G., Boadu, Kwame, Pazderka-Robinson, Hanna, & Boulianne, Shelley. (2007). Testing! testing! one, two, three - Testing the theory in structural equation models! *Personality and Individual Differences, 42*(5), 841-850.

Hayduk, Leslie A., & Glaser, Dale N. (2000). Jiving the four-step, waltzing around factor analysis, and other serious fun. *Structural Equation Modeling, 7*(1), 1-35.

Hayduk, Leslie A., & Littvay, Levente. (2012). Should researchers use single indicators, best indicators, or multiple indicators. *BMC Medical Research Methodology, 12*(159), 1-17.

Hayduk, Leslie A., & Pazderka-Robinson, Hanna. (2007). Fighting to understand the world causally: Three battles connected to the causal implications of structural equation models. In W. Outhwaite & S. Turner (Eds.), *Sage Handbook of Social Science Methodology* (pp. 147-171). London: Sage Publications.

Hayduk, Leslie A., Pazderka-Robinson, Hanna, Cummins, Greta G., Boadu, Kwame, Verbeek, Eric L., & Perks, Thomas A. (2007). The weird world, and equally weird measurement models: Reactive indicators and the validity revolution. *Structural Equation Modeling, 14*(2), 280-310.

Hayes, Andrew F. (2009). Beyond Baron and Kenny: Statistical Mediation Analysis in the New Millennium. *Communication Monographs, 76*(4), 408-420.

Herzog, Walter, & Boomsma, Anne. (2009). Small-sample robust estimators of noncentrality-based and incremental model fit. *Structural equation modelling, 16*, 1-27.

Howell, Roy D., Breivik, Einar, & Wilcox, James B. (2007). Reconsidering formative measurement. *Psychological Methods, 12*(2), 205-218.

Hu, Li-Tze, & Bentler, Peter M. . (1999). Cutoff criteria for fit indexes in covariance structure analysis: Conventional criteria versus new alternatives. *Structural Equation Modeling, 6*, 1-55.

Ilies, Remus, & Judge, Timothy A. (2003). On the heritability of job satisfaction: The mediating role of personality. *Journal of Applied Psychology, 88*(4), 650-750.

Irwin, Julie, & McClelland, Gary H. (2001). Misleading heuristics and moderated multiple regressions models. *Journal of Marketing Research, 38*(1), 100-109.

Jaccard, James, Turrisi, Robert, & Wan, Choi K. (1990). *Interaction effects in mulitple regression*. Newsbury Park, CA: Sage.

Jöreskog, Karl G., & Sörbom, Dag. (1981). *LISREL V: Analysis of linear structural relationships by the method of maximum likelihood*. Chicago: National Educational Resources.

Jöreskog, Karl G., & Yang, Fan. (1996). Nonlinear structural equation models: The Kenny-Judd model with interaction effects. In G. A. Marcoulides & R. E. Schumacker (Eds.), *Advanced structural equation modeling* (pp. 57-88). Mahwah, NJ: Lawrence Erlbaum.

Karasek, R.A. (1979). Job demands, job decision latitude and mental strain: Implications for job design. *Administrative Science Quarterly, 24*, 285-308.

Karasek, R.A. (1990). Lower health risk with increased job control among white collar workers. *Journal of Organizational Behaviour, 11*, 171-185.

Kennedy, Peter. (2002). More on Venn diagrams for regression. *Journal of Statistics Education online, 10*(1).

Kennedy, Peter. (2003). *A guide to econometrics*. Cambridge, Massachusetts: MIT Press.

Kenny, David A. , & Judd, Charles M. (1984). Estimating the nonlinear and interactive effects of latent variables. *Psychological Bulletin, 96*(1), 201-210.

Kim, Jae-On, & Mueller, Charles W. (1976). Standardized and unstandardized coefficients in causal analysis: An expository note. *Sociological Methods & Research, 4*(4), 423-438.

King, Gary. (1986). How not to lie with statistics: Avoiding common mistakes in quantitative political science. *American Journal of Political Science, 30*(3), 666-687.

Kline, Rex B. (2011). *Principles and practice of structural equation modeling* (Third ed.). New York, London: The Guilford Press.

Krosnick, Jon A. (1999). Survey research. *Annual Review of Psychology, 50*, 537-567.

Lee, Nick, & Cadogan, John W. (2013). Problems with formative and higher-order reflective variables. *Journal of Business Research, 66*(2), 242-247.

Levitt, Steven D. (1997). Using electoral cycles in police hiring to estimate the effect of police on crime. *Americal Economic Review, 87*(3), 270-290.

Levitt, Steven D. (2002). Using electoral cycles in police hiring to estimate the effects of police on crime: Reply. *Americal Economic Review, 92*(4), 1244-1250.

Lin, Chien-Hsin, Sher, Peter J., & Shih, Hsin-Yu. (2005). Past progress and future directions in conceptualizing customer perceived value. *International Journal of Service Industry Management, 16*(4), 318-336.

Little, Todd D., Bovaird, James A., & Widaman, Keith F. (2006). On the merits of orthogonalizing powered and product terms: Implications for modeling interactions among latent variables. *Structural Equation Modeling, 13*(4), 497-519.

Locke, Edwin A. (2007). The case for inductive theory building. *Journal of Management, 33*(6), 867-890.

MacCallum, Robert C. (1986). Specification searches in covariance structure modeling. *Psychological Bulletin, 100*(1), 107-120.

MacCallum, Robert C., Roznowski, M., & Necowitz, L.B. (1992). Model modifications in covariance structure analysis: The problem of capitalization on chance. *Psychological Bulletin, 111*, 490-504.

MacCallum, Robert C., Zhang, Shaobo, Preacher, Kristopher J. , & Rucker, Derek D. (2002). On the practice of dichotomization of quantitative variables. *Psychological Methods, 7*(1), 19-40.

MacKenzie, Scott B., Podsakoff, Philip M., & Jarvis, Cheryl Burke. (2005). The problem of measurement model misspecification in behavioral and organizational research and some recommended solutions. *Journal of Applied Psychology, 90*(4), 710-730.

MacKinnon, David P., Fairchild, Amanda J., & Fritz, Matthew. (2007). Mediation analysis. *Annual Review of Psychology, 58*, 593-614.

Marsh, Herbert W., Hau, Kit-Tai, & Wen, Zhonglin. (2004). In search of golden rules: Comment on hypothesis-testing approaches to setting cutoff values for fit indexes and dangers in overgeneralizing Hu and Bentler's (1999) findings. *Structural Equation Modeling, 11*(3), 320-341.

Marsh, Herbert W., Wen, Zhonglin, & Hau, Kit-Tai. (2004). Structural equation models of latent interactions: Evaluation of alternative estimation strategies and indicator construction. *Psychological Methods, 9*(3), 275-300.

Maxwell, S.E., & Delaney, H.D. (1993). Bivariate median splits and spurious statistical significance. *Psychological Bulletin, 13*(1), 181-190.

McIntosh, Cameron. (2007). Rethinking fit assessment in structural equation modeling: A commentary and elaboration on Barrett (2007). *Personality and Individual Differences, 42*(5), 859-867.

McIntosh, Cameron. (2012). Improving the evaluation of model fit in confirmatory factor analysis: A commentary on Gundy, C.M., Fayers, P.M., Groenvold, M., Petersen, M. Aa., Scott, N.W., Sprangers, M.A.J., Velikov, G., Aaronson,

N.K. (2011). Comparing higher-order models for the EORTC QLQ-C30. *Quality of Life Research, 21*(9), 1619-1621.

McKnight, Patrick E., McKnight, Katherine M., Sidani, Souraya, & Figueredo, Aurelio José. (2007). *Missing data - A gentle introduction*. New York: Guildford press.

Mowday, R.T., Porter, L.W., & Steers, R.M. (1982). *Organizational linkages: The psychology of commitment, absentism, and turnover*. San Diego: Academic Press.

Mulaik, Stanley A. (2009). *Linear Causal Modeling with Structural Equations*. Boca Raton: Chapman & Hall.

Muthén, Linda K., & Muthén, Bengt. (2002). How to use a Monte Carlo study to decide on sample size and determine power. *Structural Equation Modeling, 9*(4), 599-620.

Neisser, Ulric, Boodoo, Gwyneth, Bouchard, Thomas J. Jr., Boykin, A. Wade, Brody, Nathan, Ceci, Stephen J., . . . Urbina, Susana. (1996). Intelligence: Knowns and unknowns. *American Psychologist, 51*(2), 77-101.

Pearl, Judea. (2001). Causal inference in the health sciences: A conceptual introduction. *Health Services and Outcomes Research Methodology, 2*, 189-220.

Pearl, Judea. (2009). *Causality: Models, reasoning and inference*. Cambridge: University Press.

Pearl, Judea. (2010). *The causal foundations of structural equation modeling*. University of California. Los Angeles. Retrieved from file://S%3A%2FMethodik%2FSEM%2FPearl%20(2010)%20-%20The%20Causal%20Foundations%20of%20Structural%20Equation.pdf

Podsakoff, Philip M., MacKenzie, Scott B., Lee, Jeong-Yeon, & Podsakoff, Nathan P. (2003). Common method biases in behavioral research: A critical review of the literature and recommended remedies. *Journal of Applied Psychology, 88*(5), 879-903.

Rhemtulla, Mijke, Brosseau-Liard, Patricia, & Savalei, Victoria. (2012). When can categorical variables be treated as continuous? A comparison of robust continuous and categorical SEM estimation methods under suboptimal conditions. *Psychological Methods, 17*(3), 354-373.

Rouse, A.C., & Corbitt, B. . (2008). *There's SEM and "SEM": A critique of the use of PLS regression in information systems research*. Paper presented at the Australasian Conference on Information Systems, Christchurch.

Rozeboom, William W. (1997). Good science is abductive, not hypothetico-deductive. In L. L. Harlow, S. A. Mulaik & J. H. Steiger (Eds.), *What if there were no significance tests* (pp. 335-391). Mahwah, NJ: Erlbaum.

Sargan, John D. (1958). The estimation of economic relationships using instrumental variables. *Econometrika, 26*, 393-415.

Saris, Willem E., Satorra, Albert, & van der Veld, William M. (2009). Testing structural equation models or detection of misspecifications? *Structural Equation Modeling, 16*(4), 561-582.

Satorra, Albert, & Bentler, Peter M. (1994). Corrections to test statistics and standard errors in covariance structure analysis. In A. von Eye & C. Clogg (Eds.), *Latent variables analysis: Applications for Developmental Research* (pp. 399-419). Thousand Oaks, CA: Sage.

Schermelleh-Engel, Karin, Moosbrugger, Helfried, & Müller, Hans. (2003). Evaluating the fit of structural equation models: Tests of signifiance and descriptive goodness-of-fit measures. *Methods of Psychological Research Online, 8*(2), 23-74.

Scherpenzeel, Annette, & Saris, Willem E. (1997). The validity and reliability of survey questions: A meta-analysis of MTMM studies. *Sociological Methods & Research, 25*(3), 341-383.

Schlomer, Gabriel, Bauman, Sheri, & Card, Noel A. (2010). Best practices for missing data management in counseling psychology. *Journal of Counseling Psychology, 57*(1), 1-10.

Schmidt, Frank L., & Hunter, John E. (2004). General mental ability in the world of work: Occupational attainment and job performance. *Journal of Personality and Social Psychology, 86*(1), 162-173.

Scholderer, Joachim, & Balderjahn, Ingo. (2006). Was unterscheidet harte und weiche Strukturgleichungsmodelle nun wirklich? [What is the difference between hard and soft structural equation models?]. *Marketing, 28*(1), 57-70.

Shipley, Bill. (2000). *Cause and correlation in Biology: A user's guide to path analysis, structural equations and causal inference*. Cambridge UK: Cambridge University Press.

Sobel, M.E. (1982). Asymptotic confidence intervals for indirect effects in structural equation models. In S. Leinhardt (Ed.), *Sociological methodology 1982*. Washington DC: American Sociological Association.

Spieß, M. (1992). Missing Data Techniken: Analyse von Daten mit fehlenden Werten (Sozialwissenschaftliche Forschungsmethoden). München: Hampp

Staiger, D., & Stock, James H. (1997). Instrumental variables regression with weak instruments. *Econometrica, 65*(3), 557-586.

Steenkamp, Jan-Benedict E.M., & Baumgartner, Hans. (1998). Assessing measurement invariance in cross-national consumer research. *Journal of Consumer Research, 25*, 78-90.

Steiger, James H. (1990). Structural model evaluation and modification: An interval estimation approach. *Multivariate Behavioral Research, 25*(2), 173-180.

Steinmetz, Holger. (2010). Estimation and comparison of latent means across cultures. In P. Schmidt, J. Billiet & E. Davidov (Eds.), *Cross-cultural analysis: Methods and applications* (pp. 85-116). Taylor & Francis Group: Routledge.

Steinmetz, Holger (Ed.). (2007). *A multidimensional approach to working time*. Unpublished dissertation, The University of Giessen, Germany (http://geb.uni-giessen.de/geb/volltexte/2007/5027/).

Steinmetz, Holger, Davidov, Eldad, & Schmidt, Peter. (2011). Three approaches to estimate latent interaction effects: Intention and perceived behavioral control in the theory of planned behavior. *Methodological Innovations, 6*(1), 95-110.

Steinmetz, Holger, Schmidt, Peter, Tina-Booh, Andrea, Schwartz, Shalom H., & Wieczorek, Sigrid. (2009). Testing measurement invariance using multigroup CFA: Differences between educational groups in human values measurement. *Quality and Quantity, 43*(4), 599–616.

Sudman, S., Bradburn, N.M., & Schwarz, N. (1996). *Thinking about answers*. San Fransciso: Jossey-Bass Publishers.

Tellegen, Auke, & Waller, Niels G. (1992). *Exploring personality through test construction: Development of the multidimensional personality questionnaire*. Minneapolis, MN: University of Minnesota Press.

Tomarken, Andrew J., & Waller, Niels G. (2003). Potential problems with "well fitting" models. *Journal of Abnormal Psychology, 112*(4), 578-598.

Tomarken, Andrew J., & Waller, Niels G. (2005). Structural equation modeling: Strengths, limitations, and misconceptions. *Annual Review of Clinical Psychology, 1*, 31-65.

Vandenberg, Robert J., & Lance, Charles E. (2000). A review and synthesis of the measurement invariance literature: Suggestions, practices, and recommendations for organizational research. *Organizational Research Methods, 3*(1), 4-69.

Whitely, Susan. (1983). Construct validity: Construct representation versus nomothetic span. *Psychological Bulletin, 93*(1), 179-197.

Willis, Gordon B. . (2005). *Cognitive interviewing: A tool for Improving questionnaire design*. Thousand Oaks, CA: Sage.

Wright, Sewell. (1921). Correlation and causation. *Journal of Agricultural Research, 20*, 557-585.

Yang-Wallentin, Fan , Jöreskog, Karl G. , & Luo, Hao. (2010). Confirmatory factor analysis of ordinal variables with misspecified models. *Structural Equation Modeling, 17*(3), 392-423.

SOZIALWISSENSCHAFTLICHE FORSCHUNGSMETHODEN

Ingwer Borg, Patrick J.F. Groenen, Patrick Mair:
Multidimensionale Skalierung
Band 1, ISBN 978-3-86618-438-1, München u. Mering 2010, 102 S., € 19.80

Carolin Strobl: **Das Rasch-Modell.**
Eine verständliche Einführung für Studium und Praxis
Band 2, ISBN 978-3-86618-695-8, München u. Mering,
2. erw. Aufl. 2012, 131 S., € 19.80

Jost Reinecke: **Wachstumsmodelle**
Band 3, ISBN 978-3-86618-692-7, München u. Mering 2012, 111 S., € 19.80

Jürgen H.P. Hoffmeyer-Zlotnik, Uwe Warner:
Soziodemographische Standards für Umfragen in Europa
Band 4, ISBN 978-3-86618-827-3, München u. Mering 2013, 124 S., € 19.80

Michael Stegmann, Julia Werner, Heiko Müller:
Sequenzmusteranalyse. Einführung in Theorie und Praxis
Band 5, ISBN 978-3-86618-829-7, München u. Mering 2013, 88 S., € 19.80

Harald Klein: **Computerunterstützte Textanalysen mit**
TextQuest. Eine Einführung in Methoden und Arbeitstechniken
Band 6, ISBN 978-3-86618-831-0, München u. Mering 2013, 135 S., € 22.80

Tim Kaltenborn, Harald Fiedler, Ralf Lanwehr & Torsten Melles:
Conjoint-Analyse
Band 7, ISBN 978-3-86618-862-4, München u. Mering 2013, 143 S., € 22.80

Martin Eisend: **Metaanalyse**
Band 8, ISBN 978-3-86618-875-4, München u. Mering 2014, 104 S., € 19.80

Holger Steinmetz:
Lineare Strukturgleichungsmodelle. Eine Einführung mit R
Band 9, ISBN 978-3-95710-049-8, München u. Mering,
2. verb. Aufl. 2015, 188 S., € 24,80

Markus Burkhardt, Peter Sedlmeier:
Explorative und deskriptive Datenanalyse mit R
Band 10, ISBN 978-3-95710-044-3, München u. Mering 2015, 120 S., € 19.80